IEE Telecommunications Series 20
Series Editors: Professors J. E. Flood, C. J. Hughes and J. D. Parsons

An introduction to SATELLITE COMMUNICATIONS

Other volumes in this series

Volume 1	Telecommunications networks J. E. Flood (Editor)
Volume 2	Principles of telecommunication-traffic engineering D. Bear
Volume 3	Programming electronic switching systems M. T. Hills and S. Kano
Volume 4	Digital transmission systems P. Bylanski and D. G. W. Ingram
Volume 5	Angle modulation: the theory of system assessment J. H. Roberts
Volume 6	Signalling in telecommunications networks S. Welch
Volume 7	Elements of telecommunications economics S. C. Littlechild
Volume 8	Software design for electronic switching systems S. Takamura, H. Kawashima and N. Nakajima
Volume 9	Phase noise in signal sources W. P. Robins
Volume 10	Local telecommunications J. M. Griffiths (Editor)
Volume 11	Principles and practice of multi-frequency telegraphy J. D. Ralphs
Volume 12	Spread spectrum in communications R. Skaug and J. F. Hjelmstad
Volume 13	Advanced signal processing D. J. Creasey (Editor)
Volume 14	Land mobile radio systems R. J. Holbeche (Editor)
Volume 15	Radio receivers W. Gosling (Editor)
Volume 16	Data communications and networks R. L. Brewster (Editor)
Volume 17	Local telecommunications 2 J. M. Griffiths (Editor)
Volume 18	Satellite communication systems B. G. Evans (Editor)
Volume 19	Telecommunications traffic, tariffs and costs R. E. Farr

An introduction to SATELLITE COMMUNICATIONS

D. I. Dalgleish

Peter Peregrinus Ltd. on behalf of the Institution of Electrical Engineers

Published by: Peter Peregrinus Ltd., London, United Kingdom

British Library Cataloguing in Publication Data
Dalgleish, D. I.
An introduction to satellite communications.
1. Satellites
I. Title
523.9'8

ISBN 0-86341-132-0

Printed in England by Short Run Press Ltd., Exeter

An introduction to SATELLITE COMMUNICATIONS

D. I. Dalgleish

Peter Peregrinus Ltd. on behalf of the Institution of Electrical Engineers

Published by: Peter Peregrinus Ltd., London, United Kingdom

British Library Cataloguing in Publication Data
Dalgleish, D. I.
An introduction to satellite communications.
1. Satellites
I. Title
523.9'8

ISBN 0-86341-132-0

Printed in England by Short Run Press Ltd., Exeter

Contents

	Page
Foreword	xi
Acknowledgments	xii
1 The development of satellite communications	**1**
1.1 Introduction	1
1.2 The first satellites	2
1.3 Experimental active communications satellites	3
1.3.1 TELSTAR and RELAY	3
1.3.2 Multiple access	5
1.3.3 SYNCOM	5
1.4 Early progress in the USSR and the MOLNIYA orbit	5
1.5 International satellite communications	6
1.5.1 Introduction	6
1.5.2 INTELSAT I	7
1.5.3 INTELSAT II	9
1.5.4 INTELSAT III and global-beam antennas	9
1.5.5 INTELSAT IV and spot-beam antennas	12
1.5.6 INTELSAT IVA and frequency reuse by means of directional antennas	15
1.5.7 INTELSAT V and frequency reuse by dual polarisation	16
1.5.8 INTELSAT VI	20
1.5.9 INTELSAT VII	22
1.5.10 Competition with INTELSAT	22
1.6 Domestic and regional satellite systems	23
1.6.1 USSR	23
1.6.2 Canada	23
1.6.3 USA	24
1.6.4 Indonesia	25
1.6.5 Other countries	25
1.6.6 Regional systems	26

1.6.7 Leased transponders 26
1.7 Small earth-station antennas and specialised services 27
1.7.1 IBS 28
1.7.2 Teleports 29
1.7.3 VSATs 29
1.8 Mobile systems 31
1.8.1 Maritime 31
1.8.2 Aeronautical 32
1.8.3 Land 32
1.9 Direct broadcasting 33
1.10 Methods of modulation and multiple access 36
1.10.1 Analogue modulation 36
1.10.2 Digital transmission 39
1.11 Trends in satellite communications 45
1.11.1 Telephony 45
1.11.2 Television 45
1.11.3 Mobile services 46
1.11.4 Technical development 47

2 Satellites and launchers 49
2.1 The geostationary orbit 49
2.2 Visibility of satellites from the earth 51
2.3 Putting a satellite into orbit 54
2.3.1 General 54
2.3.2 Basic principle of launchers 54
2.3.3 Launch vehicles and launching 57
2.4 Communications satellites 65
2.4.1 General 65
2.4.2 The service module 65
2.4.3 The communications module 82
2.5 Reliability 90

3 The RF transmission path 93
3.1 The link equations 93
3.1.1 Path loss 93
3.1.2 Noise power 94
3.1.3 C/N and G/T 95
3.1.4 The link equations 95
3.1.5 C/N_o, C/T, E_b/N_o 97
3.1.6 Link budgets 97
3.2 Earth terminal characteristics 98
3.2.1 Antenna gain 98
3.2.2 Noise temperature 99
3.2.3 Solar interference 100

3.2.4 EIRP 102
3.2.5 Antenna gain off axis 102
3.2.6 Antenna beamwidth and tracking 103
3.2.7 Polarisation 103
3.3 Satellite characteristics 104
3.3.1 Antennas 104
3.3.2 *G/T* 104
3.3.3 EIRP 105
3.3.4 Transponder gain 106
3.4 Intermodulation resulting from non-linearity of power transfer characteristics 106
3.4.1 Harmonics and intermodulation products 106
3.4.2 Third-order IM products 107
3.4.3 Backoff 110
3.5 Propagation 113
3.5.1 Ionospheric effects 113
3.5.2 Atmospheric effects 113
3.6 Interference 120
3.6.1 Sources of interference 120
3.6.2 Interference between systems sharing the same frequency bands 121
3.6.3 Procedures of the ITU 124
3.7 Appendix: Examples of link budgets 127
3.7.1 Introduction 127
3.7.2 Link Budget 1 129
3.7.3 Link Budget 2 131
3.7.4 Link Budget 3 132
3.7.5 Link Budget 4 132
3.7.6 Link Budget 5 134

4 Frequency modulation **136**
4.1 Introduction 136
4.1.1 Systems using FM 136
4.1.2 Specifying system performance 138
4.2 Basic characteristics of FM 141
4.2.1 Spectrum of an FM signal 141
4.2.2 Signal-to-noise power ratio of an FM transmission after demodulation 142
4.2.3 Signal-to-noise power ratio in an FDM–FM channel 143
4.2.4 FM threshold 146
4.2.5 Bandwidth required for FDM–FM transmissions 148
4.2.6 Calculation of required carrier-to-noise power ratio for an FDM–FM transmission 151
4.2.7 Noise power ratio (NPR) 152

4.3 Distortion of FM transmissions 155
4.3.1 General 155
4.3.2 Group-delay distortion 155
4.3.3 AM–PM conversion and intelligible crosstalk 158
4.4 Energy dispersal 159
4.4.1 Spectral power density 160
4.4.2 Energy dispersal 161
4.5 Companding 163
4.5.1 Syllabic compandors 163
4.5.2 CFDM 166
4.6 SCPC–FM 167
4.7 TV–FM 169
4.7.1 Introduction 169
4.7.2 Relay of composite signals 170
4.7.3 TV sound 177
4.7.4 TV distribution 178
4.7.5 DBS and MAC 178

5 Digital satellite communications **180**
5.1 Introduction 180
5.2 Pulse code modulation 181
5.2.1 Sampling 181
5.2.2 Coding 182
5.2.3 Decoding 183
5.3 Low-rate encoding (LRE) 184
5.3.1 Introduction 184
5.3.2 Differential PCM and delta modulation 185
5.3.3 Very-low-rate encoding 188
5.3.4 Low-rate encoding of television 189
5.4 Time-division multiplexing (TDM) 190
5.5 Digital speech interpolation (DSI) 192
5.5.1 DSI gain 193
5.5.2 Freeze-out 193
5.5.3 DNI 193
5.5.4 DSI operation 194
5.5.5 SPEC 195
5.5.6 DCME 195
5.6 Modulation 195
5.6.1 Introduction 195
5.6.2 PSK 197
5.6.3 O-QPSK, MSK and 2,4–PSK 211
5.7 Error detection and correction 214
5.7.1 ARQ 214
5.7.2 FEC 215

5.8 Multiple access 221
5.8.1 CDMA 221
5.8.2 FDMA 222
5.8.3 Time-division multiple access (TDMA) 224
5.8.4 Random-access TDMA 231
5.9 Interfaces with terrestrial networks 232
5.10 Error performance objectives 233

6 Maritime and other mobile services 236
6.1 Growth of maritime radio communications 236
6.2 First steps towards an international maritime satellite service 237
6.3 MARISAT, the first maritime satellite-communications system 238
6.4 INMARSAT Standard-A system 239
6.4.1 INMARSAT becomes operational 239
6.4.2 Services available 241
6.4.3 Setting up a call 242
6.4.4 Ship earth stations (SESs) 247
6.4.5 Coast earth stations (CESs) 249
6.4.6 The space sector 252
6.5 INMARSAT second-generation and third-generation satellites 255
6.5.1 Second generation 255
6.5.2 Third generation 257
6.6 The INMARSAT Standard-C system 258
6.6.1 Introduction 258
6.6.2 Operation of the system 262
6.7 Satellite communications and safety of life at sea (SOLAS) 265
6.7.1 FGMDSS 265
6.7.2 EPIRBs 266
6.8 Aeronautical satellite communications 267
6.9 Land-mobile systems 270
6.10 Navigation and position-fixing systems 273

7 Earth stations 275
7.1 Introduction 275
7.2 Siting an earth station 276
7.2.1 Requirements 276
7.2.2 Interference 277
7.3 Antennas 284
7.3.1 Gain and noise temperature 284
7.3.2 Configuration 286

7.3.3 Radiation patterns 291
7.3.4 Polarisation 294
7.3.5 Antenna pointing and tracking 301
7.3.6 Structure, stability and safety 305
7.3.7 Radiation hazards 309
7.4 Electronic equipment 311
7.4.1 General 311
7.4.2 Earth-station noise temperature and low-noise amplifiers 311
7.4.3 High-power amplifiers (HPAs) 314
7.4.4 Signal processing equipment 319
7.4.5 Supervisory and control equipment 327

Short Bibliography **329**

Foreword

The aim of this book is to give a clear and concise exposition of the principles and practice of satellite communications. It is intended for newcomers to the design, provision, operation, management or use of communication-satellite systems and should prove useful both to engineers who have worked in other fields of telecommunications and to students.

The material is based largely on the experience of the author and his ex-colleagues in British Telecom International during more than twenty five years of designing and operating communication-satellite systems.

Many books on satellite communications include a large number of references, few (if any) of which are consulted by the reader; in this book there are only a handful of references but I have included a short list of publications which provide a wealth of additional information (including references) for those who want it.

Chapter 1 describes the development of communication-satellite systems and outlines the services which they can provide.

Chapter 2 deals in a simple but practical way with the physics and geometry of the geostationary orbit, and the construction and operation of satellites and launch vehicles.

Chapter 3 considers the factors which determine the radio-frequency (RF) carrier-to-noise power ratio at the receiving station.

Chapter 4 discusses the relation between the RF carrier-to-noise power ratio and the baseband signal-to-noise power ratio in a system using frequency modulation, particular attention being given to television transmissions.

Chapter 5 outlines the digital transmission methods used for satellite communication.

Chapter 6 is devoted to systems serving mobile stations at sea, in the air and on land.

Chapter 7 describes the most important characteristics of earth stations.

Acknowledgments

I spent twenty one happy and fascinating years (from 1962 until 1983) working on satellite communications with many good friends, first in the British Post Office and then (after the split of Posts and Telecommunications) in British Telecom. Most of the information in this book was collected during that period and, in particular, I have drawn extensively in Chapter 4 on early writing by Tony Reed of British Telecom International (BTI).

In the five years since I left BTI there have been many changes in satellite communications and I could not have kept up to date without the help of my ex-colleagues; those who have been particularly patient in answering my questions include Tony Reed, Dr. Paul Thompson and Jan Vink but my thanks go also to many others, too numerous to list. In addition I am grateful to DCC Ltd., Scientific Atlanta and Keith Thacker (of (INMARSAT), all of whom provided useful material.

In the early stages of planning, I had many useful discussions with Eric Johnston of BTI (who originally hoped to join me in the writing but had to withdraw because of other commitments) and some data which he contributed is incorporated (in Tables 3.2 and 3.3 for example). The book would never have been finished without the patience and encouragement of Professor Charles Hughes (who first suggested I should write it) and John St Aubyn of Peter Peregrinus. My good friend Stan Hill read virtually all of the manuscript, corrected a large number of editorial and other errors and made many valuable suggestions; and John Wroe of BTI saved me from making a number of mistakes (particularly in the chapter on digital systems).

Despite all the help I have received there will inevitably be some residual errors and these are entirely my responsibility. I should be very pleased if readers will let me know of any that come to their notice or of ways in which they think the book could be improved.

Last, but very far from least, I would like to thank my family for their help and forbearance while this book was being written.

D. I. Dalgleish

Chapter 1

The development of satellite communications

1.1 Introduction

'This is a great day with me and I feel I have at last struck the solution of a great problem — and the day is coming when telephone wires will be laid into houses, just like water or gas, and friends converse with each other without leaving home'. So wrote Alexander Graham Bell on the 10th March 1876, the day when he first successfully demonstrated his telephone. Bell showed extraordinary foresight in his confident prediction of the way in which the telephone network would develop, but even he might have been amazed by the size of today's worldwide system which interconnects many hundreds of millions of telephones and carries more than half a million million calls every year.

Another bold look into the future was taken by Arthur C. Clarke in a paper 'Extra-terrestrial relays — Can rocket stations give world-wide coverage?' published in *Wireless World* in 1945. Clarke not only proposed the use of satellites for the distribution of television and for other telecommunication purposes but wrote:

> 'It will be observed that one orbit, with a radius of 42 000 km, has a period of exactly 24 hours. A body in such an orbit, if its plane coincides with that of the earth's equator, would revolve with the earth and would thus appear stationary above the same spot on the planet. For a world service three stations would be required though more could be utilised.'

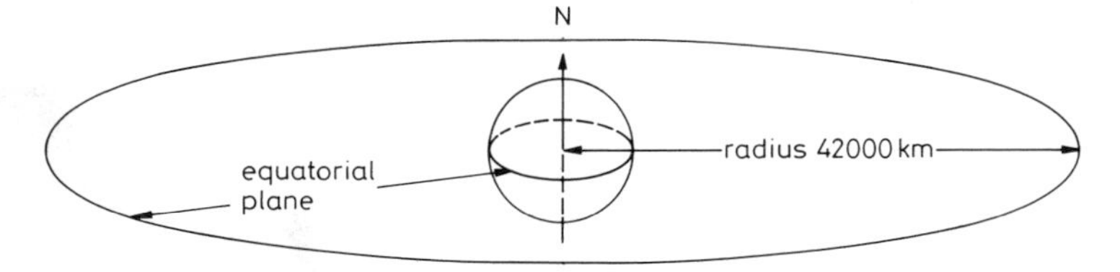

Fig. 1.1 *The geostationary orbit*

Thus twelve years before the launch of the first artificial satellite of any kind Clarke foresaw the coming of communication satellites and the use of the synchronous equatorial (or geostationary) orbit (see Fig. 1.1). It was twenty years before INTELSAT (the International Telecommunications Satellite Organisation) used a geostationary satellite to start the first commercial communications-satellite service in 1965 and another four years before it established the world service using three geostationary satellites which Clarke had envisaged. Today, in 1988, the INTELSAT system has grown to a vast enterprise, INMARSAT (the International Maritime Satellite Organisation) has transformed communication at sea and is about to start an aeronautical service, many regional and national systems have been established, satellites relay television to millions of people via cable-TV head ends and have started broadcasting direct to homes, and parts of the geostationary orbit have become uncomfortably crowded with satellites.

1.2 The first satellites

It is difficult nowadays, when the use of satellites for communications, meteorology, the survey of earth resources etc. is commonplace, to envisage the astonishment which was occasioned by the launch of the first artificial satellite (SPUTNIK I) by the USSR in October 1957. SPUTNIK I weighed 83 kg and was followed, the next month, by SPUTNIK II which was much heavier and carried a dog which lived in orbit for seven days.

The following year saw the launch of a number of satellites including SPUTNIK III (which carried about a tonne of scientific equipment), EXPLORER I and VANGUARD I (two lightweight US satellites) and SCORE which was the shell (weighing nearly 4 tonnes) of an Atlas launch vehicle. SCORE carried equipment which broadcast a tape recording of a Christmas message from US President Eisenhower and can perhaps be claimed as the first communications satellite, even though the information it communicated was recorded on tape and the message was only transmitted for twelve days.

In the next few years much progress was made in space technology. For example, in 1960 the USA launched the first meteorological satellite (TIROS I) and, in April 1961, Yuri Gagarin of the USSR Air Force made the first orbital flight by man. 1960 also saw the launch by NASA (the US National Aeronautics and Space Administration) of ECHO I, a balloon 100 ft in diameter made of plastic coated with aluminium. Telephone and television signals were successfully bounced off ECHO in its orbit at a height of about 1500 km; however, it was clear that such satellites could not be an economic means of communication because of the very high power which had to be transmitted to compensate for the huge attenuation suffered by a signal reflected from such a distant and relatively small object. Another satellite launched in 1960, COURIER IB, was able to receive and store teletype messages as it passed over an earth station and retransmit them on command; the signals were amplified as they were retransmitted and COURIER was therefore an 'active' communications satellite as opposed to ECHO which was a 'passive' satellite.

1.3 Experimental active communications satellites

1.3.1 TELSTAR and RELAY

What was needed for a convincing demonstration that satellites were a practical means of communication was an active satellite capable of the immediate relay of voice signals.

The first such satellite, TELSTAR I, was designed and built by Bell Telephone Laboratories and launched by NASA on the 10th July 1962. TELSTAR weighed 80 kg and was a sphere with a diameter of just under 0·9 m; for comparison, an INTELSAT VI satellite (the first of which should be launched by 1989) will weigh about 1800 kg and be about 12 m high. The communications package of TELSTAR comprised a single transponder (i.e. a device which receives radio signals, processes them in some way and retransmits them); this transponder had a bandwidth of 50 MHz and an output power of about 3 W. The orbit of the satellite (see Fig. 1.2) was elliptical with its perigee (nearest point to the earth) at about 1000 km and its apogee (most distant point from the earth) at about 6000 km. The plane of the orbit was inclined at approximately 45° to the equator and the period of the

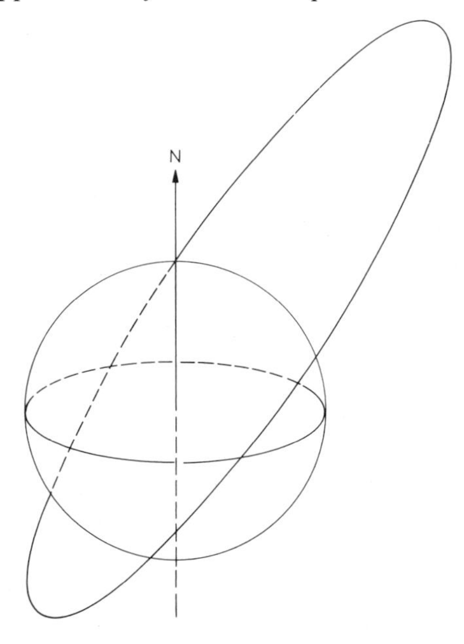

Fig. 1.2 *An inclined elliptical orbit*

orbit was about 2·5 hours. Because of the rotation of the earth, the track of the satellite as seen from an earth station appeared to be different on every successive orbit.

Three earth stations took part in the initial experiments with TELSTAR, an American Telephone and Telegraph Corporation/Bell Telephone Laboratories station at Andover, a French PTT station at Plemeur-Bodou and a British Post Office (BPO) Station at Goonhilly. As a result of the relatively low orbit, the station at Andover and the two stations in Europe were only able to see the satellite simultaneously for periods of about thirty minutes, three or four times a day.

TELSTAR produced a very low power flux density at the surface of the earth; the earth-station antennas therefore had to be very large, and yet they had to be capable of following the relatively rapid movement of the satellite. The stations in the USA and France were of very similar design; they used horn antennas 55 m long and weighing about 450 tons which were protected from wind, rain and snow by radomes 65 m in diameter. The BPO earth station used an antenna with a parabolic reflector 26 m in diameter and did not have a radome. Both types of antenna functioned well but the parabolic dish was much cheaper and virtually all subsequent earth-station antennas have been of this type.

Satellite transponders must have very high gains (of the order of 100 dB) because of the large attenuations suffered by signals on their way to and from the satellites. Satellite transponders therefore always include frequency changers to avoid the instability which might otherwise be caused by feedback from an output port to the corresponding input port. Frequencies of about 6 GHz were used for transmissions to TELSTAR and the satellite converted these transmissions to frequencies of about 4 GHz. These frequencies were also widely used by terrestrial microwave (radio-relay) systems so that many of the communication techniques required were already well developed.

The TELSTAR experiment used frequency modulation (FM) of the radio-frequency (RF) carriers. By using FM it is possible to reduce the power required to give an acceptable signal at the expense of using more RF bandwidth. With the first communication satellites it was necessary to take as much advantage as practicable of this exchange of bandwidth for power.

TELSTAR was an unqualified success. It was used for many experimental telephony transmissions but, for the public in Europe and the USA, its most impressive achievement was the immediate relay of television pictures of events on the other side of the Atlantic. A BPO report of October 1962 said: 'The results obtained from the TELSTAR demonstrations and tests to date have confirmed the expectation that active communication satellites could provide high-quality stable circuits both for television and multi-channel telephony. . . . However much remains to be done to collect the detailed information and data on which the design of future systems for commercial operation must be based'.

1.3.2 Multiple access

Over the next two years TELSTAR I was joined by RELAY I (developed by RCA and NASA), TELSTAR II and RELAY II.

All these satellites were placed in similar orbits to TELSTAR and each of them was therefore visible simultaneously to widely-separated earth stations for only a few relatively short periods each day; a large number would thus have been needed to provide full-time service. On the other hand, a single geostationary satellite can be seen 24 hours a day from about 40% of the earth's surface and this makes it possible to provide direct and continuous communication between a large number of widely separated locations. Satellites which can accept signals simultaneously from a number of earth stations are said to give 'multiple access'.

1.3.3 SYNCOM

Two questions about geostationary satellites worried those who wanted to see commercial communications-satellite systems in operation as soon as possible. Firstly, how long would it be before a satellite of reasonable traffic capacity could be launched into an equatorial orbit with a radius of 42 000 km and kept more or less stationary relative to the earth. Secondly, would the delay of approximately half a second inherent in the transmission distance of about 150 000 km from an earth station via the satellite to another earth station and back again cause difficulty in telephone conversations.

The first question was soon answered with the launch by NASA of the first geostationary satellite (SYNCOM) in July 1963. This satellite had the tragic distinction of carrying television pictures across the Atlantic, on the 22nd November 1963, of the assassination of President Kennedy (while RELAY carried the same pictures across the Pacific). A second SYNCOM, launched in 1964, was used for television transmissions from the Tokyo Olympic Games in August of that year.

The question of the likely effect on telephone calls of the delay associated with a stationary satellite continued to be hotly debated and it became apparent that the matter could only be resolved satisfactorily by a large-scale field trial.

The SYNCOM satellites marked the end of the experimental era of satellite communications because INTELSAT was founded in August 1964 and EARLY BIRD, which was to become the first commercial communications satellite and be renamed INTELSAT I, was launched in April 1965.

1.4 Early progress in the USSR and the MOLNIYA orbit

In the meantime, considerable progress in satellite communications had been made in the USSR. The first of the Soviet Union's MOLNIYA satellites was launched in the same month as INTELSAT I. These satellites were put into an elliptical orbit, but a very different elliptical orbit to those used by the early experimental satellites. What has come to be known as the MOLNIYA orbit is a highly elliptical orbit with an apogee at about 40 000 km and a perigee at about 550 km, which gives an orbital period of half a (sidereal) day; the angle of inclination of the orbit to the equator is 63·4° (at any other angle of inclination the perigee moves slowly round the earth in the plane of the orbit but at 63·4° it remains stationary). Satellites in this orbit follow the same path over the earth each day and the MOLNIYA satellites were visible for about eight hours a day from earth stations over a wide area of the Soviet territories, including stations in the far north which could not have been served by a geostationary satellite. A second advantage of the MOLNIYA orbit for the Soviet Union is that it is easier to put a satellite into this orbit, from a launch site at high latitude, than into the geostationary orbit. A disadvantage of the orbit is that three satellites are required to provide continuous coverage throughout the day.

The MOLNIYA satellites were used for telephony and facsimile services and to distribute television programmes from a central transmitting station near Moscow to a large number of relatively small (for those days) receive-only stations. The MOLNIYA system was thus the forerunner of the many systems for the distribution of television by satellite which have been established since then. Twenty-nine more MOLNIYA satellites were launched in the next ten years.

All recent communications satellites (including those of the USSR) have been put into the geostationary orbit, but in the last few years it has been proposed that satellites in highly elliptical orbits with inclinations of 63·4° and periods of 12 or 24 hours should be used for the provision of service to mobile stations at high latitudes (on land, at sea or in the air). Satellites in such orbits would be particularly suitable for use with land mobile stations at high latitudes because the satellites would appear to the stations at high elevations and obstacles (such as high ground, trees and buildings) would thus be less likely to intrude into the transmission paths.

1.5 International satellite communications

1.5.1 Introduction

The first intercontinental telephone cable (*T*rans*A*tlantic *T*elephone cable No. 1 or TAT 1) went into service in 1956. Until then intercontinental telephony was dependent on a few HF radio circuits, which were at times of such poor quality as to be almost useless. TAT 1 had an initial capacity of thirty-six circuits which it was thought would be enough to carry all the transatlantic traffic for some years ahead, but the demand for calls grew very rapidly when people discovered that they could now rely on understanding what was being said at the other end of the line; in consequence the cable was fully loaded almost from the day it went into service. TAT 1 was followed by a succession of transoceanic cables of ever-increasing capacity. Despite this there were frequent and severe shortages of intercontinental circuits (particularly transatlantic circuits) during the 1960s and 1970s.

Thus when trials with INTELSAT I established that the delay via a geostationary satellite was not a significant disadvantage for the great majority of telephone users (provided that steps were taken to suppress echoes) the satellite was quickly pressed into commercial use on the transatlantic route.

Although communications satellites were first used to provide a complementary service to long-distance cables there are important differences between the two means of transmission. For example:

(i) Cables are best suited to the provision of point-to-point circuits whereas the multiple-access property of satellite systems means that they can provide point-to-multipoint service or multipoint-to-multipoint service
(ii) The cost of cable circuits increases with their length whereas the cost of satellite circuits is independent of the distance between earth stations
(iii) Satellite transmissions can leap over physical and political barriers that are impassable to cables
(iv) Satellites can provide service to mobile stations

In the long term it is these differences between satellites and cables which have exercised most influence on the way in which satellite systems are used. Thus satellites are proving to be the only way in which a full range of communications services can be provided to many smaller countries and to isolated communities in larger countries, satellites are playing a major part in the distribution of sound and television programmes, and satellites are providing the first reliable maritime communications service.

The rest of Section 1.5 is a survey of the technical development of the INTELSAT system and INTELSAT satellites from INTELSAT I (1965) to INTELSAT VII (the first of which should be launched around 1993). Major advances were of course being made in other satellite-communications systems while the INTELSAT system was developing, and these advances and those made by INTELSAT interacted in many ways. Nevertheless, we will concentrate on the INTELSAT system because it is the major international communications-satellite system and because its evolution (over two decades) illustrates many of the problems fundamental to satellite communications.

1.5.2 INTELSAT I

INTELSAT was established in August 1964 and appointed COMSAT (the US Communications Satellite Corporation) as its first manager. COMSAT had already placed a contract for a geostationary satellite and this satellite became INTELSAT I. It was launched in April 1965 by NASA and was used for commercial service from June of that year.

INTELSAT I (see Fig. 1.3) was cylindrical, 0·7 m in diameter and 0·6 m long and weighed about 39 kg. It was spin stabilised, i.e. it was spun about its axis of symmetry to prevent it tumbling as it moved in its orbit. The spin axis was parallel to the axis of the earth and the microwave antenna, which projected from the end of the drum, turned with the satellite; the radiation pattern of the antenna therefore had to be symmetrical about the spin axis and this meant that only a small fraction of the power delivered to the antenna was radiated towards the earth. The curved surface of the satellite was covered in solar cells which generated about 40 W for operation of the electronic and electrical equipment.

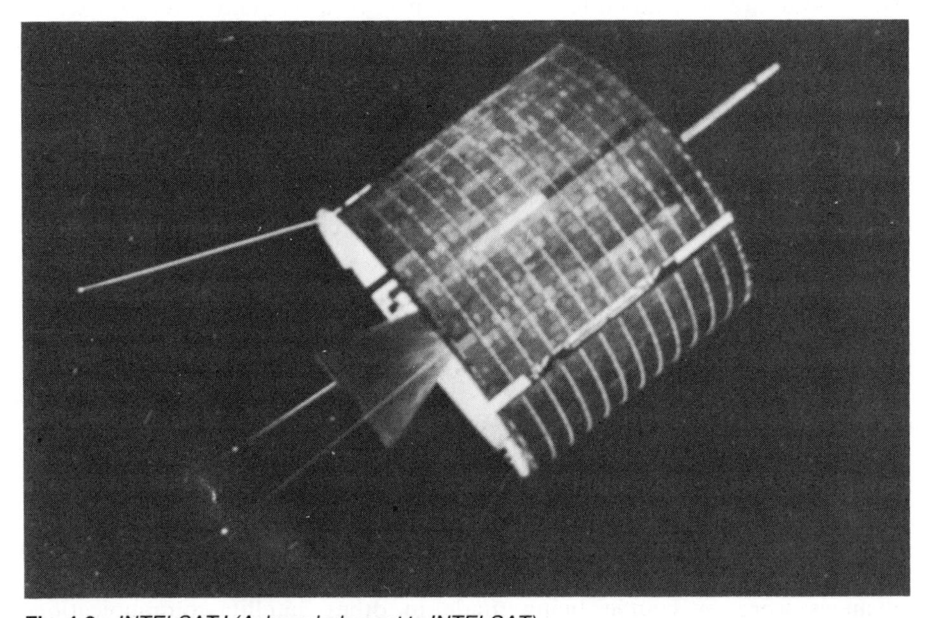

Fig. 1.3 *INTELSAT I (Acknowledgment to INTELSAT)*

The satellite had two transponders (each with a bandwidth of 25 MHz) which received transmissions at 6 GHz and converted them to 4 GHz for re-transmission. One transponder carried signals from Europe for North America and the other carried signals from North America for Europe. The total traffic capacity was 240 telephone circuits or one television transmission. Each transponder could transmit only one radio-frequency carrier at a time; all the telephone channels in each direction had therefore to be multiplexed together and modulated on to a single carrier. Thus although there were now four earth stations in Europe, the two originals at Plemeur-Bodou and Goonhilly having been joined by a large antenna at Raisting in Germany and a smaller one at Fucino in Italy, only one of these stations could be used with INTELSAT I at any one time. Because multiple access was not possible, each European earth station took it in turns to work to the satellite (Fucino being the smallest station usually worked on Saturdays and Sundays when traffic was lightest) and a network of terrestrial links was used to carry traffic to the working earth station. Similarly, on the other side of the Atlantic, stations at Andover (USA) and Mill Village (Canada) took it in turns to work to INTELSAT I. Satellite communications on this basis would not be attractive today, but the arrangement was acceptable while those concerned were learning how to operate and maintain the first commercial system.

During the early months of operation a number of surveys were made to establish what users thought of the service and, in particular, whether they were bothered by the time delay inherent in communication via the geostationary orbit. These surveys confirmed that very few users were worried by the delay and showed that the great majority thought that the quality of the service was good.

1.5.3 INTELSAT II

INTELSAT I was only designed to operate for eighteen months but gave full-time service for more than three-and-a-half years. In 1967 it was joined in orbit by three satellites of the next generation. The main differences between INTELSAT II and INTELSAT I were that:

(i) The two transponders of INTELSAT I were replaced by a single transponder (with a bandwidth of 130 MHz) which was designed to carry signals from several earth stations simultaneously (i.e. it allowed multiple access)
(ii) INTELSAT II was designed for a life of three years.

The first of the INTELSAT IIs provided service to the Pacific Ocean Region, the second provided additional traffic capacity for the Atlantic Region and the third became a spare in orbit for the Pacific Ocean satellite.

1.5.4 INTELSAT III and global-beam antennas

With the next generation of satellites INTELSAT established the world-wide service that Arthur Clarke had envisaged. In January 1969 the first operational INTELSAT III began service to the Atlantic Ocean Region (AOR) and by mid-1969 it had been joined by two more INTELSAT IIIs over the Indian and Pacific Oceans. Fig. 1.4 shows the earth stations which were in service (or planned) by this time; many more stations have, of course, been added to the INTELSAT system in later years.

It was estimated that 500 million people saw the television pictures,

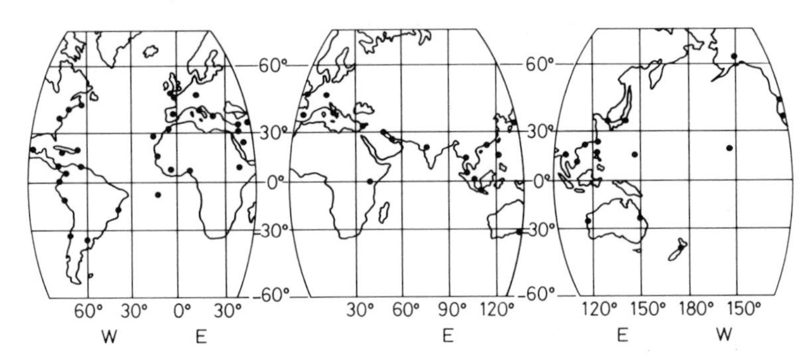

Fig. 1.4 *INTELSAT III: Global coverage*

relayed by the INTELSAT III satellites, of the landing on the moon in July 1969 by US astronauts.

The INTELSAT III satellites had significant advantages relative to their predecessors. They had two transponders, each with a bandwidth of 225 MHz and an output power of about 6 W and their capacity was 1200–1500 telephone circuits or about 700–900 telephone circuits plus a television channel (the exact capacity depended on the pattern of the traffic through the satellite).

Like the previous INTELSAT satellites, INTELSAT III received transmissions at 6 GHz and downconverted them to 4 GHz for transmission back to the earth. Fig. 1.5 shows how each transmission received and transmitted by the satellite occupied a different frequency band; this method of operation is known as frequency-division multiple access (FDMA). The telephony

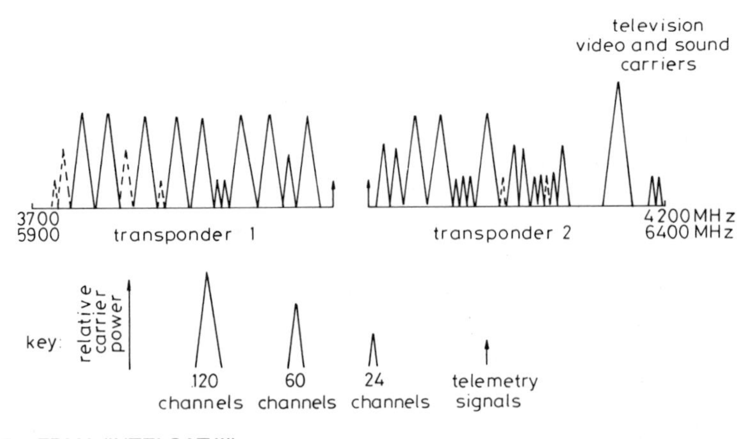

Fig. 1.5 *FDMA (INTELSAT III)*

Not to scale
broken lines show spectrum reserved for new transmissions

carriers were frequency modulated with frequency-division-multiplex (FDM) assemblies of 24, 60 or 120 telephony channels and the corresponding RF transmissions occupied 5, 10 or 20 MHz.

This type of operation is known as FDM–FM–FDMA (see Section 4.1.1 for further details) and over half the INTELSAT telephony traffic is still carried by this method, although most new satellite-communications equipment uses digital techniques.

Possibly the most significant innovation on INTELSAT III was the new type of communications antenna. Previous satellites had radiation patterns which were symmetrical about the axis of spin and most of the RF power they transmitted was therefore wasted by being radiated into space. On the other hand, virtually all the power transmitted by INTELSAT III was radiated towards the surface of the earth by means of a 'global-beam' antenna with a beamwidth of 19° (i.e. just wide enough to subtend the earth from the geostationary orbit and allow for a small pointing error). Fig. 1.6 is an artist's impression of an INTELSAT III stationed over the Atlantic Ocean.

As INTELSAT III was spin stabilised (about a N–S axis) it was necessary to spin the antenna at the same speed as the satellite but in the opposite direction of rotation in order to keep it pointing at the earth; this was done by 'despin' motors under the control of infra-red sensors which detected the earth's horizon.

The INTELSAT III programme had more than its fair share of

FIG. 1.6 *INTELSAT III Over the Atlantic Ocean (Acknowledgment to INTELSAT)*

difficulties. Three satellites failed to reach the geostationary orbit (two of them because of failure of the Delta launch vehicle, which otherwise had an enviable reputation for reliability) and two satellites failed in orbit (because the bearings between antenna and satellite seized up in the hostile environment of space). Thus out of eight satellites only three survived for their design life of five years.

1.5.5 INTELSAT IV and spot-beam antennas

Although virtually all the RF power transmitted by an INTELSAT III satellite was directed towards the earth by the global-beam antenna the resultant power flux density at the surface of the earth was still extremely low. The very large earth-station antennas in use at that time had collecting areas of over 500 m^2 but received less than one hundredth of a picowatt per telephone channel. With this power it was only possible to operate a commercial service by stretching technology to its limits and by using virtually all the available bandwidth of 500 MHz for the transmission of 1500 (or less) telephony channels.

With the demand for satellite circuits increasing rapidly it was essential that the next generation of satellites should produce much higher power flux densities than INTELSAT III. Power flux density may be increased by using higher transmitter power and by using antennas with higher gain. In practice both methods are used to increase the capacity of communication satellites and, for a satellite of given mass, the optimum balance between the two usually depends on the pattern of the traffic flow.

An antenna with a higher gain than a global-beam antenna has a narrower beamwidth and cannot cover the whole of the area from which the satellite is visible; it may therefore be necessary to equip the satellite with more than one antenna in order to serve the whole of the required coverage area and this introduces the problem of interconnecting a number of input and output antennas.

The INTELSAT IV satellites, the first of which went into service in 1971, used both a global-beam antenna and two high-gain 'spot-beam' antennas, with beamwidths of about 4·5°; the global-beam antenna was used both to receive and to transmit and it provided multiple access between all the stations which could see a given satellite; the spot-beam antennas were used only to transmit the largest streams of traffic and were steerable so that they could be pointed, under ground command, at the areas to which these streams of traffic flowed. In the case of a satellite over the Atlantic, for example, one spot beam was directed at Europe and one at North America (see Fig. 1.7).

INTELSAT IV had twelve transponders, each of which had the same output power as one of the two INTELSAT III transponders. The bandwidth of each transponder was 36 MHz and there were guard bands of 4 MHz between adjacent transponders.

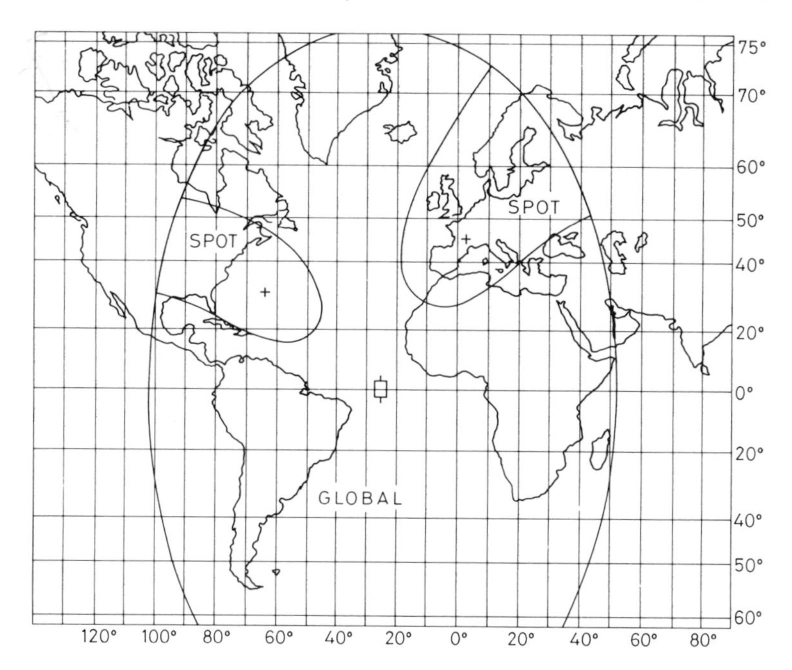

Fig. 1.7 *Spot-beam and global-beam coverage areas (INTELSAT IV)*

All the transponders were fed from the global-beam antenna and the outputs of four of the transponders were connected permanently to the global antenna. The outputs of the other eight transponders could be switched, under ground control, to either the global antenna or one of the spot-beam antennas.

The division of the frequency band between twelve transponders and the ability to switch transponders between global and spot beams gave flexibility which made it possible to use the satellite efficiently in each of the three regions despite differing patterns of traffic. The provision of twelve transponders also reduced the number of carriers to be amplified simultaneously by each satellite transponder and thus reduced the number of intermodulation products generated, which increased the efficiency of the satellites.

Horn antennas were used to provide the global beams, and antennas with parabolic reflectors were used to form the two spot beams. The antennas were mounted on a mast and the overall height of the satellite was 5·3 m. The mass in orbit of INTELSAT IV was 730 kg (nearly five times that of INTELSAT III) and the power generated by the solar array was about 500 W.

The main body of INTELSAT IV was, like that of INTELSAT III, a spinning cylinder which carried the solar array on its outside surface and the batteries (and other ancillary equipment) inside. The communications equipment was mounted on the despun platform which carried the antennas; this had the advantage of avoiding the need for rotating joints carrying microwave signals between the antennas and the transponders. The mechanical and electrical interface between the spinning cylinder and the despun platform comprised two large ball-race bearings, slip rings for transferring electrical power and a rotary transformer for carrying the telemetry and command signals*. Such an interface is known as a BAPTA (bearing and power transfer assembly).

The capacity of INTELSAT IV was about 4000 telephone circuits plus one or two television carriers (the actual capacity depending both on the number of carriers per transponder and the number of transponders connected to the spot-beam antennas). Eight INTELSAT IV satellites were launched between 1971 and 1975 of which seven reached orbit.

At this time INTELSAT was buying satellites as fast as it could get them because:

(i) It was necessary to replace the INTELSAT III satellites with the more powerful INTELSAT IV satellites as quickly as possible
(ii) It was decided that a spare satellite in orbit must be provided in each of the three regions so that service could be restored without delay if an operational satellite failed
(iii) The traffic in the AOR was rising so fast that it became necessary to provide an additional operational satellite, and eventually two additional operational satellites, for this region.

1.5.5.1 Primary and major-path satellites

One of the operational satellites over the Atlantic was designated the primary (P) satellite and this was used to provide multiple-access communication between virtually all the earth stations in the region; the other two satellites were designated major-path satellites (MP1 and MP2) and were used to carry large groups of circuits between countries with traffic sufficiently heavy for them to be willing to build second (and in some cases third) earth terminals for Atlantic Ocean service. The capacity in the primary satellite released by the transfer of traffic to the major-path satellites allowed new earth stations to enter the system and smaller users to expand their services without the expense of building new terminals. The use of primary and major-path satellites also provided diversity of traffic routing for users with two (or more) terminals and thus gave some protection against failure of a terminal (and some additional protection against failure of a satellite).

* Telemetry signals carry information about the state of the satellite to a monitor station on the earth. Command signals are sent from the earth to the satellite to operate switches, steer antennas etc.

At a later stage the traffic in the Indian Ocean Region also grew to the point where it was necessary to use a primary and a major-path satellite.

1.5.6 INTELSAT IVA and frequency reuse by means of directional antennas

Traffic was growing so fast in the Atlantic Region in the early 1970s that INTELSAT found it difficult to keep pace despite the introduction of INTELSAT IV and the major-path satellites. The introduction of still more operational satellites would have required the provision of many additional earth terminals and would have been uneconomic. INTELSAT therefore decided in early 1972 to take up an offer by the manufacturers of INTELSAT IV to modify the satellite design to make it capable of carrying about 6000 telephone circuits and two television channels. This could be done in far less time than was needed to develop a completely new satellite and the first of the modified (INTELSAT IVA) satellites was completed in less than two years from contract and was launched in 1975.

It will be recalled that in the early days of satellite communications there was plenty of frequency spectrum to spare and bandwidth was willingly exchanged for power. By the mid-1970s this was no longer the case; for example, in the AOR the 500 MHz available in the 6/4 GHz bands was being used three times (i.e. by the P, MP1 and MP2 satellites) and additional bandwidth was still needed. There was more spectrum available for satellite communications at higher frequencies but use of these frequencies had to await the development of the next generation of satellites. It was therefore necessary to make still more use of the 6/4 GHz spectrum, but without increasing the number of operational satellites, and this was done by means of directional antennas on the satellites.

Fig. 1.8 shows the basic principle. Transponders A and B are allocated the same 36 MHz band of frequency spectrum; A is fed by a directional antenna which receives signals only from the west of the region while B is fed from a directional antenna which receives signals only from the east (for example, in the case of the AOR the west receive antenna collects signals only from N. and S. America while the east receive antenna collects signals only from Europe, Africa and the Middle East). When the signals have been down-converted from 6 GHz to 4 GHz and amplified they are fed to directional transmitting antennas.

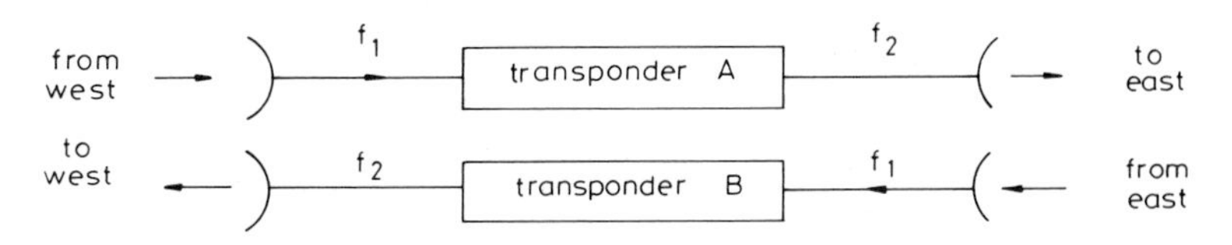

Fig. 1.8 *Frequency reuse by spatial discrimination*

The directional antennas on INTELSAT IVA were offset parabolic reflectors fed by an array of horns and the beam could be tailored to the required shape by selecting the appropriate horns by means of switching matrices. The west and east receive beams, which served about half of the global coverage area, were known as hemispheric (or hemi) beams. If desired, all the power from a transponder could be concentrated in the northern or southern zone of a hemispheric beam.

The satellite had eight pairs of transponders connected to the directional receive and transmit antennas. There were also four transponders connected to global antennas. Thus the total bandwidth available was $[(2 \times 8) + 4)] \times 36 = 720$ MHz whereas with INTELSAT IV it was only $(12 \times 36) = 432$ MHz.

Extensive use of switches gave considerable flexibility in routing signals; thus, for example, signals received from the west could be transmitted back to earth stations in either the eastern or western coverage area.

Subsequent INTELSAT satellites have developed the concept of hemispheric beams and zone beams still further.

1.5.7 INTELSAT V and frequency reuse by dual polarisation

INTELSAT IVA gave much-needed time for the development of the INTELSAT V satellites, the first of which was launched in 1980. INTELSAT V is designed to use most of the 6/4 GHz frequency spectrum four times over by a combination of the directional-antenna techniques introduced on INTELSAT IVA and orthogonally-polarised transmissions. Additional capacity is also provided by the use of frequency spectrum at 14 and 11 GHz.

1.5.7.1 Orthogonal polarisation

Discrimination between two or more signals occupying the same frequency band may be achieved by the use of directional antennas except where the signals follow the same path; in the latter case it is still possible to discriminate between two signals provided that they are orthogonally polarised.

No energy is transferred from a vertically-polarised wave to a horizontally-polarised antenna or from a horizontally-polarised wave to a vertically-polarised antenna. Thus, by using both vertically-polarised and horizontally-polarised transmitting and receiving antennas it is possible to transmit two signals on the same frequency and over the same path without interference between them (i.e. they are isolated from one another). Moreover the antennas do not need to be vertically polarised and horizontally polarised; any two orthogonal states of polarisation will do including right-hand and left-hand circular polarisation. It is important to note that whatever initial polarisation state is chosen there is only one orthogonal state; thus it is only possible to reuse the frequency spectrum once by means of polarisation isolation.

The isolation obtainable in practice is limited by the physical and electrical imperfections of the antennas and by the inhomogeneity of the transmission medium.

For further information on polarisation isolation see Section 7.3.4.

1.5.7.2 Communications facilities

INTELSAT V uses global beams, hemispheric beams, zone beams and spot beams and examples of hemi-beam, zone-beam and spot-beam coverage areas are shown in Fig. 1.9. The global-beam transmissions to and from the

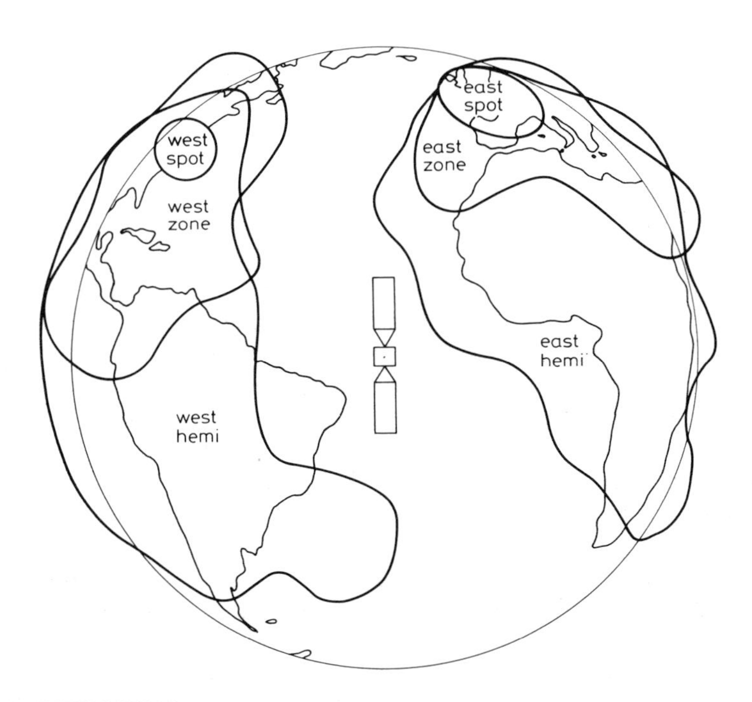

Fig. 1.9 *INTELSAT V Coverage patterns*

satellite are left-hand circularly polarised (LHCP) and right-hand circularly polarised (RHCP), respectively; the hemi-beam transmissions have identical polarisations to the global-beam transmissions but there is no interference between them because they use different parts of the 6/4 GHz frequency bands. However, the zone beams lie either totally or partly within the hemispheric beams and, to a large extent, use the same frequency spectrum;

the zone-beam transmissions are therefore polarised orthogonally to the hemi-beam transmissions (i.e. zone-beam transmissions to and from the satellite are RHCP and LHCP, respectively). Although frequency reuse by orthogonal polarisation had been employed by satellites of other systems before its use on INTELSAT V this was the first time that it had been combined with frequency reuse by means of directional antennas.

The spot-beam antennas receive at 14 GHz and transmit at 11 GHz. As can be seen from Fig. 1.9, one of these antennas points east and one points west. The east and west beams are widely separated and there should therefore be little interference between them; nevertheless, orthogonal linearly-polarised transmissions are used to give additional isolation.

Rain causes a reduction in the isolation between cross-polarised transmissions. The reduction is less for linear polarisation than for circular polarisation and linear polarisation is therefore used for the spot beams. On the other hand, circular polarisation is used for the 6/4 GHz transmissions because circularly-polarised transmissions are unaffected by Faraday rotation (see Section 3.5.1) which might otherwise cause problems at the lower frequencies.

A single transmit antenna generates hemi-beam and zone-beam transmissions from the satellite; similarly, a single antenna receives zone-beam and hemi-beam transmissions from the earth stations. Each hemi/zone antenna consists of a reflector fed by an array of 88 horns, each of which can handle RHCP and LHCP transmissions simultaneously.

The hemispheric beams give a coverage pattern which is suited to operation in all three ocean areas, but the zone coverage pattern required in the Indian Ocean Region is very different to that required in the Atlantic or Pacific Ocean Regions. Switches are therefore provided so that the array of feed horns can be configured to suit the different regions. These switches can be operated by command from the ground if the coverage pattern needs to be changed because the satellite is moved from one region to another.

The satellite has seven working receivers (one each for the global beam, the east and west hemi beams, the east and west zone beams and the east and west spot beams). The transmissions received on each beam are split into channels by means of an array of bandpass filters. After the frequencies of the transmissions have been changed from the 6 and 14 GHz bands to the 4 and 11 GHz bands, each channel is amplified by a separate power amplifier (twenty seven working amplifiers being needed in all). The outputs from these amplifiers are then recombined for retransmission via the seven down beams. Channels are directed from the receive beams to the required transmit beams by means of switching matrices which can be operated from the ground to suit the pattern of traffic. The total bandwidth provided by INTELSAT V is over 2000 MHz and the traffic capacity is (typically) 12 000 telephone circuits plus two TV channels. Further details of the communications module are given in Section 2.4.3.2.

1.5.7.3 Body stabilisation

INTELSAT V is very different in shape from all the previous INTELSAT satellites. The body is a box about 1·7 × 2 × 1·8 m. The antennas are mounted on a tower and the total height of the satellite is about 6·5 m. Two flat solar arrays extend nearly 8 m north and south of the body; these arrays are rotated to face the sun and are designed to produce at least 1200 W throughout the seven-year design life of the satellite. Both the solar arrays and the antenna complex have to be folded to get them into the nose fairing of the launcher.

A spin-stabilised satellite is defined as one in which the main body of the satellite is spun. INTELSAT V is not spin-stabilised because its form makes it impracticable to spin the body. Stabilisation is therefore provided by a mass spinning normal to the plane of the orbit (i.e. parallel to the axis of the earth) within the body. This mass is called a momentum wheel and satellites like INTELSAT V are called (perhaps rather confusingly) body-stabilised satellites.

INTELSAT V has a mass of nearly two tonnes before it is launched but about half of this is the mass of the apogee motor (the solid rocket motor which performs the final stage of the launch process).

1.5.7.4 Earth stations

The introduction of dual polarisation and the 14/11 GHz band produced problems for INTELSAT earth stations.

Antennas fitted with dual-polar feeds became available several years before the first INTELSAT V was launched but it was necessary to modify antennas of earlier design so that they could receive and transmit both polarisations (or, if they were required to use only one polarisation, so that their transmissions did not interfere with transmissions by stations using the opposite polarisation).

The operators with the largest traffic requirements had to build additional terminals for use at 14/11 GHz. These new terminals had antennas with a diameter of about 17 m and were known as Standard-C earth stations to distinguish them from the two groups of stations working at 6/4 GHz, i.e. Standard-A stations (which had antennas with diameters of up to 30 m) and the later Standard-B stations (having antennas with diameters of about 10 m).

As there are relatively few Standard-C stations, it is essential that INTELSAT satellites have the ability to interconnect stations working at 6/4 GHz and stations working at 14/11 GHz; this is known as cross-strapping. The satellites therefore convert all transmissions received (whether at 6 GHz or 14 GHz) to the 4 GHz band; any transmissions intended for Standard-C stations are subsequently up-converted to 11 GHz.

1.5.7.5 INTELSAT VA

The main difference between INTELSAT V and INTELSAT VA satellites is the addition to the latter of down-path spot-beam antennas working at 4 GHz; these antennas use the same reflectors as the 14/11 GHz spot-beam antennas and produce beams with a diameter of about 5°. The traffic capacity of an INTELSAT VA satellite is about 15 000 (4 kHz) channels, i.e. about 3000 channels greater than that of an INTELSAT V satellite.

1.5.7.6 INTELSAT VB

INTELSAT VB satellites are a variation of INTELSAT V intended to be particularly suitable for the provision of business services.

The satellites operate at 6/4 GHz and at 14/11 GHz or 14/12 GHz (the choice of 11 or 12 GHz working can be made by command from the ground). The advantage of using 12 GHz instead of 11 GHz is that the frequency band 12·5–12·75 GHz is allocated exclusively to satellite-communications down links in ITU Region 1 (i.e. Europe, Africa and parts of the Middle East); earth stations receiving these frequencies can therefore be sited in urban areas of Region 1 without any risk of interference from terrestrial stations.

INTELSAT VB has also been modified to give greater flexibility in the connections which can be made between the input and output beams; a received signal can be transmitted in any or all of the spot, hemi or zone beams.

1.5.7.7 MCS

Some INTELSAT V satellites are provided with a maritime communications system (MCS) package which is leased to INMARSAT (see Section 6.4.1).

1.5.8 INTELSAT VI

Planning for international communications is a long-term business and planning for the post-INTELSAT V era commenced in 1975 with long-range forecasts of traffic and service requirements in the three ocean regions. A large number of different system configurations were analysed, and work on specification of the INTELSAT VI satellite commenced in 1979. A request for proposals (RFP) was issued in 1981 and a contract for development and provision of five satellites (with an option for eleven more) was placed in 1982. The 1982 price for the initial five satellites was $525 M (excluding incentive payments). It was intended that the first INTELSAT VI should be brought into operation in the Atlantic Ocean Region (AOR) in 1987 but shortages of launch vehicles resulting from problems with the Shuttle and Ariane programmes prevented this and it is now likely to be 1989 before the satellite is carrying traffic.

Each INTELSAT VI satellite should be able to carry over 30 000 telephone circuits and four television transmissions. This requires a high degree of reuse of the frequency spectrum. Orthogonal polarisation allows only one reuse and all other reuse must therefore be achieved by spatial isolation.

The satellite employs hemi beams and zone beams on opposite polarisations in the 6/4 GHz band (as does INTELSAT V) and four spatially-isolated zone beams are provided (whereas INTELSAT V only has two); INTELSAT VI thus makes sixfold use of most of the 6/4 GHz band and needs even more complex antenna feed arrays than does INTELSAT V. The zone and hemi beams are formed by selecting groups from an array of 146 dual-polarised horns in the focal planes of a 3·2 m-diameter reflector and a 2 m-diameter reflector (for the 4 and 6 GHz transmissions, respectively). The footprints (i.e. coverage areas) of the four zone beams can be reconfigured in orbit to match the differing requirements of the three regions (AOR, IOR and POR) by selecting one of three preset groups of horns. The east and west spot beams of INTELSAT VI reuse the 14/11 GHz frequencies by spatial isolation in the same way as the spot beams of INTELSAT V; however, the spot beams of INTELSAT V are not fully steerable whereas those of INTELSAT VI are.

INTELSAT VI employs parts of new frequency bands which were allocated to satellite communications at the World Administrative Radio Conference (WARC) held in 1979; these new bands are contiguous with the original 6/4 GHz bands. The satellite also uses 500 MHz of the 1000 MHz available at 14/11 GHz (whereas INTELSAT V used only 420 MHz). The result of the combination of spatial isolation, orthogonal polarisation and additional frequency bands is that the total bandwidth available via the INTELSAT VI transponders is 3200 MHz.

Each of the ten received beams (two global, two hemi, four zone and two spot) is connected to a broadband receiver. Transmissions in the up-beams and down-beams are interconnected by means of filter banks and switching matrices just as in INTELSAT V. However, in INTELSAT VI dynamic switches are provided in two of the transponder frequency bands in addition to the static switches. These dynamic switches permit zone and hemi beam-to-beam interconnection patterns to be changed about every 30 μs when the transponders are being used with time-division multiple-access (TDMA) transmissions (see Section 5.8.3); this method of operation is known as space-switched time-division multiple access (SS–TDMA) and it allows more efficient use to be made of transponders than does simple TDMA.

INTELSAT V requires a power of over 1 kW; an important factor favouring the choice of a body-stabilised configuration for this satellite was that it was considered difficult to produce such a high power from a solar array on the body of a conventional spin-stabilised satellite of practical dimensions. Nevertheless INTELSAT VI, which is spin-stabilised, produces a power of over 2 kW. It does this by using two cylindrical solar arrays which are concentric at launch; an inner fixed panel (2·2 m in length) is surrounded by an outer panel (3·8 m in length) which slides down after the satellite is in orbit. The mass of the satellite in orbit is over two tonnes.

1.5.9 INTELSAT VII

For reasons which are explained in Section 1.11, communications satellites are now less likely to be used for point-to-point routes carrying heavy traffic but are finding increasing use for communications with mobile terminals, for broadcast and other point-to-multipoint services, and for routes which are not easily served by cable.

This change in usage is taking place quite slowly but is already reflected in the provisional requirements for INTELSAT VII satellites, the first of which will be launched, at earliest, in 1993. INTELSAT VII satellites are expected to be larger than INTELSAT V but smaller than INTELSAT VI; they will thus be the first INTELSAT satellites since INTELSAT II which will not be very much bigger than their predecessors.

INTELSAT's present (1988) plans are to specify a satellite which is primarily geared to the requirements of the Pacific Ocean Region but can easily be adapted to the needs of the Atlantic Ocean Region. The communication package will probably be not dissimilar to those of INTELSAT V and INTELSAT VI but with a rather smaller total bandwidth than the latter. It is intended that the package shall be associated with an off-the-shelf space platform in order to minimise costs and it is hoped that the lifetime in orbit will be as much as twenty years.

1.5.10 Competition with INTELSAT

The founders of INTELSAT agreed that its purpose was the provision of the space segment of an international communications-satellite system primarily intended to carry public telecommunications services, i.e. services available to anyone who wishes to use them; this has not, of course, prevented INTELSAT from providing the space segment of communications-satellite systems for the private use of Government departments, international corporations etc.

It was also agreed that members of INTELSAT might co-operate in the establishment and operation of other communications-satellite systems where it could be shown that these systems would not cause interference or economic harm to the INTELSAT system, and many such systems (e.g. the US and Canadian domestic systems, and regional systems such as EUTELSAT and ARABSAT) have been successfully co-ordinated with INTELSAT.

The founders of INTELSAT wanted to create a truly global communications system which would provide for the needs of developing countries as well as highly-industrialised countries. They therefore agreed that the price charged for a service would be independent of the route; this means that routes carrying heavy traffic subsidise those carrying light traffic (although cheaper rates are charged for channels carried by more efficient transmission methods such as TDMA). Until the 1980s this philosophy went virtually unchallenged and there was no serious competition to INTELSAT on long-distance communications-satellite routes between its member nations; however, with the deregulation of telecommunications (first in the USA and later in other countries) it was inevitable that private companies would offer service on the more profitable routes. The Federal Communications Commission (FCC) of the USA has now authorised a number of private companies to provide specialised services in competition with INTELSAT, and one company (PanAmSat) already has a satellite in orbit. Proponents of deregulation believe that competition will bring greater efficiency and cheaper communications. Opponents say that only the largest users will derive benefit, that everyone else will pay more and get worse service, and that the development of an efficient global system of communications will be delayed. There are many members of INTELSAT who now consider that, if the organisation is to compete successfully with other systems, it can no longer afford to continue its policy of cross-subsidisation.

1.6 Domestic and regional satellite systems

1.6.1 USSR

Because the cost of a circuit via a communications satellite is independent of the distance between earth stations it is not surprising that satellites were at first used almost entirely for international circuits, except in the USSR where the vast distances between some of the cities provided an incentive to use satellites for national service as soon as possible. The USSR used communications satellites for relaying television programmes from central studios to distant broadcasting stations as early as 1964; telephone circuits were later added to the system and for nearly ten years the USSR was the only country to have a 'domestic' satellite system.

However, by the early 1970s satellites had been developed to the point where it was practicable to use them to provide national communications for countries smaller than the USSR.

1.6.2 Canada

Canada is the second largest nation in the world (in terms of its geographical area) and it was the second nation to use satellites for national communications. The Canadian Government Agency TELESAT put the first of its

ANIK satellites into service in 1973. Anik is the Eskimo word for brother and one of the first uses of the ANIK system was to bring telephone and television service to remote communities in the far north of Canada.

The earth stations associated with small remote communities must clearly be robust, simple to maintain and relatively cheap. Fortunately, it is usually practicable to serve all of a national territory with one or more satellite antennas with relatively narrow beams and high gains, and this allows the use of small earth-station antennas. The antennas of the first ANIK stations serving communities in the far north of Canada had diameters of 4·5 m (which was very much smaller than most of the earth-station antennas in use when the system was established, although larger than many antennas used with more recent systems).

In addition to serving remote communities the ANIK system:

(i) provides telephony and data circuits on heavily-loaded routes of the Trans-Canada Telephone System
(ii) is used for the distribution of television and radio programmes for the Canadian Broadcasting Corporation
(iii) can be used for other services such as video conferencing.

Antennas with diameters from 4·5 m to 30 m are used for these purposes.

1.6.3 USA

The Federal Communications Commission (FCC) ruled in June 1972 that anyone in the USA might apply for a licence to run a domestic communications-satellite system.

The first company to offer service was RCA which hired transponders on an ANIK satellite in December 1973 but later developed and built its own (SATCOM) satellites, the first of which was launched in 1975. The SATCOM system, like the ANIK system, provides a wide range of services including service to remote communities. In this case the remote locations are in Alaska where, for the commencement of service, the State authorities bought 100 small earth stations which were installed in communities which had hitherto had no telephones. SATCOM also brought television to these communities for the first time and this provided educational programmes as well as entertainment. The latest SATCOM satellite now serves more than 200 community stations in Alaska.

Other early brands of US communications satellite were WESTAR (the first of which was launched by Western Union in 1973) and COMSTAR (ComSat General, 1975). These were followed in the 1980s by ASC (American Satellite Company), FORDSAT (Ford Aerospace), GALAXY (Hughes), GSTAR (GTE), SPACENET (GTE Spacenet), SBS (Satellite Business Systems) and TELSTAR (AT&T). There are now several hundred satellite transponders available for US domestic purposes.

An application of domestic satellites which grew very fast in North America was the distribution of television and sound programmes to the head ends of cable-TV systems, local broadcasting stations, hotels etc. The first transmission of pay-TV by satellite was made in 1975 and by 1982 there were over 10 000 TV receive-only (TVRO) stations feeding cable-TV systems in the USA. This rapid growth in the size of the market brought an equally rapid reduction in the cost of TVRO earth stations. Thus when, in 1984, it became legal for private individuals in the USA to receive any signal (provided it was not used for commercial purposes) the stage was set for another burst of demand for TVRO stations. By mid-1985 it was possible to buy a TVRO station (comprising a 3 m-diameter antenna, LNA, downconverter and receiver) for around $2000 and well over a million were in use. Many of these stations were being used by private households for free (legal) reception of programmes intended for paying customers such as cable-TV networks and hotels, and some were being used (illegally) by 'pirates' who were supplying television services to blocks of flats and other customers. As a result, the use of encoding (to prevent unauthorised reception) quickly became widespread and, for a short time, there was a downturn in the sales of TVRO stations.

Another fast-growing sector of domestic satellite communications in the USA has been that of VSATs (very-small-aperture terminals, see Section 1.10).

1.6.4 Indonesia

The next country, after the USA, to use satellite communications for domestic purposes was Indonesia. Indonesia comprises about 13 000 islands which span a distance of over 5000 km. In the early 1970s inter-island communication relied on HF and troposcatter radio links (and some cables) while communication within islands was mainly by terrestrial microwave systems. When it became necessary to plan a major improvement in national communications it was clear that a communications-satellite system was ideally suited both for the provision of a network linking the islands and the distribution of TV programmes for entertainment and education. The first Indonesian satellite (PALAPA–A) was brought into service in 1976 and the earth sector of the system now comprises well over 200 earth stations using antennas with diameters of 5 or 10 m.

1.6.5 Other countries

Other countries which own communications or direct-broadcast satellites (or both) include Australia (AUSSAT), Brazil (BRAZILSAT), France TELE-SAT and TDF), the Federal Republic of Germany (TV-SAT), India and Japan.

1.6.6 Regional systems

Regional systems such as EUTELSAT and ARABSAT provide communications between a group of countries with common interests and offer a way of spreading the very large investment required to establish a satellite system.

In an area such as Europe, which already has a well-developed telecommunications network and where many of the most heavily-loaded routes are relatively short, a communications-satellite system cannot expect to attract more than a small fraction of the mainstream communications traffic.

European governments were nevertheless prepared to subsidise the development of the space sector of the EUTELSAT system in order to support the European aerospace and communications industries. EUTELSAT has been lucky because the demand for the relay of TV programmes has risen much more rapidly than was forecast and this has, to some extent, compensated for the relatively small demand for telephony circuits.

Although Europe does not include vast, sparsely populated areas (such as are found in Russia, North America and Australia) there are remote communities and thin routes which are difficult to serve satisfactorily by the use of terrestrial systems. Thus the Nordic countries are developing a high-powered geostationary satellite (TELE-X) which will be used for communications and direct broadcasting to homes, and Italy is showing particular interest in the use of satellites for communications with the most southerly part of the country.

1.6.7 Leased transponders

When there is insufficient domestic or regional traffic to justify the expense of putting dedicated satellites into orbit, it is usually possible to lease a transponder (or part of a transponder) on an existing satellite. Costs can be reduced by leasing pre-emptible capacity, that is capacity (e.g. on a spare satellite) which the lessor has the right to reclaim under specified circumstances (such as a failure affecting a service with higher priority). Because of the high reliability of most satellite systems it is very rare for the lessor to exercise the right of pre-emption. Over thirty countries now lease capacity on INTELSAT satellites either for domestic communications or for the international relay of television programmes.

For circuits forming part of its global network INTELSAT specifies in detail the transmission standards to be used; this is essential to ensure satisfactory interworking between the many nations using the system. For leased capacity INTELSAT specifies only those conditions of use which are necessary to ensure that there is no interference with other users of its satellites; this allows lessees maximum freedom in designing their system.

1.7 Small earth-station antennas and specialised services

The first generation of earth stations used very large antennas (with diameters of the order of 30 m) in order to maximise the number of channels obtainable via the early satellites. The price of each antenna was many millions of dollars but the cost per channel was acceptable because these stations were intended to carry a great deal of traffic.

The use of smaller earth-station antennas reduces the capacity of a satellite system and it is therefore necessary to charge the operators of smaller antennas more per satellite channel; nevertheless, for stations carrying relatively few channels, the saving in earth-station costs can more than offset the increased space-sector charges. Thus, for example, INTELSAT introduced the Standard-B earth station which requires an antenna with a diameter of about 10 m and many thousands of antennas of this size are now in operation. This was one of the first steps on a long road which has led to the availability and use of earth-station antennas in a very wide range of sizes.

Modern satellites have much greater power and bandwidth than the early generations so that it is no longer necesary for even those earth stations with the largest traffic loads to use antennas with diameters of 30 m and the biggest earth-station antennas made today have diameters of about 18 m. At the other end of the scale are the antennas used with microterminals or very-small-aperture terminals (VSATs) which have diameters of the order of 1 m; VSATs make inefficient use of satellite power and bandwidth but are very cheap and this makes it economic to use them in systems where a large number of terminals (say 100 or more) are needed.

Small earth terminals may be located at or near to the premises of those who need communications services and this has a number of important potential advantages:

(i) Direct interconnection of two (or more) sites via a satellite may be cheaper than interconnection via local exchanges and the toll and trunk networks; direct interconnection can also provide channels with bandwidths which are impracticable using existing local networks.

(ii) Service is available, between as many sites as required, as soon as earth stations are installed; there is no need to wait while complex routes are established by means of connections to and between existing terrestrial networks. In situations where a number of different organisations would be involved in setting up long-distance national or international terrestrial channels, the use of small earth terminals gives the customer the added advantage of dealing with a single organisation both during procurement and subsequently when maintenance is required or if there is cause for complaint.

(iii) Many of the faults in terrestrial communications systems arise in the connections between users and local exchanges; by contrast, satellites virtually never fail and small earth stations at customers' premises can be made very reliable.

(iv) Digital channels at virtually any required bit speed can be provided and rerouted without difficulty, whereas switched 64 kbit/s terrestrial channels will not be widely available until the 1990s (at earliest) and switched terrestrial channels at higher bit rates are unlikely to be widely available in this century.

These advantages are of particular value in the market for what are variously called 'private-user systems', 'business systems' or 'specialised services'. Specialised-service systems can provide all the communications requirements (national, regional and international) of users such as business organisations and government offices including high-speed data transfer, video conferencing (using data rates of 1·5 or 2 Mbit/s), high-speed facsimile, electronic mail, digital channels at 56 or 64 kbit/s (or multiples of these bit rates) and all forms of communication available via other networks.

Specialised services were originally conceived as being provided via domestic systems and earth stations with small antennas situated at the users' premises but are now available via national, regional and international systems, and via earth stations using a wide range of antenna sizes.

1.7.1 IBS

International specialised services are provided by INTELSAT and EUTELSAT amongst other organisations. INTELSAT Business Services (IBS) can be provided via:

(i) A country gateway which uses an earth station with a large antenna (e.g. a Standard-B station working at 6/4 GHz); country gateways usually serve all, or a large part, of a nation.

(ii) An urban gateway which uses an earth station with a medium-sized antenna and serves a major city or industrial region.

(iii) A user gateway comprising a small earth station at the customer's premises used solely by that customer (or an earth station at a nearby site shared by a number of customers).

INTELSAT VB satellites are the best suited of the INTELSAT V family to the provision of business services. These satellites can use the 12 GHz frequency band (part of which is allocated exclusively to satellite communications in some parts of the world, e.g. Europe and Africa) as well as the 14/11 and 6/4 GHz bands and they have been modified so as to give great flexibility in the connections which can be made between input and output beams.

INTELSAT earth stations type E2 (14/12/11 GHz) and F2 (6/4 GHz) use antennas with diameters of about 5·5 and 7·5 m, respectively, and are suitable for urban gateways; earth stations type E3 (14/12/11 GHz) and F3 (6/4 GHz) with antenna diameters of about 3·5 and 4·5 m, respectively, are suitable for use as user gateways.

INTELSAT offers two types of business service: closed network and open network. For closed network systems INTELSAT specifies only those RF transmission characteristics required to ensure compatibility with the satellite and avoid interference to other users or systems; this affords maximum freedom in the design and use of the system. Closed network systems work well when all the users belong to one organisation or when there are only two parties who have to reach agreement but they can be difficult (if not impossible) to set up when, say, a carrier in Europe wants to start a business service to the USA and finds that he has to reach agreement on the characteristics of the system with more than 20 carriers in the USA, many of whom are already operating similar domestic services with disparate equipments. The INTELSAT open network system avoids this type of difficulty because the terminals are specified by INTELSAT in sufficient detail to ensure that they can be interconnected with other terminals to the same specification and will give an assured end-to-end performance. The standardisation of terminals can both reduce costs and offer new opportunities by facilitating interconnection between systems; INTELSAT and EUTELSAT have therefore harmonised their specifications for open-network earth terminals.

1.7.2 Teleports

The full advantages of specialised services are not realised unless earth terminals are provided at or close to the users' premises; on the other hand the cost of procuring, installing and maintaining earth terminals, and in some cases the difficulties of siting them so as to avoid interference, tend to prevent all but the largest organisations acquiring their own earth stations. One way of bringing customers and earth terminals together is the 'teleport'. The term 'teleport' originally meant a site large enough to accommodate a number of earth stations and commercial buildings, and close to a big city (but chosen so as to be as free from interference as practicable). Companies occupying offices or factories at such a teleport can provide their own earth stations or share an earth terminal provided by the teleport administration. Buildings and earth terminals are connected by optical-fibre cable, and optical-fibre cable is also used to connect the teleport with the main business centres in the city. (The first teleport of this type was in New York). However, the term teleport is now often used for any urban earth-station site providing specialised services to the local business community and a large number of such teleports have been established.

1.7.3 VSATs

Earth stations capable of carrying services requiring medium-speed or high-speed bit streams have turned out to be more expensive to manufacture, site and install than was expected and the growth of business services has therefore not been as fast as had been hoped. On the other hand VSATs (using antennas with diameters of less than 2 m) are cheap and can be installed almost anywhere; the market for these terminals has, in consequence, been growing rapidly in North America and may be about to take off in Europe.

VSATs are usually used in conjunction with a hub station with a comparatively large antenna (say about 8 m in diameter) which transmits to or receives transmissions from all the other stations in the system. VSAT systems may be interactive, may have a selective-addressing facility and often use packet switching. Typical applications are credit-card verification at petrol stations, stock control for large chains of retail premises, distribution of stock exchange prices and collection of geophysical data.

The use of very small antennas at earth stations requires the use of high satellite powers for the transmission of relatively low bit rates but this is economic because the cost per station is small. VSAT systems frequently use error-correction or spread-spectrum techniques to achieve the required performance and to reduce interference to and from other systems; they therefore tend to use a lot of bandwidth, as well as a lot of satellite power. It has, however, been suggested that the progressive transfer of heavy traffic routes from satellite systems to optical fibres (see Section 1.11) may mean that more bandwidth will become available for the use of VSAT systems.

INTELSAT offers two VSAT services: INTELNET I which is a data-distribution service and INTELNET II which is a data-gathering service. Many services require communication in both directions but the information flow in one of the directions may be very small and it may be more economical to use ordinary terrestrial communications systems in this direction. An example is the APOLLO system, the development of which is being funded by the European Space Agency (ESA), the European Commission and a number of European telecommunications carriers. The APOLLO system is intended for the distribution of information from very large central libraries (e.g. the British Museum Library) to small local libraries all over Europe. A request for information may be passed to the central library by telephone or telex and a facsimile of the required documents is then transmitted by satellite to the local station. The system is equally suited to the distribution of documents or data from any large central office to outlying offices and particular packages can be addressed to specific offices or groups of offices.

The development of VSAT systems is in danger of being stifled in many countries (particularly in Europe) by licensing and regulatory procedures. These procedures, devised in the days when satellite systems comprised only a few earth stations, may prove very slow and restrictive when applied to networks comprising hundreds or even thousands of stations.

1.8 Mobile systems

Only radio systems can provide communications with mobile stations. VHF transmissions give good service over line-of-sight paths (i.e. over distances up to about 40 miles for ships, and up to about 100 or 200 miles for aircraft flying at 10 000 or 30 000 ft, respectively). For longer distances the choice is between HF radio (which has been in use since the beginning of the century) and satellite communications. Maritime satellite communications have been available for over a decade and global aeronautical satellite communications will be available by about 1990.

1.8.1 Maritime

Voice communication with ships via HF radio is unreliable and subject to delays because of the variable nature of the ionosphere and the limited bandwidth available in the HF spectrum; Morse telegraphy, which requires narrower bandwidths and is less prone to fading, is therefore widely used but has the disadvantages that it requires trained operators and is slow. Satellites can, on the other hand, provide immediate and reliable communications using voice, telex or data channels.

The first maritime communications-satellite system was the US MARISAT system which started operation in 1976. This was succeeded in early 1982 by the system operated by the International Maritime Satellite Organisation (INMARSAT). INMARSAT started by leasing the MARISAT satellites and quickly acquired additional capacity by leasing satellites from ESA and maritime communications subsystems (MCSs) on INTELSAT V satellites; it is now awaiting delivery of its own, second-generation, satellites. About 7000 ships are already fitted with MARISAT/INMARSAT Standard-A ship earth stations (SESs); this type of SES can provide a wide range of telephony, telegraphy and data services but the equipment is relatively complex and expensive, and can be difficult to install. By 1990 INMARSAT will be operating a store-and-forward data service; the Standard-C SESs required for this service will be only a fraction of the cost of a Standard-A station and very much smaller, and it is expected that many tens of thousands of these stations will be installed in the early 1990s. INMARSAT is also going to introduce a Standard-B system which will use all-digital communications methods and will be matched to the rapidly-developing terrestrial integrated-system digital networks (ISDNs); Standard-B terminals may eventually displace Standard-A terminals.

Ship earth stations are used by offshore oil/gas exploration rigs and production platforms as well as by ships. On the rigs they are used, *inter alia*, for the rapid transmission of data back to shore (bit rates up to 56 kbit/s and, with special arrangements, up to 1 Mbit/s are available).

Satellite communications make an important contribution to safety at sea. Every INMARSAT SES is fitted with a distress facility which gives immediate access to a Rescue Coordination Centre (RCC); other services which contribute to safety and are available via the INMARSAT system are navigational information, meteorological information and medical advice. Another safety facility is provided by satellites launched by the USA and USSR which relay signals from emergency position-indicating radio beacons (EPIRBs) which are carried on board ships and aircraft, and are credited with the rescue of about 700 people in the period 1982–87. The valuable role which satellites can play in promoting safety at sea has been recognised by the International Maritime Organisation (IMO) which has included both SESs and EPIRBs in its plans for the Future Global Maritime Distress and Safety at Sea System (FGMDSS).

1.8.2 Aeronautical

The success of maritime satellite communications together with the rapid growth of the VHF air-to-ground public telephony service in the USA has encouraged the development of aeronautical satellite communications. INMARSAT is carrying out a programme of tests, demonstrations and trials which will lead to a global service in 1990. The Aviation Satellite Corporation (AvSat), a new organisation based in the USA, has also announced that it intends to provide a global service (starting in 1989) but must inevitably face much greater difficulties than INMARSAT in achieving its aim since the latter already has satellites, a world-wide organisation and a vast amount of experience at its disposal.

1.8.3 Land

Many land vehicles, unlike long-distance aircraft and ocean-going ships, spend nearly all their time in areas where there is easy access to a terrestrial communications system and thus only a small proportion of vehicles are likely to use a communications-satellite system; nevertheless, this small proportion of a vast number of vehicles represents a significant market. North America is an example of a region where it is expected that land mobile satellite communications will prove profitable. There are a very large number of cellular systems in the USA and Canada but these systems are concentrated on cities and towns and, in the vast areas between these centres of population, there is virtually no service available to long-distance travellers. The USA and Canada are therefore planning a joint system intended to provide service to mobile stations on land (this system will also offer maritime and aeronautical communications).

Paging services offer a cheap way of conveying a limited amount of information and BTI has started a paging service via the INTELSAT Atlantic satellite which is intended to serve travellers in Europe, Africa and the Middle East.

1.9 Direct broadcasting

In 1945 Arthur Clarke suggested that television programmes would some day be broadcast via satellites. However, although the first uses of communications satellites included the relay of television programmes for rebroadcast by terrestrial stations or for distribution by cable-TV systems, it is only now (more than 40 years after Clarke suggested it) that satellites are about to come into widespread use for broadcasting direct to private homes.

The reason for this is, of course, that no matter whether satellite broadcasts are paid for by advertisers, sponsors or subscribers they must attract a mass audience if they are to justify their existence. This means that home receiving equipment must be cheap and easy to install whereas the terminals needed to receive transmissions from the earlier satellites were very expensive and far from suited to home installation.

There are now two schools of thought on the best way to introduce broadcasting via satellites. On the one hand there are those who intend to use high-power satellite transmissions at frequencies around 12 GHz allocated by ITU to the broadcast satellite service (BSS); i.e. frequencies intended specifically for broadcasting direct to homes. On the other hand there are those who intend to use lower-power transmissions from satellites of the type already in use for the relay of satellite programmes to cable-TV networks (and general telecommunications); these satellites use frequencies in the 11/12 GHz bands allocated to the fixed satellite service (FSS), i.e. frequencies not specifically intended for broadcasting.

The high-power (BSS) satellite transmissions generally have EIRPs of the order of 60 dBW. Because a direct-broadcast satellite (DBS) must serve a large number of homes scattered over a wide area it cannot employ a narrow-beam antenna with a high gain; the high EIRP must therefore be obtained by using output amplifiers which can deliver hundreds of watts to the antenna (whereas the output amplifiers of most other satellites are rated at tens of watts). Even a large satellite can only develop sufficient power to support three or four such high-power amplifiers and the cost per channel of BSS satellites is therefore considerable. However, with modern equipment it should be possible for many viewers to get pictures of acceptable quality using antennas with a diameter of about 0·3 m and this type of antenna will be cheap and easy to install. The exact size of antenna required depends on whether the receiver is in the centre of the satellite coverage area or towards its edge; thus, for example, it is claimed that satisfactory reception from the West German satellite TV–SAT will be possible in the main service area when using an antenna with a diameter of not more than 40 cm but that antenna sizes of 50 and 70 cm will be required in Paris and London, respectively.

Transmissions from the lower-power satellites will have EIRPs of the order of 50 dBW and these satellites can support many transmissions (e.g. the Astra satellite, which is being constructed for a consortium based in

Luxembourg, will provide 16 channels). Thus a low-power television-relay channel can be rented at about one-quarter to one-fifth of the cost of a high-power BSS channel (say about $3M p.a. compared with $15M p.a.). On the other hand, it will be necessary to use an antenna with a diameter of at least 0·8 m to get satisfactory picture quality when receiving transmissions from the lower-power satellites. This will increase the cost of the receiving terminal and antennas of this size may be significantly more difficut to install than antennas with a diameter of around 0·3 m, not least because of the greater wind loading. At the moment no one knows for sure what the public will be willing to pay or how many channels the market will support.

Direct broadcasting has already made one false start. In 1982, the FCC (Federal Communications Commission of the USA) issued permits to eight companies to provide DBS systems. One company, which decided that it would start service by using low-power transmissions in order to be first in the market place, began operations in late 1983 and ceased trading by the end of 1984. This failure may seem surprising occurring as it did at a time when the number of backyard stations was growing rapidly (see Section 1.6.3). However these backyard stations were owned by enthusiasts who were willing to pay several thousand dollars for terminals with antennas 2 to 4 m in diameter and who constituted only a minute fraction of the market needed to support a direct broadcasting system.

Western Europe will be the first region with a number of DBSs in operation. These will include:

(i) TV–SAT 1 (West Germany) which was launched in late 1987; this satellite is only partly operational because one of its two solar panels failed to open and, as a result, only two of its four transponders can be used. (It is intended to launch TV–SAT 2 in 1990)
(ii) The French satellites TDF–1 (1988) and TDF–2 (1990) (the TDF satellites are virtually the same design as the TV–SATs)
(iii) An experimental satellite (OLYMPUS), which is being constructed for the European Space Agency (ESA) and will be launched in early 1989; this will carry two DBS channels — one for a pilot service by Italy and the other for general experimental use by European countries
(iv) TELE–X, a satellite which is being procured by the Nordic countries to provide broadcast and other services (launch is planned for late-1988)
(v) A satellite which is being procured by British Satellite Broadcasting (BSB), a UK consortium which expects to be offering three channels in 1990.

All of these satellites are high-power (BSS) satellites. There will also be a number of low-power satellites (e.g. ASTRA 1 and ASTRA 2) broadcasting channels intended for direct reception and, of course, there are many channels intended for distribution via cable networks.

The USSR also plans to introduce an extensive DBS system (with up to five satellites) to serve Eastern Europe.

It should be mentioned that Japan, in the forefront of satellite communications as ever, launched an experimental BSS satellite in 1984; unfortunately the high-power amplifiers on this satellite failed within a month of launch, thus demonstrating once again the problems of introducing new technologies into the space environment. A second Japanese direct-broadcast satellite was successfully launched and put into operation in 1988.

The ITU held a World Administrative Conference on Broadcasting Satellites (WARC–BS) in 1977. The aims of this WARC were to decide on suitable characteristics for transmissions from satellites in the BSS service and to plan the allocation of the BSS spectrum among the nations of the world in such a way as to minimise the probability of interference between systems. Up to this time radio frequencies and positions in the geostationary orbit were allocated on a first come, first served basis. Some members of the ITU felt that this system was giving all the easily-exploited regions of the radio spectrum and the best orbital positions to the larger and richer countries, thus leaving the poorer nations at a disadvantage when they wanted to start service. ITU Regions 1 and 3 therefore made plans based on a radical change of policy. The broadcast frequency band of 11·7–12·5 GHz was divided into 80 channels and orbital positions spaced at intervals of 6° were selected. Each nation was allocated at least one unit comprising four or five TV channels (each with a bandwidth of 27 MHz) and a position in the geostationary orbit. Nations with very extensive territories were allocated more than one unit; the USSR, for example, received six units.

Brazil, Canada and the USA were not willing to commit themselves to these principles; planning for Region 2 (of which they form an important part) was therefore postponed to a Regional Administration Radio Conference (RARC) which was held in 1983. The demand for channels in Region 2 was much greater than that in Regions 1 and 3; in particular, the USA wanted 64 channels for each of a number of service areas and the total number of entries in the region (i.e. channels multiplied by areas) was nearly 2000. The problem was solved by making extensive use of computers to develop a plan which met the stated requirements of each country as closely as possible. Although not every requirement was met in full, the plan did give an average of about 16 channels per service area compared to about four channels per service area in an equivalent bandwidth in Regions 1 and 3. The more efficient use of the spectrum associated with the Region 2 plan results partly because the planners were able to take account of improvements in technology between 1977 and 1983 (e.g. reductions in the noise figure of cheap receivers) and partly because the plan meets a particular set of traffic requirements and does not attempt to distribute the resources of orbit and bandwidth in accordance with any principles of equity.

1.10 Methods of modulation and multiple access

1.10.1 Analogue modulation

1.10.1.1 FM

Virtually all the early experimental and operational communications-satellite systems used frequency modulation (FM) and a large proportion of the telephony traffic on the INTELSAT system is still transmitted on frequency-modulated carriers. However, most new earth-station equipment (with the exception of equipment for transmitting and receiving broadcast-standard television transmissions) uses digital techniques and in another five years digital transmissions will probably account for about 90% of all telephony and data traffic in the fixed satellite service (FSS). Nevertheless, the high cost of replacing earth-station equipment and retraining staff in new operational and maintenance procedures will ensure that many stations continue using their FM equipment for as long as is practicable; thus there will probably still be a significant residue of FM transmissions in ten or fifteen years time.

1.10.1.2 FDM–FM–FDMA

The transmission method most commonly used during the first twenty years of commercial satellite communications was FDM–FM–FDMA, i.e. the grouping of voice channels into frequency-division-multiplex (FDM) assemblies, frequency modulation of the multiplexes on to RF carriers and the use of carriers of different frequencies to give a number of stations simultaneous access to the satellite. This method is still in widespread use on routes carrying moderate to heavy traffic loads but little new FDM–FM equipment is being installed.

The increase in the equivalent isotropically-radiated power (EIRP) available from successive generations of satellites made it possible to reduce the frequency deviation of FM transmissions (and therefore the RF bandwidth required per channel) and the traffic capacity of transponders carrying a single FDM–FM carrier soon became very high. For example, a transponder of INTELSAT IV working with a spot-beam antenna and earth-station antennas with a diameter of about 30 m was able to carry an 1800-channel FDM multiplex in a bandwidth of 36 MHz (which corresponds to an RF bandwidth of less than 20 kHz per telephone channel). With the introduction of frequency reuse (by spatial discrimination and orthogonal polarisation) the maximum number of channels in some 36 MHz transponders was reduced to about 1000 because of the effects of interference between carriers using the same frequency.

1.10.1.3 Intermodulation

The non-linearity of transponder output amplifiers results in intermodulation (see Section 3.4). In order to limit this intermodulation to acceptable levels it is usually necessary to back off the amplifier (i.e. use it at less than its full output power) and this may cause a serious reduction in traffic capacity.

The power of the intermodulation products increases rapidly with the number of signals being amplified per transponder; this power can therefore be reduced by decreasing the total number of carriers through a satellite and increasing the number of transponders. The number of carriers in an FDM–FM–FDMA system is usually minimised by using multi-destination carriers (i.e. carriers which bear channels to more than one earth station). The number of transponders is limited primarily by the mass of the satellite but also by the minimum RF bandwidth required; for example, most INTELSAT satellites use transponders with a minimum bandwidth of 36 MHz for the transmission of FDM assemblies, TDM assemblies and broadcast-quality TV programmes, and transponders with a bandwidth of 72 MHz for the transmission of TDMA.

1.10.1.4 CFDM

Considerable increases in capacity can be achieved by the use of companded FDM (CFDM) assemblies, i.e. assemblies comprised of companded speech channels (see Section 4.5); typically, the use of companding halves both the satellite power and the bandwidth required to transmit a given number of channels.

1.10.1.5 SCPC

FDM multiplexes comprise groups of twelve channels. Thus earth stations operating in FDM–FM–FDMA systems must provide equipment for the transmission and reception of packages of 12 channels. This can be wasteful, particularly for small stations which operate to several routes (each carrying only two or three circuits).

The use of multi-destination carriers also means that a station may have to provide a large amount of demultiplexing equipment to separate out the channels intended for it from those intended for other stations, and this demultiplexing equipment may have to be reconfigured (and possibly supplemented) when it becomes necessary to make changes in the pattern of traffic flow through the satellite. The expense of maintaining and modifying the demultiplexing equipment may, once again, be particularly burdensome for the smaller stations.

These stations may therefore prefer single-channel-per-carrier (SCPC) working. In SCPC systems, as the name implies, each channel is modulated on to a separate carrier. There are SCPC systems using both analogue and digital modulation but all SCPC systems are also, necessarily, FDMA systems.

A typical FM–SCPC system is the INTELSAT VISTA system which provides 1200 channels spaced at intervals of 30 kHz across a transponder with a usable bandwidth of 36 MHz. Each carrier is voice activated, i.e. it is switched on only when speech is present. Because the two parties to a telephone conversation do not usually speak simultaneously, and because there are silent intervals between sentences, words and syllables, voice activation reduces the average number of voice carriers by 60% or more. The satellite power available to each carrier is therefore more than doubled and, at the same time, the reduction in the number of carriers abates the intermodulation noise power. Virtually all FM–SCPC systems use companded speech channels.

FM–SCPC earth-station equipment comprises a channel unit for each pair of (outgoing and incoming) channels together with IF and RF equipment which is common to a group of channel units. Earth-station operators may thus minimise their initial capital investment by commencing operation with one set of common equipment and just enough channel units to carry the expected traffic.

1.10.1.6 SSB

Some US terrestrial systems and domestic communications-satellite systems use single-sideband (SSB) modulation. In these systems an RF carrier is amplitude modulated with an FDM assembly of telephone channels and the residual carrier component is suppressed, together with one of the sidebands. The resultant transmission comprises as many RF signals as there are telephone channels; SSB thus produces an assembly of SCPC transmissions. The spacing of the RF carriers is the same as that of the FDM channels (i.e. 4 kHz); 9000 channels could thus, in theory, be transmitted via a 36 MHz satellite transponder if it were practicable to equip the satellite with transmitters with sufficient high output power.

However, because SSB transmissions are amplitude modulated they are very susceptible to the distortion caused by amplifiers with a non-linear relation between output power and input power. In SSB systems it is therefore usually necessary to back off the output of TWT or klystron amplifiers by about 10 dB. It may also be necessary to avoid the use of a frequency band at the centre of the transponder (where the number of intermodulation products is a maximum).

SSB requires relatively high power per channel but the total power available from a backed-off satellite transmitter is low; it is therefore necessary to use companded telephone channels (see Section 4.5) with SSB systems in order to avoid either a severe restriction of the number of channels available or an unacceptable decrease in the quality of the transmissions.

Other practical problems with SSB are:

(i) It is very susceptible to interference

(ii) Coherent demodulation is necessary; this means that extremely stable oscillators and an effective automatic-frequency-control (AFC) system are required.

Despite continuing progress in developing more linear amplifiers, SSB has not so far found widespread use in satellite systems. It is, however, receiving renewed attention as a modulation method for use with land mobile satellite systems, which have rather different requirements to most other communications-satellite systems. The power of a signal arriving at a land mobile station may vary over a very wide range mainly because (as the vehicle moves) trees, buildings and other obstacles pass through the transmission path; under these circumstances it is better to use a modulation method like SSB in which the baseband signal-to-noise ratio is proportional to the RF carrier-to-noise ratio than to use a method like FM where the baseband signal degrades rapidly when the RF signal-to-noise ratio falls below a certain threshold value. A disadvantage of SSB for mobile systems is that the mobile station must be equipped with a linear output amplifier with relatively high output power (say about 20 W mean power for operation to a global beam transponder).

1.10.2 Digital transmission

One of the main advantages of digital transmission methods for terrestrial systems is that, by interposing regenerators at sufficiently frequent intervals along the transmission path, the quality of a received signal can be made virtually independent of the distance it has travelled. Regeneration is not of equal advantage to satellite systems because the only point in a satellite link where a regenerator can be placed is at the satellite itself. Commercial communications satellites have not, up to now, included regeneration, but regeneration at the satellite could give a significant improvement in performance (see Section 2.4.3.3) and may be in use in commercial systems by the latter half of the 1990s.

Other advantages of digital methods are first that they are well suited to transmission of the ever-increasing amounts of data being generated by the use of information technology and secondly that it is now practicable (because of the rapid advances in the development and manufacture of large-scale integrated circuits) to process digital signals in complex ways which greatly increase the efficiency with which they can be transmitted and used. These advantages are already increasing the efficiency and flexibility of satellite communications and, looking to the future, it seems likely that virtually all satellite systems will adopt digital transmission methods in order to be compatible with digital terminal equipment and integrated-service digital-network (ISDN) operation.

Nearly all digital transmissions via communications satellites are modulated onto carriers using phase-shift keying (PSK) or methods which are closely related to PSK (see Section 5.6).

FDMA can be used with digital transmissions just as with analogue transmissions. However, the ease with which digital signals can be stored and manipulated makes it practicable for a number of earth stations to use the same frequency by interpolating their transmissions in time; this is time-division multiple access (TDMA).

1.10.2.1 SPADE and DA

The first commercial digital communications-satellite system was SPADE, which was brought into service by INTELSAT in 1971. SPADE is an acronym for *S*ingle-channel-per-carrier, *P*ulse-code-modulation, multiple-*A*ccess, *D*emand-assignment *E*quipment.

In systems where a fixed number of satellite channels are pre-assigned to each earth station there are times when channels at some stations remain unused while other stations have to refuse traffic because they have insufficient channels. In a demand-assignment system each earth station requests the allocation of just sufficient satellite capacity to satisfy its immediate needs; as soon as any of this capacity is no longer required the station returns it to a pool where it becomes available to any other station which needs it. Demand assignment (DA) thus makes more efficient use of satellite capacity than pre-assignment (PA) particularly if the system serves a number of routes with different busy hours. DA can be particularly advantageous for stations with very small traffic loads because it is possible to gain access to any station in the system with (in the limit) a single channel and because space-segment charges can be related to the time for which a satellite channel is used.

SPADE terminals are computer controlled and linked to other terminals in the system by a signalling channel. When a demand for a call arrives at an earth station the SPADE terminal:

(i) Selects a pair of carrier frequencies (one for each direction of transmission) from a list of available frequencies
(ii) Advises all other terminals that the pair of frequencies it has selected is in use
(iii) Sets up a circuit, in co-operation with the SPADE terminal at the destination earth station, via carriers on the chosen frequencies.

At the end of the call the originating station notifies all other stations that the pair of carrier frequencies it was using is now free.

SPADE terminals comprise common equipment and a number of channel units. The common equipment includes a signalling and switching processor, an IF subsystem and timing circuits. Each channel unit includes a voice detector (the output of which is used to turn the carrier on and off), a PCM codec, a QPSK modem and a channel-frequency synthesiser. Each time a carrier is turned on and a speech burst is transmitted the receiving station has to synchronise the demodulator with the carrier frequency, and recover the bit timing and framing information before it can decode the speech burst. Each speech burst must therefore be preceded by an additional sequence of bits; this is supplied by a sub-unit of the channel unit known as the transmit synchroniser.

As SPADE is essentially intended to provide for light traffic streams to many destinations it requires a global transponder. The transponder bandwidth of 36 MHz supports 400 pairs of channels.

Although over 30 countries in the INTELSAT system bought SPADE terminals, the service did not grow as fast or to the extent that had been hoped. The main reason for this was that the system was advanced for its time and terminals were quite expensive to buy and to maintain. The cost of the terminals could have been reduced by vesting control of channel assignment in a central station but, in the early 1970s, many nations would not use a system in which control over any part of their communications traffic was exercised by a station in another country; however, the concept of centralised control stations now seems to be more widely accepted. In addition, advances in manufacturing techniques have made this type of equipment cheaper and more reliable.

1.10.2.2 PCM–QPSK–SCPC

INTELSAT PCM–QPSK–SCPC equipment was derived from SPADE equipment by omitting the control processors. The initial expenditure required to start service with SCPC equipment is minimal and this, plus more simple operation and better reliability than was associated with the early SPADE terminals, offsets the additional space-sector costs resulting from the loss of demand assignment. As a result, the INTELSAT PCM–SCPC service expanded very quickly and the equipment was manufactured in large quantities (which brought reductions in capital cost and fuelled further expansion of the system). Many other SCPC systems, including delta-modulation SCPC and FM–SCPC systems, soon followed.

The INTELSAT SCPC system carries both voice and data transmissions. Voice channel units are very similar to SPADE channel units. Data channel units accept incoming bit rates of either 48 kbit/s or 56 kbit/s; in the former case the unit applies rate 3/4 FEC (forward error correction) and in the latter case rate 7/8 FEC; the output rate is thus 64 kbit/s in both cases.

1.10.2.3 TDM–PSK–FDMA

The increasing use of time-division-multiplex (TDM) assemblies of digital signals on terrestrial links has led to the introduction of TDM–PSK–FDMA operation in satellite systems. The TDM assemblies are modulated directly (by phase-shift keying) onto RF carriers at the earth stations and are assembled at the satellite in FDMA. Like any other FDMA system, TDM–PSK–FDMA suffers from intermodulation between carriers passing through a common high-power amplifier in the satellite (or earth station). Its advantages are that the earth-station equipment is relatively simple and cheap and that TDM–PSK–FDMA equipment is usually easy to incorporate into an earth station which has previously been working to an FDM–FM–FDMA system.

1.10.2.4 IDR

The INTELSAT TDM–PSK–FDMA system is known as the intermediate data rate (IDR) digital-carrier system. The IDR system uses QPSK modulation and rate 3/4 FEC. A digital signal at any information rate from 64 kbit/s to 44 Mbit/s can be transmitted but INTELSAT defines a set of recommended rates which include the channel rate of 64 kbit/s, the primary (24-channel and 30-channel) multiplex rates of 1·544 Mbit/s and 2·048 Mbit/s, higher multiplex rates such as 34·368 Mbit/s and 44·736 Mbit/s and a number of intermediate rates. The allocated satellite bandwidth is 67·5 kHz per 64 kbit/s information channel corresponding to a capacity of about 500 channels in a 36 MHz transponder.

1.10.2.5 TDMA

The problem of intermodulation between carriers can be avoided by using time-division multiple access (TDMA). In a TDMA system a number of earth stations transmit in succession on the same carrier frequency. Each earth station must therefore store the information it receives from the terrestrial network until it is time for it to transmit; the stored information is then sent to the satellite as a burst of high-speed data (which must be precisely timed so as to avoid overlap with the preceding and succeeding bursts). The receive side of the TDMA terminal at each station must demodulate the bursts of transmission from other stations, extract the information addressed to the station and feed it to the appropriate terrestrial channels at the correct speed.

Digital methods make it easier to store large amounts of information and to process it in complex ways; TDMA is therefore almost always associated with digital signals. TDMA terminals usually comprise a common TDMA terminal equipment (CTTE) and a number of terrestrial interface modules (TIMs).

In FDM–FM–FDMA systems significant changes in the pattern of traffic usually require the production of a new frequency plan; this comprises a statement, for each transmission, of the carrier frequency and the number of channels, and also (either explicitly or implicitly) the FM deviation, the allocated bandwidth and the EIRPs required on the up-path and down-path. Implementation of a new plan may thus require significant changes in the configuration of the equipment at many earth stations.

In a TDMA system the burst time plan (BTP) is the equivalent of the frequency plan in an FDMA system, and that part of the BTP relevant to each station is held by the station as data which is used by the program controlling the TDMA equipment. A change of traffic pattern and BTP may thus be implemented simply by feeding new data into the earth-station control processors and the only hardware change required is the addition of TIMs when the existing TIMs become overloaded.

TDMA is well suited to systems requiring a high degree of multiple access and can easily respond to changes in the pattern of traffic. The equipment is, however, more expensive to procure and complex to maintain than TDM–PSK–FDMA equipment; TDM–PSK–FDMA is therefore often preferred where the flexibility of TDMA is not required.

1.10.2.6 DSI

TDMA terminals usually include facilities for digital speech interpolation (DSI).

Almost all long-distance telephone calls are made over a ‘four-wire’ circuit, i.e. a circuit comprising a separate channel for each direction of transmission. We have already noted (in connection with SCPC transmission in Section 1.10.1.5) that voice channels are occupied, on average, for less than 40% of the time. It is therefore possible for n channels to serve more than $5n/2$ customers (provided n is reasonably large). It is of course necessary to allocate a channel to a customer only when he starts to speak and to recover it as soon as he stops (if it is needed by someone else). The satellite channel associated with a given talker may thus be continually changing and it is necessary to give immediate notice of these changes to the distant terminals.

Speech interpolation has been used on submarine-cable systems for many years; the earliest of these systems, used on transatlantic cables, was time-assignment speech interpolation (TASI) which operated on analogue signals. DSI, as its name suggests, interpolates speech signals after they have been digitally encoded.

The capacity of the INTELSAT and EUTELSAT TDMA systems without DSI is about 1700 telephone channels in 72 MHz. The capacity with DSI depends on the proportion of channels in the system to which DSI cannot be applied (i.e. the number of non-speech channels) but is about 3000 channels with the present composition of INTELSAT traffic.

DSI can be used with other types of digital transmission such as TDM–QPSK–FDMA.

1.10.2.7 LRE

DSI is a means of multiplying the number of voice channels which a digital system can carry and DSI equipment is therefore also known as digital-channel multiplication equipment (DCME).

Another way of improving the capacity of digital systems is to use more efficient coding methods so that channels can be transmitted using lower bit rates; low-rate encoding (LRE) is thus another form of digital-channel multiplication.

LRE takes advantage of the fact that voice, video and music signals are not random and that it is therefore possible, by studying the past form of the signals, to make an intelligent guess at their future behaviour. This is obvious in the case of a television signal because large sections of the picture often remain the same from frame to frame; it is therefore possible to make big savings in the amount of information which has to be transmitted by comparing each picture with the previous one and updating only those sections which have changed. This is equivalent to guessing that the picture will remain unchanged and sending a signal only when this guess proves to be wrong. The redundancy in speech and music signals is not as obvious as that in video signals but it is there nevertheless. Most data signals are not redundant and thus their bit rate cannot be reduced.

One of the most widely-used methods of LRE (see Section 5.3.2) is adaptive differential PCM (ADPCM). CCITT is close to agreeing on a standard for an ADPCM codec which will reduce the bit rate required for a voice channel from 64 kbit/s to 32 kbit/s. The use of 32 kbit/s ADPCM in conjunction with DSI could, in theory, give a DCM factor of 5 (relative to unprocessed 64 kbit/s PCM) but in practice the multiplication factor is smaller because of the presence of signals, such as data channels, to which DCM methods cannot be applied. Further improvements in LRE are on the way (e.g. a bit rate of 16 kbit/s has already been adopted for the pan-European cellular system).

1.10.2.8 FEC

LRE reduces the number of bits used to represent a signal. On the other hand, it is possible to add bits to a signal in such a way that the decoder can recognise, by the form of the received bit pattern, if errors have occurred during transmission. If errors are detected the receiver can ask for the signal to be retransmitted; this is known as ARQ operation. Alternatively it is possible for the decoder to compare what it knows to be an incorrect pattern with all of the allowable patterns and to decide which of the allowable patterns was most probably transmitted; this is forward error correction or forward error control (FEC). FEC can give a large reduction in the bit error ratio (BER) of a digital signal at the expense of either an increase in bandwidth or a reduction in the information bit rate; this enables power to be exchanged for bandwidth and provides a flexible way of matching a signal to the characteristics of a satellite system. Systems using FEC include the INTELSAT PSK–SCPC, TDMA and TDM–PSK–FDMA systems, the INMARSAT Standard-C and aeronautical systems, and many others.

1.11 Trends in satellite communications

1.11.1 Telephony

INTELSAT's first income came from supplying transatlantic telephone channels at a time when cables could not keep pace with the rapid growth of traffic. Since then satellite systems and cables have shared the traffic on many heavily-loaded point-to-point routes, and satellites have provided the first effective service on thousands of routes not served by cable.

Optical fibre has now become such a cheap means of providing large numbers of channels that it is being used in preference to satellites to add capacity on major routes. It is worth noting that circuits via a communications-satellite system specifically designed to serve busy long-distance routes would very probably prove no more expensive than those provided via optical fibre, but customers prefer the shorter transmission delay associated with the latter. Thus although the delay inherent in communication via the geostationary orbit is not seen as a major disadvantage when there is a shortage of circuits (see Sections 1.5.1 and 1.5.2) perceptions change on routes where cable circuits are in plentiful supply.

The situation on long-distance routes carrying moderate or light traffic is very different. Many of these routes can only be served economically by means of multiple-access satellite systems and some of them cannot be served effectively by any other means because of geographical barriers such as mountains or deserts. This type of route will therefore continue to provide a major role for satellites.

Jefferis* estimates that 'the proportion of international traffic carried by satellite other than between neighbouring countries will fall only modestly from its present level, of probably around 60% to perhaps 50% in 10 years time and 40% in 20 years'; Jefferis also notes that even with traffic growth rates as low as 10% this means that the number of international satellite voice circuits required will double in ten years.

1.11.2 Television

Prior to the advent of satellites it was impracticable to relay television programmes across the oceans because of the limited bandwidth of submarine

* JEFFERIS, A.K.: 'Commercial satellite communications — The next twenty years', *Electron. & Power*, Sept. 1986, pp. 657–660.

coaxial cables. Present-day optical-fibre submarine cables have more than enough bandwidth but are unlikely to displace satellites for this purpose because the latter can distribute the programmes directly to many widely-separated stations. This point-to-multipoint capability has also led to the widespread use of satellites in domestic and regional systems for the distribution of television programmes to remote communities, terrestrial broadcasting stations and the head ends of cable-TV networks. The demand for this type of service continues to grow but may eventually be offset to some extent by the use of satellites to distribute television programmes direct to homes.

The era of direct broadcast by satellite is only just beginning but it seems likely that the service will grow quite quickly. For example, the first direct-broadcast satellite for Western Europe was launched in November 1987 but it is expected that more than twenty transponders will be serving this region by end-1988. It is, however, too soon to guess at the optimum balance between satellite EIRP and the size of domestic antennas. There are various inexpensive ways in which it is possible to add data transmissions to the services carried via broadcasting satellites and this may make it practicable to provide cheap information services and electronic mail for both domestic and business purposes.

1.11.3 Mobile services

Radio is the only practicable means of communication with mobile stations and satellites are the only satisfactory means of communication with mobile stations remote from terrestrial stations.

The INMARSAT system is already making a profit and, with the introduction of the standard-C SES, the number of ships working to satellites should increase rapidly over the next few years.

The requirement for better communications with cockpit and cabin crew on long-distance flights, together with the known demand for an aeronautical public telephony service, seems to offer a good chance that aeronautical satellite communications will become profitable, although it may take longer to reach the break-even point than did maritime satellite communications. Much will depend on the future level of air travel.

A large section of the market for land mobile communications is already served by cellular networks, which are spreading rapidly and becoming more numerous. It therefore seems probable that (at least in the short term) satellites will provide land mobile communications mainly in regions where there are a significant number of vehicles which are temporarily or permanently out of range of terrestrial stations, and where specialised services are required.

Finally, the suitability of satellites to the provision of mobile communications can be exploited in two other ways. First in the use of temporary transportable stations to provide emergency communications or on-the-spot news coverage. Secondly in the rapid establishment of communications links for specialised services; these services are often difficult to supply via those sections of telecommunications networks which still rely on analogue equipment and copper wire.

1.11.4 Technical development

Over the past 25 years the most obvious trend in satellite communications has been the growth in size of satellites and their antenna arrays together with the proliferation of smaller and smaller earth-station antennas. In the early days, when a satellite system included only a few earth stations, the costs of the space sector were dominant and it made sense to spend an extra million dollars, say, on each station if this gave a large increase in the capacity of the system. On the other hand a reduction of, say, $1000 in the cost of TVRO stations for domestic reception may lead to a very large increase in the number of stations sold and this may correspond to the difference between success and failure of broadcast systems; it may thus make sense to increase the cost per channel of the satellite by millions of dollars in order to reduce the cost of the earth stations. The EIRP per satellite transmission can be increased in three main ways:

(i) By reducing the number of transmissions per satellite
(ii) By increasing the total power available from the transmitters
(iii) By increasing the gain of the antennas.

Increasing the transmitter power requires larger satellites and/or more efficient transmitters. NASA is considering the development, in the long term, of platforms built in space and capable of carrying transmitters radiating tens of kilowatts of RF power but it is more realistic to count on modest developments (at least in this century) such as an increase in satellite transmitter power of not more than about 10 dB and the use of solid-state transmitters which will be lighter, more efficient and more linear than the present TWT amplifiers.

The extent to which satellite antennas will undergo further development is debatable; on one hand Jefferis (*op. cit.*) considers that 'progress beyond INTELSAT VI to even more numerous coverage areas is unlikely because of the connectivity problems introduced' while, on the other hand, NASA has plans for a first-generation mobile communications satellite using two to eight beams (to be launched before 1990) leading to a second-generation satellite with up to 24 beams (to be launched around the mid-1990s) and a third-generation geostationary platform generating 100 beams (to be launched before the end of the century).

The use of satellite antennas with narrow beams raises particular problems in the case of communication with mobile stations firstly because it is less easy to define geographical areas likely to generate heavy traffic and secondly because of the problems arising when the mobile station moves out of a beam. INMARSAT has avoided these problems so far by specifying first-generation and second-generation satellites which rely exclusively on global antennas, but it is clear that the third generation of maritime satellites will need to make some use of antennas with higher gains.

Communications satellites serving mobile stations may well use on-board processing to assist in the interconnection of multiple beams, the tracking of mobile stations from beam to beam and the setting up of calls; the latter application could result in a considerable increase in efficiency by reducing the number of (or even eliminating) the messages between earth stations which are required when setting up a demand-assigned circuit via a transparent satellite.

Another subject on which there is considerable disagreement is the extent to which the higher-frequency bands will be used for satellite communications. Japan launched the first satellite making operational use of the 30/20 GHz bands at the end of 1977 but, in the ensuing decade, no other country has followed suit. This is because the high propagation losses occurring during heavy rain make it difficult to achieve high availability at reasonable cost. Nevertheless, NASA believes that there are so many satellites operating at 6/4 GHz and 14/12/11 GHz in the part of the geostationary orbit used by North and South America that it will be essential to make extensive use of the 30/20 GHz bands for satellite communication in the early 1990s. A considerable amount of thought is now being given to economic ways of improving availability. For example, as a signal fades it would be possible to keep the error rate constant by introducing more powerful FEC; in some systems this would mean that the rate of transfer of information would vary with the depth of fade but in TDMA systems the rate could be kept constant by allocating additional frame time to the fading link. It is planned to use the European OLYMPUS satellite (to be launched in 1989) to compare a number of different fade countermeasures.

Low-bit-rate encoding can be expected to make considerable strides in the next few years and this will increase the efficiency of all types of telephony system. However, satellite systems and mobile systems (and particularly mobile systems using satellites) will feel the benefits more than most because they tend to be the systems with the lowest carrier-to-noise ratios. 64 kbit/s encoding of voice is already giving way to 32 kbit/s, and 16 kbit/s has been adopted for the Pan-European cellular system and will undoubtedly be in widespread use before many years have passed. Lower rates such as 9·6 kbit/s and 4·8 kbit/s are proposed for specialised use and the planned North American mobile communications-satellite system may even use 2·4 kbit/s for emergency voice channels.

Chapter 2

Satellites and launchers

2.1 The geostationary orbit

A satellite is a body which moves in an orbit around another body (of greater mass) under the influence of the gravitational force between them. The force (in newtons) required to keep a satellite in a circular orbit is given by:

$$m\omega^2 r$$

where m = mass of the satellite, kg
ω = angular velocity of the satellite, rads/s
r = radius of the orbit, m.

The gravitational force acting on a satellite of mass m at distance r from the centre of the earth is:

$$mgR^2_e/r^2$$

where g = acceleration due to gravity at the surface of the earth (9·807 ms^{-2})
R_e = radius of the earth (6378 km at the equator)

Thus, for a satellite in a stable circular orbit round the earth:

$$mgR^2_e/r^2 = m\omega^2 r$$

or

$$r^3 = gR^2_e/\omega^2 \quad (2.1)$$

The Period of the orbit T, i.e. the time taken for one complete revolution of 2πrads, is

$$T = 2\pi/\omega \quad (2.2)$$

From eqns. 2.1 and 2.2:

$$r^3 = gR^2_e T^2/4\pi^2 \quad (2.3)$$

Now, for a synchronous orbit, T is equal to the period of a complete rotation of the earth relative to the fixed stars (the sidereal day), which is 23 hours 56 minutes and 4 seconds.

Substituting the numerical values of g, R, T and π in eqn. 2.3 we get

$$r = 42\ 176\,\text{km}$$

This is the distance of a synchronous satellite from the centre of the earth; its height above the surface of the earth is therefore 35 798 km or approximately 36 000 km (the latter being the figure usually quoted). The synchronous orbit in the equatorial plane is called the geostationary orbit. A satellite in the geostationary orbit does not appear to be moving when seen from the earth but its velocity in space is over 3000 m/s. (*Note:* Slightly different values for the radius of the geostationary orbit may be found in other books and papers. The differences are the result of 'rounding-up' errors and of minor inconsistencies in the values taken for g and R_e.)

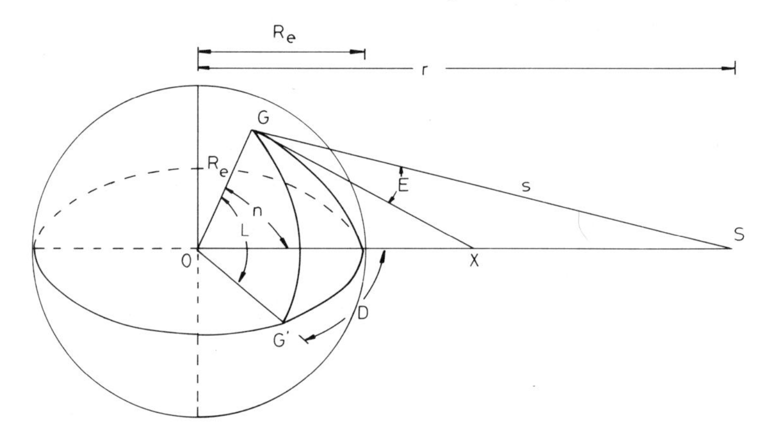

Fig. 2.1 *Distance between a geostationary satellite and an earth station*

S = position of satellite
G = position of earth station
GG′ = line of longitude through G
GX = local horizontal through G
SGX = elevation angle of satellite from G
GOG′ = latitude of G
SOG′ = difference in longitude of satellite and G

The distance s between a satellite in an equatorial orbit and an earth station is given by (see Fig. 2.1):

$$s^2 = r^2 + R^2_e - 2rR_e\cos n \qquad (2.4a)$$

where $\cos n = \cos L \cos D$

r = radius of the orbit (approximately 42 200 km for a geostationary satellite)

L = latitude of the station

D = difference in longitude between the station and the satellite

The distance is also given by (see Fig. 2.1 again):

$$s = R_e([(1/k)^2 - \cos^2 E]^{1/2} - \sin E) \qquad (2.4b)$$

where k = ratio R_e/r (= 0·1512 for a geostationary satellite)
E = elevation of the satellite from the earth station

The distance from an earth station to a geostationary satellite seen at an elevation angle of 5° is 41 130 km (compared with approximately 36 000 km from an earth station on the equator to a satellite overhead).

2.2 Visibility of satellites from the earth

One of the most important pieces of information about a communications satellite is whether it can be seen from a particular location on the earth. Fig. 2.2 gives the azimuth and elevation E of a satellite from an earth station in terms of the latitude of the station L and the difference in longitude D between the station and the satellite. Fig. 2.2*a* gives the azimuth as an acute angle A to the local meridian and Fig. 2.2*b* shows the relation between A and the azimuth of the satellite expressed, in the usual way, as degrees east of north. Thus, for example, if D = 50° and L = 40° then the elevation of the satellite is 21° and A = 62°. Now, if the satellite is east of the station and the station is south of the equator the azimuth of the satellite from the station is equal to A, i.e. 62° E of N; on the other hand if the station is north of the equator then the azimuth of the satellite is (180 – A) = 118° E of N. Similarly if the satellite is west of the station, then the azimuth of the satellite is (180 + A) = 242° E of N for a station north of the equator and (360 – A) = 298° E of N for a station south of the equator. Fig. 2.2*a* is based on the expressions:

$$\tan A = \tan D/\sin L \qquad (2.5a)$$

$$\tan E = (\cos n - k)/\sin n \qquad (2.5b)$$

where n and k have the same meanings as in Section 2.1

The allowable longitudinal separation between earth station and satellite decreases rapidly with increasing latitude at the higher latitudes. At an elevation angle of 5°, for example, the maximum allowable longitudinal separation between an earth station and a satellite is about 55° for a station at a latitude of 65°, about 20° for a station at a latitude of 75° and zero for a station at a latitude of 76°. Geostationary satellites do not therefore give good coverage at high latitudes. Fortunately only a very small proportion of the world's population lives at latitudes greater than about 65°. The poor coverage at high latitudes can, however, be a disadvantage when geostationary satellites are used for maritime or aeronautical communication systems.

By following a line of constant elevation on Fig. 2.2*a* it is possible to plot the maximum possible coverage area of a satellite at that elevation angle (the

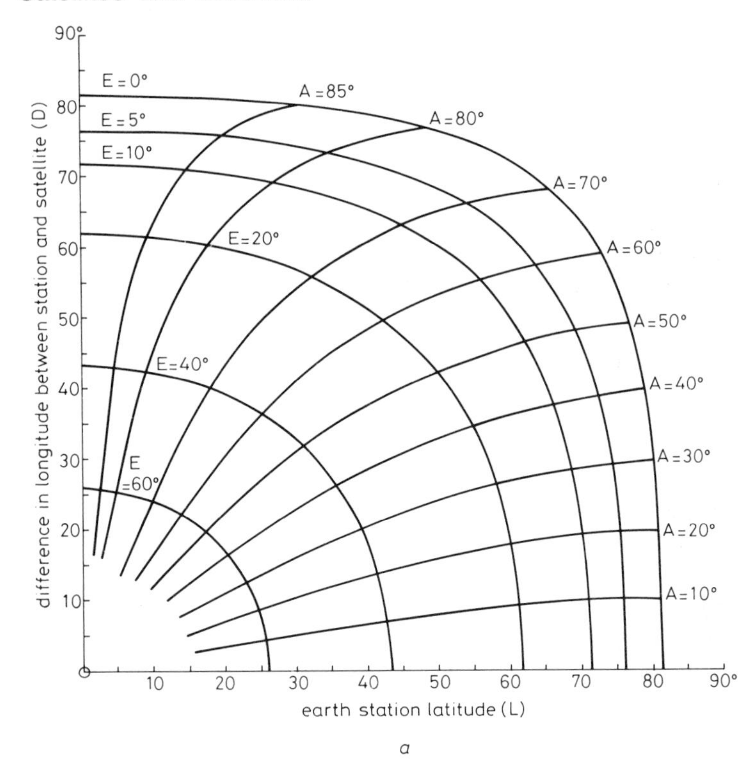

a Elevation and angle *A* as functions of *D* and *L*
E = elevation angle
For relation of *A* and azimuth, see Fig. 2.2*b* opposite

Fig. 2.2 *Elevation and azimuth of a geostationary satellite as seen from an earth station*

usable coverage area may be less than the maximum because the satellite antennas may only illuminate part of the area). By following a line of constant latitude on the same Figure it is possible to derive the elevation and azimuth from an earth station at that latitude to points in the visible portion of the goestationary orbit.

It is not possible to put a satellite into a perfect geostationary orbit. Any practical orbit:

(i) is slightly inclined to the equatorial plane
(ii) is not exactly circular
(iii) does not have exactly the same period as that of the earth's rotation.

Moreover, disturbing forces (such as the attraction of the sun and moon) are constantly trying to change the orbit and the effects of these forces have to be

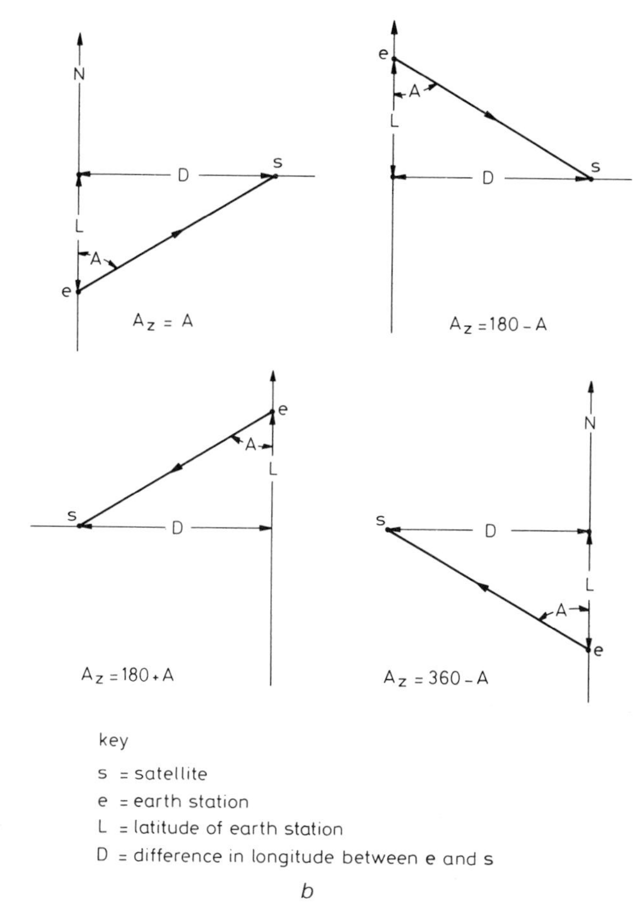

b Relation of angle A and azimuth

counteracted from time to time by operating thrusters on the satellite (see Section 2.4.2.5). Thus practical geostationary satellites do not appear completely stationary when seen from the surface of the earth. The combination of eccentricity of the orbit with inclination of the orbital plane causes the daily track of the satellite relative to the earth to be a Lissajous figure (typically an ellipse), and the difference between the periods of the satellite and the earth's rotation causes the Lissajous figure to drift slowly in longitude.

The orbital errors of modern satellites are usually small; for example INTELSAT V satellites are generally kept to within ±0·1° of the equator and their nominal longitudes, and it is intended that INTELSAT VI satellites shall be kept to within ±0·02° of the equator and ±0·06° of their nominal longitudes.

2.3 Putting a satellite into orbit

2.3.1 General

Launching and controlling satellites is a very expensive business. The cost of the launchers and support services required for the safe and effective use of satellites can exceed the cost of the satellites themselves. For example, in mid-1987 INMARSAT quoted typical costs as:

	\$M
Satellites (each)	50
Launch (per satellite)	45—55
Services (per annum)	70

Satellites are usually designed to be compatible with more than one model of launch vehicle in order that operators can take advantage of competition for the provision of launch services and so that there is an alternative if one of the chosen methods of launch proves unreliable. The chosen launch vehicles will place constraints on the size, shape and mass of the satellite and will influence some of the other characteristics such as ability to withstand vibration and acceleration.

2.3.2 Basic principle of launchers

The basic principle of any launcher is that of the rocket; i.e. the vehicle is propelled by reaction to the momentum of hot gas ejected through exhaust nozzles. Thus, suppose:

(i) A mass of propellant dM is burned in the rocket motor of a launcher and the resulting hot gas is ejected with velocity v

(ii) the mass of the launcher is M and the change in its velocity is dV

The change in momentum of the launcher must be equal to the momentum of the gas which is ejected, i.e.

$$v\ (dM) = M\ (dv) \tag{2.6}$$

Note that, for the moment, we assume that the mass of propellant burned dM is so small that the mass M of the launch vehicle remains virtually constant. Note also that the change in velocity of the launcher is in the opposite direction to the velocity of the ejected gas and that the launcher is steered by swivelling the nozzles through which the gas is ejected.

Most engines use the oxygen in the atmosphere to burn their fuel but the rocket motors used for launchers must work in space where there is no atmospheric oxygen. The propellant for a launcher must therefore comprise both a fuel and an oxidising agent. Propellants may be either solid or liquid. Solid propellants are a more or less homogenous mixture of fuel and oxidiser whereas the fuel and oxidiser comprising a liquid propellant are carried in separate tanks on the launch vehicle.

Rockets using liquid propellants tend to be more efficient, reliable and controllable; however, liquid propellants may be difficult to handle and more dangerous than solid propellants, especially if they are hypergolic (that is they ignite spontaneously on contact with each other). Large launch vehicles nearly always use liquid propellants for the main engines. For example, Ariane (the European launcher) uses UDMH (unsymmetrical dimethyl hydrazine) as the fuel and nitrogen tetroxide as the oxidiser in its first two stages, and liquid hydrogen and liquid oxygen for its third stage. (Liquid hydrogen and liquid oxygen are the most effective combination of fuel and oxidiser but they are not used in the first two stages because of engineering problems.) Although liquid propellants are used for the main engines of launch vehicles, rockets using solid propellants are often employed as strap-on units (boosters) to give extra thrust.

Expression 2.6 would allow the change in velocity of a launch vehicle to be calculated directly from the amount of propellant burned if the mass of the launch vehicle were virtually constant. However, the propellant may constitute around 90% of the total mass of a launcher at lift-off and, in a large launcher, it may be burned at a rate of several tonnes per second. It is therefore necessary to integrate expression 2.6 in order to derive the total change in velocity; the result, sometimes called 'the rocket equation', is:

$$V = v\log_e(M/m) \qquad (2.7)$$

where V = velocity increment
v = exit velocity of the gases
M = initial mass (fuel tanks full)
m = final mass (fuel tanks empty)

(Note that the rocket equation ignores the effects of gravity and the drag of the atmosphere).

2.3.2.1 *Multi-stage vehicles*

Suppose that the mechanical structure of a launcher (including the engines and all the control equipment) accounts for one-eighth of its mass (the other seven-eighths of the mass being fuel) and that the mass of the payload is 1% of the initial mass of the launcher. The mass ratio M/m is $101/[(100/8)+1] = 7{\cdot}48$ and V (when all fuel is burnt) is therefore $(\log_e 7{\cdot}48)v = 2{\cdot}01v$. Not much is gained by doubling the mass of the launch vehicle because the mass ratio is then $201/[(200/8)+1)] = 7{\cdot}73$, which gives $V = 2{\cdot}05v$, an increase in velocity increment of only 2%. Fortunately it is possible to get very worthwhile increases in velocity increment by building the launch vehicle in stages which are fired sequentially, the shell of each stage being jettisoned before the next stage is fired. Table 2.1 gives a comparison of the velocity increment of single-stage and two-stage launchers having the same total mass, the same amount of fuel and the same payload.

Table 2.1 *Comparison of single-stage and two-stage launch vehicles*

Launch vehicle	Mass (tonnes)			M/m	$v\log_e(M/m)$
	Fuel	Structure	Payload		
A (Single stage)	175	25	2	202/27	2.01v
B Stage 1	154	22	2	202/48	1.44v
Stage 2	21	3	2	26/5	1.65v
Total	175	25	2	—	3.09v

Vehicle A has a mass ratio of 202/27 and a final velocity of $2{\cdot}01v$. The total mass of vehicle B at the end of stage 1 has been reduced from the initial mass of 202 tonnes to 48 tonnes by burning 154 tonnes of fuel; i.e. the mass ratio for the first phase is 202/48 and the corresponding velocity increment is $1{\cdot}44v$. The structure of stage 1 is now jettisoned, thus reducing the initial mass for the second phase of operations from 48 tonnes to 26 tonnes. The mass ratio for the second stage is 26/5 with a corresponding further velocity increment of $1{\cdot}65v$. Thus launcher B reaches a velocity of $3{\cdot}09v$, more than 50% greater than that achieved by launcher A. Further advantage can be derived by splitting the launch into three or more stages and, as will be seen later, launch into geostationary orbit usually involves four stages.

2.3.2.2 *Exhaust velocity and specific impulse*

Up to now we have expressed velocity increment only as mass ratio multiplied by the exhaust velocity of the hot gases without considering what exhaust velocity can be realised.

The specific impulse I of a propellant is the ratio of the force developed to the rate at which the propellant is consumed. If the rate of consumption is expressed in kilogrammes per second then the force developed should be expressed in newtons (i.e. kg m s^{-2}) and the unit of I is then N s/kg which is equivalent to m/s; I therefore has the dimensions of velocity and it can be shown that this velocity is the effective exhaust velocity v for the propellant. However, in calculating I it is common practice to take the unit of force as kilogrammes and the unit of I then becomes [kg/(kg/s)], i.e. seconds; in this case the exhaust velocity is related to the specific impulse by the expression

$$v = gI \tag{2.8}$$

where g = acceleration due to gravity at the surface of the earth (i.e. $9{\cdot}807\ \mathrm{m/s^2}$)

I is dependent mainly on the propellant used but also on such factors as local ambient pressure (which varies with altitude) and combustion temperature.

Table 2.2 gives values of specific impulse *in vacuo* for some liquid bi-propellants; values at sea level would be about 10% less. (The values are given both as seconds and as a velocity).

Table 2.2 *Specific impulse of some liquid-propellant systems*

Oxidiser	*Fuel*	*Specific Impulse I* (s)	(m/s)
Liquid oxygen	+ hydrogen	391	3835
Liquid oxygen	+ UDMH	310	3040
Liquid oxygen	+ kerosene	300	2942
Nitrogen tetroxide	+ UDMH	292	2864
Nitrogen tetroxide	+ UDMH	285	2795

The first stage of the powerful proton launcher (USSR) carries over 400 tonnes of propellant (UDMH/nitrogen-tetroxide) which burns for 130 s. Thus the total consumption of fuel is over 3 tonnes/s and this produces a thrust of about 9 MN.

It must again be emphasised that this very elementary treatment of launchers has ignored the effects of gravity and atmospheric drag which significantly reduce the performance of the first stage of any launcher.

2.3.3 *Launch vehicles and launching*

2.3.3.1 *Types of launch vehicle*

Three types of launch vehicle can be used to put satellites into geostationary orbit, namely:

(i) Expendable three-stage vehicles
(ii) Expendable direct-injection vehicles
(iii) Reusable (or partially reusable) vehicles

The great majority of communications satellites have been launched by expendable vehicles and this is likely to continue to be the case for many years to come.

2.3.3.2 Launch by expendable vehicles

Satellites are launched in an easterly direction, if possible, because this allows advantage to be taken of the rotation of the earth to increase the velocity of the vehicle; for geostationary satellites it is essential to launch in an easterly direction as the satellite must move in the direction of the earth's rotation.

The first and second stages of an expendable launch vehicle are usually designed to lift it clear of the earth's atmosphere and to accelerate it horizontally to a velocity of about 8000 m/s; this places it in a parking orbit at a height of about 200 km. The plane of the parking orbit will be inclined to the equator at an angle not less than the latitude of the launch site.

The most efficient way of getting from the parking orbit to a circular equatorial orbit at 36 000 km is to convert the parking orbit into an elliptical transfer orbit (with the perigee at the height of the parking orbit and the apogee at 36 000 km) and then to convert the transfer ellipse to the geostationary orbit (see Fig. 2.3*a*).

The transfer orbit, which is also called the Hohmann transfer ellipse, is in the same plane as the parking orbit. The third stage of an expendable vehicle

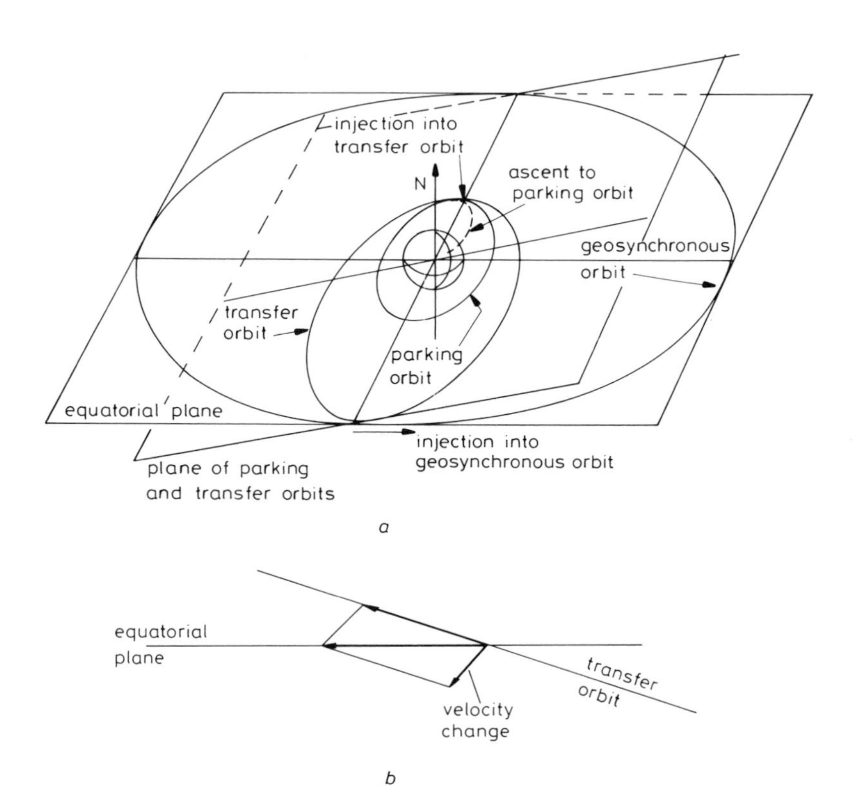

Fig. 2.3 *Launch sequence*

is fired as the satellite crosses the equator; this ensures that the apogee of the geostationary transfer orbit (GTO) is in the equatorial plane.

When the satellite is in the GTO, the third stage of a three-stage launch vehicle has completed its mission and is jettisoned. The final phase of the launch sequence is carried out by means of an apogee kick motor (AKM) built into the satellite. This motor is required to provide (at the apogee of the GTO) a velocity increment of such a magnitude and in such a direction as to reduce the inclination of the orbit to zero (see Fig. 2.3*b*); the AKM must also provide the change in velocity required to make the orbit circular. As can be seen from the Figure, the satellite will only be put into the geostationary orbit (which is an equatorial orbit) if the apogee of the transfer orbit is in the equatorial plane. The period of the GTO is not, of course, the same as the period of the earth; successive apogees therefore occur at different longitudes. To ensure that the AKM is fired at exactly the right moment, the satellite should be within sight of two control stations; the satellite is therefore allowed to complete as many transfer orbits as are required to bring it to the right longitude (at apogee) relative to the control stations.

Once the satellite is in the geostationary orbit, the attitude is corrected, the antennas (and solar panels, if any) are deployed and the satellite is drifted to the correct longitude for operation.

The higher the inclination of the transfer orbit, the greater is the energy required to convert it to an equatorial orbit; a launch from a site at high latitude therefore takes more energy than a launch from a site at low latitude. The latitudes of the NASA launch site at Cape Canaveral and the Ariane launch site at Kourou are 28·5° and 10°, respectively, and the corresponding velocity changes required at apogee are approximately 1800 m/s and 1500 m/s.

Solid-fuel apogee motors may account for up to about half the total mass of a satellite. The use of a liquid-fuel apogee motor (with a higher specific impulse) can result in a significant saving in mass.

Typical three-stage vehicles used to launch communications satellites into geostationary transfer orbit are:

(i) Delta 3914 (USA) and Long March 3 (China) which can put payloads of about one tonne into geostationary transfer orbit (this is sufficient for a single large satellite with its AKM)
(ii) Atlas–Centaur (US) and Ariane 3 (Europe) which can put about 2·5 tonnes into transfer orbit (Ariane 3 can launch two satellites at a time).

Three-stage launchers with still greater payload capacity are being developed; for example some versions of Ariane 4 (which should be available by 1989) will be able to put over 4 tonnes into transfer orbit and Ariane 5 (1995) will be able to put over 8 tonnes into transfer orbit.

If a satellite is launched by means of a direct-injection launch vehicle it does not need an AKM because direct-injection launchers have a fourth stage which converts the transfer orbit to the geostationary orbit. Typical direct-injection launchers are Titan 34D (USA) and Proton (USSR).

2.3.3.3 Reusable space vehicles

NASA's Space Transportation System (STS), or Space Shuttle as it is more often called, is the only reusable vehicle to have seen regular service but the USSR is expected to launch a reusable vehicle in late 1988. A Shuttle has three main elements:

(i) The orbiter which carries the payload and crew
(ii) A very large external tank containing over half a million litres of liquid oxygen and one and a half million litres of liquid hydrogen (the propellant for the main engines of the orbiter)
(iii) Two solid-propellant boosters.

The boosters are dropped two minutes after lift-off and recovered from the sea; the orbiter's main engines are shut down about 8·5 min after lift-off and the main fuel tank, which is not recovered, is then jettisoned. The orbiter itself goes into a parking orbit and returns to earth when its mission is achieved. An orbiter can carry about 30 tonnes into parking orbit but cannot by itself deliver a payload into geostationary orbit or even into a transfer orbit. Communications satellites which are to be launched by Shuttle must therefore be associated with both a perigee stage, or perigee kick motor (PKM), which will put the satellite into the transfer orbit, and an apogee stage.

Shuttles made over twenty successful operational flights between 1981 and end-1985 but in January 1986 the Shuttle *Challenger* was destroyed and all the crew killed in a tragic accident. This accident was the result of a fault in the design of the solid-propellant boosters which, in turn, caused the rupture of the external tank and the explosion of the liquid propellant.

Launches were resumed in mid-1988 after redesign of the booster but it is unlikely that NASA will now use the Shuttle for the regular launch of commercial satellites: firstly because most Shuttles will, for the next few years at least, be required for the backlog of US military payloads and secondly because the US Government wants to encourage its aerospace industry to enter the commercial launch market.

The Shuttle disaster has put the emphasis back, at least temporarily, on the use of expendable vehicles for the launch of commercial satellites. However, as has been noted, the reusable vehicle developed in the USSR is nearly ready for flight; in addition, studies of a number of smaller reusable vehicles are under way, but it is questionable whether any of the latter could carry sufficient payload to be able to launch communications satellites.

Amongst the smaller reusable vehicles being studied are:

(i) The Hermes orbiter: Work on this 'mini-shuttle' was initiated by France and has now been taken under the wing of the European Space Agency. Hermes, which would be launched on an Ariane 5, is designed to take a crew (of two to four people) and cargo into near-earth orbit but is not intended, in its present form, for launching satellites.
(ii) HOTOL (HOrizontal Take Off and Landing craft): This 'spaceplane' proposed by British Aerospace and Rolls–Royce, differs from the Space Shuttle and Hermes in several important ways. First it would comprise only a single stage intended to reach a parking orbit at an altitude of about 300 km and to be completely reusable, secondly it would have a hybrid engine which would use atmospheric oxygen to burn liquid hydrogen fuel at low altitudes and then switch to normal rocket propulsion (using liquid oxygen and liquid hydrogen) at higher altitudes, and thirdly it would be able to take off from (and land on) runways intended for large conventional aircraft.
(iii) TAV (TransAtmospheric Vehicle) and the Sanger Spaceplane: These are similar vehicles to HOTOL which are being studied in the USA and West Germany, respectively.

2.3.3.4 Choosing a launcher

Published information on the performance and cost of launchers sometimes is, or appears to be, inconsistent. Some of the reasons for this are that:

(i) Launch vehicles are constantly being modified and improved, and there are often many variants of a basic vehicle. For example, Table 2.3 shows six proposed versions of Ariane 4 and the payload which each will put into geostationary transfer orbit.

Table 2.3 *Variants of Ariane 4 and the corresponding payloads into transfer orbit*

Variant	Payload (tonnes)
Basic	2·0
+2S	2·7
+4S	3·1
+2L	3·3
+2S+2L	3·8
+4L	4·2

+2S = with two solid boosters
+4L = with four liquid boosters

(ii) It is not always made clear whether the payload quoted is the mass which can be put into parking orbit, or transfer orbit or geostationary orbit (or what apogee and perigee stages are being assumed, if these are necessary).

(iii) The development of most launchers (and in some cases production and support services as well) may be directly or indirectly subsidised by governments for either military or trade reasons (or both). Launch prices are not therefore determined on normal commercial criteria and may be subject to sudden variation. As a consequence, widely differing costs may be quoted for launches using similar vehicles or even for launches using the same vehicle.

It is obviously important to avoid dependence on a single launch vehicle or organisation, but the additional constraints on satellite design imposed by the requirement for compatibility with two or more types of launch vehicle can prove expensive.

Organisations planning to put satellites into orbit are often faced with difficult decisions, especially when major development of launchers is taking place. For example, when INTELSAT was planning the INTELSAT V programme both the Space Shuttle and Ariane were being developed, but there was no certainty that either of them would prove to be reliable nor was it possible to forecast which would prove the cheapest to use. It was therefore specified that INTELSAT V should be compatible with three launch vehicles: Ariane 3, the Shuttle and Atlas–Centaur (the latter being an existing launcher of proven reliability). As things turned out, the Shuttle option had to be dropped because it proved impracticable to interface the perigee stage with the satellite; all INTELSAT Vs have therefore been launched by Atlas–Centaur or by Ariane. More recently, INMARSAT specified that its second-generation satellites should be compatible with at least two of Ariane, Atlas–Centaur, Proton, Thor–Delta, Titan and the Shuttle.

The importance of providing for as many options as possible was emphasised by a string of launch failures in the first half of 1986 and, in particular, by the history of the Shuttle. When the latter was being developed NASA claimed that it would dramatically reduce the cost of launching satellites. The expected reduction in costs was never realised but the Shuttle appeared to be reliable. As a result, NASA accepted a large number of bookings for the launch of satellites by Shuttle and started to phase out the use of expendable launch vehicles. When the disastrous accident to *Challenger* occurred in January 1986 and shuttle launches were suspended, many space-sector operators were left desperately seeking other means of launch.

An obvious alternative for many of these satellites would have been Ariane, but Arianespace (the company which markets Ariane) had a full order book for about three years ahead. Moreover, the Shuttle disaster was followed by a number of other launch failures which, for a time, shook the confidence of the Western aerospace industries. In April 1986 a Titan rocket exploded on lift-off, in May a Delta rocket veered off course and had to be destroyed and, on the 30th May, the third stage of an Ariane 3 failed and all Ariane launches were suspended until late 1987. Owners of commercial satellites therefore began to consider seriously, for the first time, what launch facilities were available from sources other than the USA and Arianespace.

After the USA, the country with by far the greatest experience of launching satellites is the USSR. One of its most successful launchers is Proton. Proton is a very large four-stage rocket with the ability to place a payload of over two tonnes into geostationary orbit. It was designed over twenty years ago, there have been well over 100 launches of the vehicle and it is claimed that there were no failures in over 35 flights between January 1983 and early 1986 (but there has since been a failure in 1987). Proton has been offered to (but, for political reasons, has so far not been used by) INMARSAT, ESA and INTELSAT; it will, however, be used for the launch, in 1988, of an earth-resources satellite for India. It is said that the price asked for a launch by Proton is very much less than that for corresponding launches by US vehicles or Ariane, and that the USSR is prepared to offer a second launch at half price if the first launch fails. The USSR is now planning to improve its marketing of launch services.

The Peoples Republic of China had launched about 18 satellites by mid-1986 including two which were put into geostationary orbit using the Long March rocket. Long March 3 is a three-stage rocket capable of putting 1·3 tonnes into geostationary transfer orbit and China has said that it could provide up to twelve commercial launches per year. Since early 1986 China has discussed the use of Long March 3 for the launch of commercial communication satellites with both Swedish and US companies. Once again it is said that launches by Long March are being offered at a much lower price than launches by Ariane or US launch vehicles.

Japan has been developing expendable launch vehicles since 1970. The first Japanese launchers (N–I and N–II) used US (Thor–Delta) technology. The first of a new all-Japanese vehicle (H–I) was launched in August 1986; this can put just over one tonne into geostationary transfer orbit (or over half a tonne into geostationary orbit); the second generation of this launcher (H–II) should be available in 1992 and this will be able to carry about four times the payload of H–I.

India has launched several small satellites and has plans for a new launch vehicle which should (by 1992) be capable of carrying 2·5 tonnes into GTO.

Some modern launch vehicles can carry more than one satellite. The Shuttle can carry two satellites into parking orbit; Ariane 3 can carry two satellites into GTO. Special adaptors are needed to mount two satellites on one rocket and ensure successful separation, without collision, at the correct time. The adaptors for Ariane 3 and Ariane 4 are called SYLDA (Système de

Lancement Double Ariane) and SPELDA (Structure Porteuse Externe Pour Lancement Ariane), respectively.

2.3.3.5 Support services

A launch vehicle is useless without the complex and expensive support services needed to mate it with the payload, prepare it for firing, monitor its progress while it is being launched, and transmit the commands necessary to guide it into the correct orbit and keep it there.

The launch site must be chosen so that there is no danger to life if the vehicle has to be destroyed shortly after take-off because of a malfunction; sites immediately to the west of an ocean or an area of uninhabited land are therefore most suitable. As has already been noted, the energy required to put a given payload into equatorial orbit can be reduced by using a launch site at a low latitude.

The site must be equipped not only with launch pads but with buildings (including areas with high roofing and overhead cranes) which will provide a clean, air-conditioned environment where vehicle and satellite can be mated and prepared for launch. Some of the preparations, especially those involving the handling of liquid propellants, are very hazardous and must be carried out by highly trained operators wearing special protective clothing and working in isolated areas under the surveillance of safety officers using remote-control cameras. The preparation and launch areas must be connected by a transport system capable of carrying a launcher weighing hundreds of tonnes. The mating of satellite and launch vehicle, followed by testing and preparation for launch, requires a large team and a period of many weeks.

To monitor and control the launch vehicle and satellite during the launch phase it is necessary to have a worldwide network of stations to receive telemetry signals from the spacecraft, to keep track of its position and to send control messages to it; these functions are usually grouped together as telemetry, tracking and command (TT&C). Once a geostationary satellite has been established in its nominal operational position a worldwide network of tracking stations is no longer required. However, it is still necessary to receive telemetry messages from and send commands to the satellite, and to measure any changes in its position; these functions are often carried out via one or more communications earth stations (under contract to and under the control of the managers of the space sector).

The orbit and position of a satellite may be determined by making accurate measurements of its distance and angular direction from an earth station, or by measurements of distance from a number of earth stations.

When a satellite is in the correct orbit but before it becomes operational it is necessary to check that its performance still meets the specification. During the operational phase transmissions from the satellite must be continually monitored (e.g. to check that earth stations are transmitting at the correct frequencies and at the correct EIRPs and that the performance of the satellite is not deteriorating). The monitoring functions and the TT&C functions are sometimes linked together under the portmanteau term 'telemetry, tracking, command and monitoring' (TTC&M). Some further discussion of the TT&C functions is given in Section 2.4.2.6.

2.4 Communications satellites

2.4.1 General

A communications satellite may be regarded as comprising two main modules:

(i) The communications module (the 'payload')
(ii) The service module (the 'bus' or 'space platform')

An apogee motor also forms an integral part of the satellite if it is to be put into orbit by a three-stage launcher.

The service module provides all the support services, such as power and thermal control, required to enable the communications module to work satisfactorily. Important design requirements common to all payloads and service modules are minimisation of mass and power consumption, and maximisation of reliability.

The total mass (at launch) of an INTELSAT V satellite is about two tonnes; about 50% of this is the mass of the apogee motor and about 10% is the mass of the communications module. The power consumption of the communications module is about 70% of the total power consumption (which is approximately 1 kW).

The high cost of developing and manufacturing satellites can be reduced by designing a bus which can be adapted for use with a number of different communications packages; for example, the same bus was used with both the European communications satellite (ECS) and the first-generation maritime-communications satellite (MARECS). Additional advantages accruing from the use of a bus which has already been proven in space are a reduction in the time required for development and increased confidence in reliability (because tests under simulated conditions can never give complete confidence that equipment will prove effective and reliable in the environment of space).

2.4.2 The service module

2.4.2.1 General

The main subsystems of the service module are:

(i) Structural subsystem

(ii) Thermal-control subsystem
(iii) Power subsystem
(iv) Attitude and orbit-control subsystem
(v) Telemetry and command subsystem
(vi) Apogee motor

None of these subsystems can be designed independently of the others; nor can the design of the bus make progress without continual reference to the characteristics of the launch vehicle and the payloads with which it is to be associated. Some spacecraft carry several payloads; e.g. the Indian satellite INSAT I carries payloads for communications between fixed points, for broadcasting services and for meteorological services.

2.4.2.2 The structural subsystem

The structural subsystem has to hold together the other subsystems of the bus and the communications module, provide protection from the environment and facilitate connection of the satellite to the launcher.

Structural subsystems comprise a skeleton on which the equipment modules are mounted, and a skin which provides protection from micrometeorites and helps to shield the equipment from extremes of heat and cold. Most communications satellites are either cylindrical or box shaped (see Figs. 2.4 and 2.5); the cylindrical satellites are stabilised by spinning the whole of the main body whereas the box-shaped satellites are stabilised by means of inertia wheels spinning within the body (see Section 2.4.2.5).

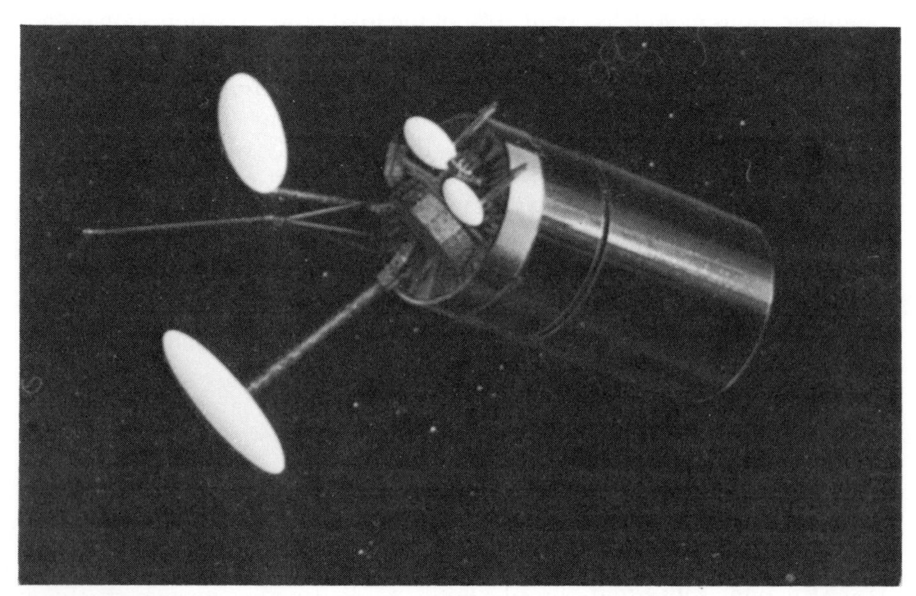

Fig. 2.4 *INTELSAT VI (Acknowledment to INTELSAT)*

Fig. 2.5 *INTELSAT V (Acknowledgment to INTELSAT)*

A spacecraft is virtually free of gravitational stress when it is in orbit and it is therefore possible to use structures, such as large deployable arrays of solar cells, which would collapse under their own weight on earth if special precautions were not taken. However, large stresses are developed during launch as a result of acceleration and the intense vibration generated by rocket motors. The satellite structure must be sufficiently strong to withstand these stresses and it is usually connected to the launcher by a robust thrust tube or cradle.

Large structures, such as the solar arrays of three-axis stabilised satellites, usually have to be folded so that they will fit into the launcher; this folding may be used to give the structures additional stiffness during launch but every folded structure requires a reliable mechanism to unfold it and lock it open once the satellite is in orbit.

Although materials in space are not subject to gravitational stress or atmospheric corrosion, the effects of the space environment are not by any means all benign. The high vacuum of space causes some materials to sublime or evaporate and some metals to weld together on contact; the latter behaviour means that special attention has to be given to the materials used for bearings.

Materials used for spacecraft structures must have a high ratio of strength to mass; those commonly used include magnesium and aluminium alloys, titanium, graphite-fibre-reinforced plastic (GFRP) and sandwich structures with a honeycomb filling. The structural subsystem of INTELSAT V accounts for about 10% of the total mass of the satellite.

During launch the satellite is covered by a fairing which is part of the launcher. This fairing protects the satellite from pressure effects and frictional heating as it travels through the atmosphere at low altitudes.

2.4.2.3 The thermal-control subsystem

Many of the components of a satellite will operate satisfactorily only over a limited temperature range and may be damaged if they are subjected to extremes of heat or cold; for example, waveguide multiplexers will usually operate satisfactorily only over a temperature range of, say, −5°C to +45°C (even when made of materials with low coefficients of expansion such as Invar) and nickel–cadmium (Ni–Cd) batteries may be damaged if they are subjected to temperatures greater than about 40°C or less than about −30°C. However, parts of a satellite are usually in direct sunlight (with a flux density of over 1 kW/m^2) while other parts are facing cold space (at a temperature of about −270°C); a thermal-control subsystem is therefore required to protect the equipment from these extremes of heat and cold, and from thermal shocks such as those produced when the apogee motor is fired or when the satellite moves into the shadow of the earth (in the latter case the temperature of parts of the surface of the satellite may drop by as much as 200°C). The thermal-control subsystem must also get rid of the heat produced by the communications equipment; as there is no convection in space this heat must be conducted to some part of the satellite where it can be radiated away.

Active and passive thermal control

Both passive and active means of controlling the temperature inside a spacecraft may be used. The active means include louvres and blinds (operated by bimetallic strips) and electrical heaters. The passive means, such as surface finishes and insulating blankets, are cheaper and more reliable and are preferred where practicable.

Surface finishes can be chosen to absorb a high or low proportion of the incident solar energy or the albedo (i.,e. the solar energy reflected from the earth); they may also be chosen to have a high or low capability to radiate heat from the satellite to the cold sky. Insulating blankets are usually made of many layers of plastic film coated with aluminium.

Any equipment generating heat, such as the output amplifiers of the communications module, must be situated close to some part of the surface of the satellite which is good at radiating heat into space (i.e. a surface which has a high emissivity) while absorbing only a small proportion of solar energy (i.e. having a low absorptivity). Mirrors have a low ratio of absorptivity to emissivity and are used as surface finishes on parts of many satellites. The mirrors may be optical solar reflectors (OSRs) made of quartz backed with silver, or second-surface reflectors (SSRs) made of plastic backed with silver or aluminium.

There are major differences between the thermal control problems posed by spin-stabilised and body-stabilised satellites. A body-stabilised satellite rotates only once a day relative to the sun whereas a spin-stabilised satellite usually rotates at one or two revolutions per second. Thus on a body-stabilised satellite there is time for large temperature differences to build up between the surfaces of the satellite facing the sun and the rest of the body whereas the temperature round the rotating drum of a spin-stabilised satellite tends to be uniform.

INTELSAT VI, a spin-stabilised satellite, will have two cylindrical solar-cell arrays (see Fig. 2.4). One cylinder is housed within the other during launch but when the satellite is in orbit the outer cylinder will slide down to expose the inner cylinder. The outer cylinder is covered with high-efficiency solar cells; these cells have a high absorptivity but this does not matter because the cylinder does not enclose any equipment when it is in the operational position. Cells with a lower absorptivity are used on the inner cylinder in order to keep the temperature of the equipment compartment within acceptable limits; these cells have a lower efficiency but this is a disadvantage that has to be accepted. Above the inner array of solar cells there is a cylindrical band of quartz mirrors which are the primary means by which the heat from the equipment is radiated into space.

Because a body-stabilised satellite carries its solar cells on outriggers much of the main body of the spacecraft can be enclosed in a thermal blanket. The north and south faces of the body are not usually insulated, because each of them receives no solar radiation for six months of the year and is illuminated only obliquely for the other six months; these surfaces are therefore frequently used to radiate into space the heat developed by the equipment within the satellite.
Two methods of active control are:

(i) The substitution of a resistive load for a transponder when the latter is turned off; the total heat generated is thus kept constant.
(ii) The use of heaters during eclipse to maintain the temperature of critical items such as batteries, fuel lines to thrusters, and thruster valves; these heaters may be operated by command from the ground or (preferably) by thermostats.

2.4.2.4 The power subsystem

(a) Solar arrays

The primary source of power for a communications satellite is the sun. Solar cells are used to convert energy received from the sun into electrical energy.

Each cell can deliver about 150 mA at a few hundred millivolts and an array of cells must be connected together (in series and parallel) to give the required voltage and current for operating the equipment. The solar array is mounted on the curved surface of spin-stabilised satellites or, in the case of body-stabilised satellites, on solar panels which are deployed in orbit; these panels are usually rotated so that they intercept the maximum practicable amount of solar energy.

If the solar panels of a body-stabilised satellite were always to intercept the maximum possible amount of energy (i.e. they were always to be normal to the sun's radiation) they would have to be rotated about two axes so as to compensate for both the daily rotation of the satellite relative to the sun and the apparent movement of the sun from solstice to equinox. In practice only movement about a single axis is provided because the additional complication required to provide rotation about two axes is not worth the resulting gain in power. The output from the solar array is therefore about 10% greater at equinox than it is at solstice.

The amount of solar energy intercepted by a cylindrical solar array on a spin-stabilised satellite is proportional to the projected area of the array dl, where d is the diameter of the cylinder and l is its length, whereas the actual area of the array is πdl. Thus the solar array on a spin-stabilised satellite must have a considerably greater area than that required on a body-stabilised satellite to produce the same output power. It might be expected that the ratio of the two areas would be π. However, the solar cells on a body-stabilised satellite get hotter and suffer more degradation from bombardment by high-energy particles than the cells on a spin-stabilised satellite; this means that the cells on a body-stabilised satellite are rather less efficient at converting solar energy to electrical energy at the beginning of their life and that their efficiency decreases more rapidly with time. As a consequence, the ratio of the areas of the equivalent solar arrays is usually between 2 and 2·5.

The additional mass of the solar-cell array on a spin-stabilised satellite, relative to that on a body-stabilised satellite, is partly offset by a reduction in the mass of the thermal-control subsystem, and the attitude and orbit-control subsystem (which are usually simpler in a spin-stabilised satellite). The large mass of the solar-cell array situated on the perimeter of the satellite has the advantage that it increases the angular momentum and therefore the stability of the satellite (see Section 2.4.2.5). Until recently the power available to a spin-stabilised satellite was limited by the size of the cylinder which could be fitted into the nose cone of the launch vehicle but the concentric solar arrays on INTELSAT VI, and other similar satellites, have loosened this constraint.

It is fortunate that solar energy is free because solar cells are not very efficient. Cells working under the best conditions convert about 15% of incident radiation to electrical power; the efficiency of the solar cells on satellites is usually significantly less than this at the start of working life and can drop by about one-third during seven to ten years in orbit.

The major factor causing a drop in the efficiency of solar cells in orbit is bombardment by high-energy particles, which are mainly protons from hydrogen plasma (the solar wind). Unfortunately the solar wind can be very strong in the region of the geostationary orbit at times of solar flares and a single severe solar flare can cause a sudden drop in efficiency of as much as 5%; although such flares are not frequent there is a significant probability that a satellite will meet one during its working life. Solar cells are fitted with cover slips which do something to limit the damage from high-energy radiation and also from micrometeorites.

The efficiency of solar cells decreases as their temperature increases. Any radiation which is not converted to electrical power or re-radiated raises the temperature of the cells (and also the temperature inside the satellite). It is therefore essential that the surface of a solar cell shall have as high an emissivity as is practicable; the cover slips are usually designed specifically to reject the longer-wavelength infra-red radiation (which would not contribute much to electrical output but would, if absorbed, result in a significant rise in temperature).

In practice the output from the solar panels on a modern body-stabilised satellite is about 60–70 W/m^2 and the panels weigh between 3 and 5 kg/m^2. Power outputs of around 1–2 kW are usually needed so that arrays of 15–30 m^2 are required on body-stabilised satellites. Considerable effort is being put into improving the efficiency of solar cells and reducing their mass.

(b) Eclipse

When a satellite passes through the earth's shadow the solar arrays stop producing power. If the earth's axis were normal to the plane of its orbit round the sun then all satellites in equatorial orbits would pass through the earth's shadow for a period, around midnight at the longitude of the satellite, every day. However, because the earth's axis is tilted relative to its orbital plane a geostationary satellite is always in sunlight at the summer and winter solstices but suffers a period of eclipse each day for a number of days around the equinoxes (see Fig. 2.6). Part of the geosynchronous orbit begins to move into the earth's shadow 21 days before an equinox and the period of eclipse increases each day until it is a maximum at the equinox. The shadow of the earth subtends an angle of 17·4° at the geostationary orbit so that this maximum period of eclipse is (17·4/360) × (23 h 56 min) = 69·4 min (Note: 23 hrs 56 min is the period of the sidereal day). After the equinox the period of eclipse declines again until, after another 21 days, the geostationary orbit is once again completely in sunlight. As there are two equinoxes, there are a total of 84 days each year when a geostationary satellite suffers some period of eclipse.

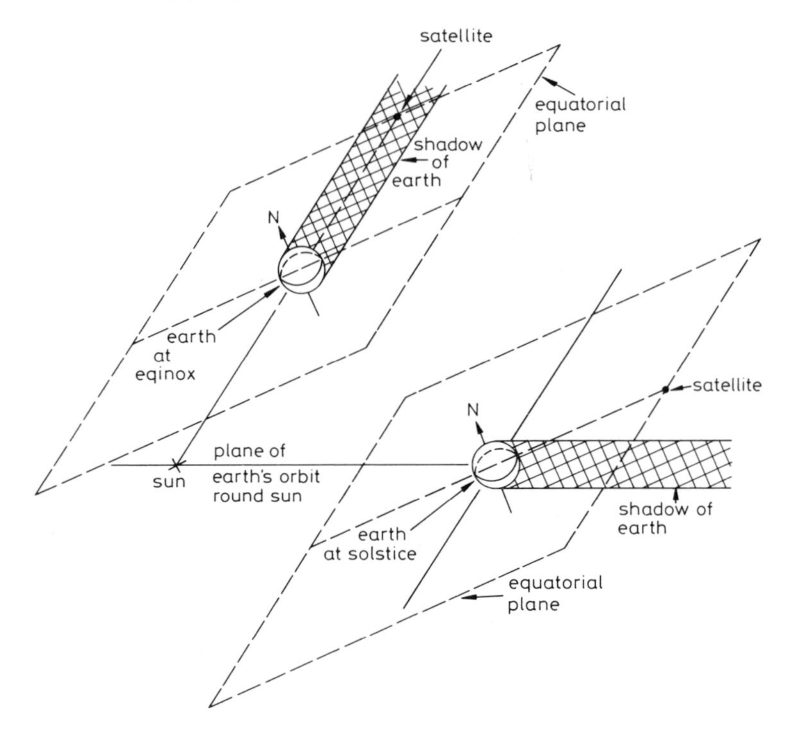

Fig. 2.6 *Eclipse of an equatorial satellite*

(c) Batteries

With some satellite systems (e.g. those used for the relay of television programmes) it may be possible to accept the outages resulting from eclipse, especially if the period of eclipse is delayed until after midnight (local time) by using a satellite to the west of the coverage area. With other systems (e.g. INMARSAT) an outage is quite unacceptable and in these cases it is necessary to equip the satellites with batteries.

It is essential that the batteries shall have a long life and that the ratio of electrical energy stored to the mass of the battery shall be as high as possible (this ratio is called the specific energy of the battery and is usually expressed in watt hours per kilogramme). Until recently virtually all satellites used batteries of nickel–cadmium (Ni–Cd) cells; these have a relatively low specific energy (30—40 Wh/kg) but have a long life if used properly. The later INTELSAT Vs used (and INTELSAT VI will use) nickel–hydrogen (Ni–H) cells which can store at least 50% more energy per kilogramme.

The lives of most types of rechargeable cell depend on the number of charge–discharge cycles, the depth of discharge and the temperature at which the cells are stored and operated. For Ni–Cd batteries on communications satellites the capacity is usually chosen so that the maximum depth of discharge of the cells in normal use is about 60% of their capacity and average discharge is about 40%. With Ni–H batteries the depth of discharge can be up to about 70% without reducing life. Ni–Cd batteries are normally reconditioned prior to the eclipse season, that is the cells are fully discharged at a low rate and then recharged; this helps to keep the plates of the cells in good condition. On most satellites the batteries can be switched to charge or to load by command from the ground.

The satellite solar array must be capable of supplying all the power demanded by the communications and bus equipment, and charging the batteries at the same time. The batteries must have sufficient initial charge to operate equipment such as the telemetry and telecommand systems, thruster valves and heaters during the launch phase; i.e. before the satellite has reached geostationary orbit and the solar array has been deployed (if necessary) and is properly aligned with respect to the sun.

Because the power available from the solar array is greatest at the beginning of life and the voltage of the cells is very dependent on temperature it is necessary to provide some means of regulating the supplies to the battery and electronic equipment. Central regulators or regulators at the input to each subsystem may be used.

For INTELSAT V the mass of the power subsystem (including batteries) is about 8% of the total mass of the satellite at launch. Charging the batteries takes up to 8% of the total power generated by the solar arrays.

2.4.2.5 Attitude and orbit control

Suppose that a satellite has been placed in the geostationary orbit, that it has been positioned at the correct longitude and that its attitude has been adjusted so that its antennas can be pointed in the right directions. The position and attitude of the satellite will immediately begin to change under the influence of disturbing forces. The most significant of these forces are:

(i) The gravitational attractions of the sun and moon; these result in the orbit of the satellite becoming inclined to the equatorial plane at a rate of about 0·9° per annum.

(ii) The non-uniformity (called tri-axiality) of the earth's gravitational field. The earth is not a perfect sphere and the gravitational field therefore has maxima and minima; all geostationary satellites would, if left to themselves, drift to one of the two points of maximum gravitational attraction on the equatorial orbit (at longitudes of about 75°E and 255°E).

(iii) The pressure of radiation from the sun which tends to cause changes in attitude (because the centre of solar pressure is usually offset from the centre of mass of the satellite) and also eccentricity of the orbit (in the case of satellites with a large surface area).

Other factors which can cause disturbance are the impact of meteorites and interaction of the satellite with the earth's magnetic field.

Inclination of the orbit, drift in longitude and change of attitude must all be corrected before they cause significant deterioration in the performance of the system or cause the satellite to interfere with other radio systems. INTELSAT I was allowed to drift for about ten months before a correction was applied, by which time it was about 5° from its nominal longitudinal position. Such a large drift was only acceptable because the few earth stations working to the satellite were well within the satellite-antenna coverage area and there were no other satellites nearby to suffer interference. The situation is very different now that the geostationary orbit is crowded and satellite antennas with narrow beamwidths are in use; in consequence, the Radio Regulations (RRs) of the International Telecommunications Union (ITU) require that satellites be kept to within 0·1° of their nominal longitude (a correction every two or three weeks is needed to achieve this) and that errors in pointing satellite antennas shall be not more than 0·3° or 10% of their half-power beamwidth (whichever is the greater). Modern satellites such as INTELSAT VI can do much better than this (see Section 2.2).

Small rocket motors called thrusters are used for station keeping (i.e. reducing the inclination of the orbit and correcting longitudinal position), for attitude control and also (in spin-stabilised satellites) for spinning up the satellite; the same thrusters may be used for both station keeping and attitude control if they are suitably located on the satellite. For correction of the orbital inclination and the longitudinal position of the satellite (without changing its attitude) the resultant force exerted by the thrusters must act through the centre of mass of the satellite. On the other hand, if the attitude of the satellite is to be altered without changing its location the resultant of the forces acting on the satellite must constitute a couple (i.e. two equal and opposite parallel forces) and there must be no resultant force through the centre of mass.

The attitude of the satellite is usually stabilised by spinning the satellite body or by means of a spinning (momentum) wheel within the satellite. Spinning wheels may also be used to exert torques on the satellite by the reaction to a change in their speed of rotation.

(a) Thrusters

Thrusters work on the same principle as launchers, i.e. force is exerted on the satellite by the reaction to gas expelled through a jet. However, the force exerted by thrusters of the type used for in-orbit manoeuvres is tiny compared with that developed by the rocket motors of launchers; for example, INTELSAT V has ten working thrusters (and ten standbys) with thrusts varying from 0·3 N to 22 N whereas the thrust developed by the first stage of a large launcher may be several meganewtons.

In the past, virtually all satellites used completely separate propulsion systems for apogee boost and in-orbit adjustments, apogee boost being provided by a solid-propellant motor whereas the in-orbit system commonly used hydrazine thrusters. In a hydrazine thruster the hydrazine is decomposed by passing it over a heated catalyst; in an electrothermal hydrazine thruster (EHT) the gases are heated electrically after decomposition by the catalyst; this raises the exhaust velocity and thus reduces the amount of hydrazine required for a given work load. INTELSAT V carries both ordinary (catalytic) hydrazine thrusters and EHTs. The latter are used for north–south (N–S) station keeping (i.e. correction of inclination) which uses much more energy than other types of manoeuvre.

Some modern satellites (including INTELSAT VI) are now using a common bi-propellant propulsion system to provide apogee boost and in-orbit adjustments; the use of a single propulsion system for the two purposes and the high specific impulse of bipropellants result in a useful saving in mass.

Fine control of thrust is required in order to allow accurate control of attitude and station, and the flow of fuel is usually controlled by pulsed solenoid valves.

(b) Station keeping

In order that the appropriate station-keeping corrections can be applied it is essential that the orbit and position of a satellite are accurately determined. This may be done by making measurements of the angular direction and distance of the satellite from an earth station, or by making measurements of distance from a number of earth stations (angular measurements are not usually, by themselves, sufficiently accurate for the purpose). When the orbit and position of the satellite have been determined it is possible to calculate the velocity increments required to keep the N–S and E–W excursions of the satellite within the allowable limits.

The frequency with which N–S corrections must be made depends on the maximum allowable value of the orbit inclination but the total increment required each year to cancel out the attraction of the sun and moon is 40—50 m/s (the exact value changes with periodic variations in the orbit of the moon).

E–W station keeping is usually achieved by allowing the satellite to drift towards the nearest point of equilibrium until it reaches the maximum allowable error in longitude. It is then given a velocity increment just sufficient to carry it to the point of maximum allowable error on the other side of the nominal longitude. From here the satellite drifts back once more towards the point of equilibrium and the process is repeated. The frequency and magnitude of the velocity increments required depend on the angular distance between the satellite and the points of equilibrium and on the allowable error. The maximum total velocity increment required per annum is about 2 m/s.

(c) *Attitude control*

The motion of a ship about its centre of mass is described in terms of roll, yaw and pitch, which are rotations about three orthogonal axes. The same terms are used to describe the changes in attitude of a satellite and the three axes are shown in Fig. 2.7. The axis of roll is the direction of motion, the axis of yaw is the local vertical and the axis of pitch is parallel to the N–S axis of the earth. It will be seen that the directions in space of the roll and yaw axes change as the satellite moves in its orbit.

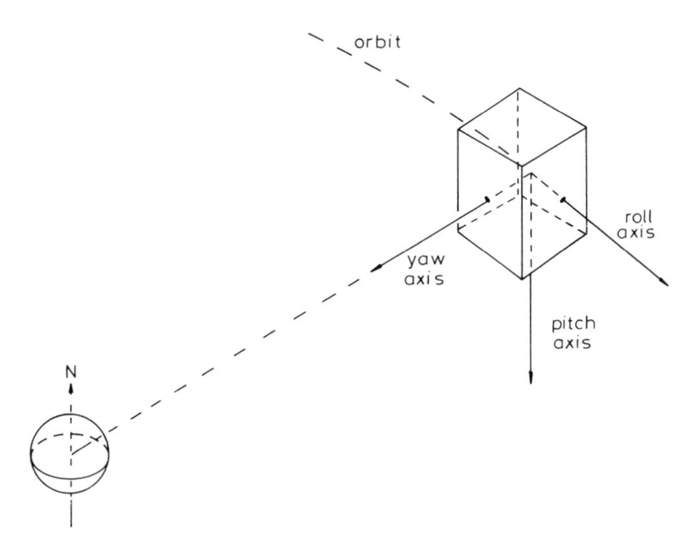

Fig. 2.7 *Pitch, yaw and roll*

The essential components of an attitude control system are:

(i) A means of determining the attitude of the satellite
(ii) A logic system to compare the attitude with the required attitude and to determine what errors exist, and what action is needed to correct them
(iii) A method of rotating the satellite about its axes.

The sensors most commonly used on satellites for determining attitude are sun sensors and earth-horizon sensors. The latter are usually infrared sensors which will work even when the earth is in darkness. However, the boundary between earth and space is not well defined (because of the earth's atmosphere) and earth sensors are not therefore suitable for use if the satellite has spot beams subtending an angle of less than about 1·3°. A better method of steering narrow-beam antennas on satellites is to use an RF beacon transmitted by an earth station. Inertial sensors (gyroscopes) can also be used.

(d) Spin stabilisation and body stabilisation

Satellites may be divided into two classes as far as attitude control is concerned. In the first class ('spin-stabilised' satellites) the body of the satellite is spun about the pitch axis; in this case the antenna (or the antenna platform if there are several antennas) must be spun in the opposite direction so that the antennas remain pointing at the earth. INTELSATs III, IV and VI are examples of spin-stabilised satellites. In the second class ('body-stabilised' satellites) the body of the satellite is maintained in a fixed attitude relative to the earth. (Body-stabilised satellites are also known as 'three-axis stabilised' satellites.) INTELSAT V is an example of a body-stabilised satellite. Body-stabilised satellites are often stabilised about the pitch axis by means of the angular momentum of a spinning mass but, in this case, the mass is a momentum wheel within the body of the satellite not the body of the satellite itself. Body-stabilised satellites may also use reaction wheels on one or more axes to control attitude.

(e) Angular momentum

Fig. 2.8 shows a wheel spinning, with angular velocity ω, in plane YOZ (the spin axis is OX). If the inertia of the wheel is I then its angular momentum is ωI. The angular momentum may be represented by the vector OA drawn along the axis of rotation; the length of the vector represents the magnitude of ωI and the direction of the vector is conventionally taken as the direction in which a right-handed screw would move if it were rotated in the same sense as the wheel.

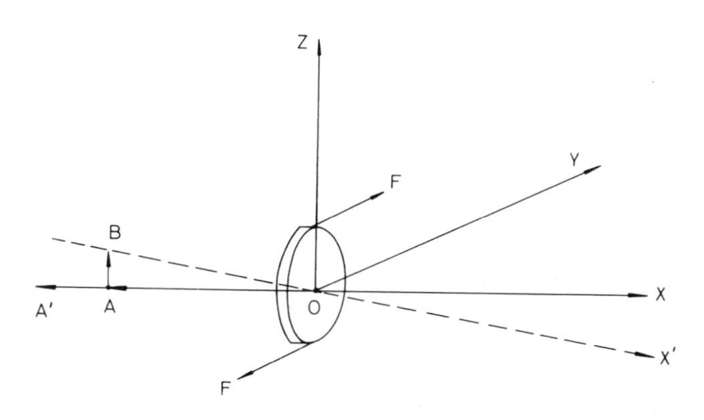

Fig. 2.8 *Change of angular momentum*

Consider the case where the angular velocity of the wheel changes but the direction of the axis of rotation remains unchanged. Everyday experience tells us that such a change of angular velocity is produced by applying a couple in the plane of the wheel; thus, for example, the two forces F in Fig. 2.8 constitute a couple which would increase the momentum of the wheel (from OA to OA′) without changing the direction of the spin axis. The rate of change of angular momentum is proportional to the couple (or torque) T, i.e.

$$T = I(d\omega/dt) \tag{2.9}$$

Any change in the angular momentum of the wheel is, of course, accompanied by a reaction which is equal and opposite to the couple causing the change. Thus when a reaction wheel mounted on, say, the roll axis of a satellite is accelerated or decelerated (by a torque applied by an electric motor) the wheel exerts an equal and opposite torque on the satellite via the motor (and the satellite therefore rotates about the roll axis unless some other forces stop it doing so). If a reaction wheel is continually used to apply torque in the same sense then the speed of the wheel will continually increase; in this case the energy of the wheel must be dumped before the maximum acceptable speed is reached. To dump energy the wheel is decelerated by its motor and the torque generated by doing this is cancelled by an equal and opposite torque applied to the body of the satellite by means of thrusters.

Returning to Fig. 2.8, now consider the case where the angular velocity of the wheel remains constant but the axis of spin rotates in the plane XOZ with angular velocity ω_p. After a short time δt the spin axis has moved to OX′ and the angular momentum is now represented by vector OB. The change of momentum is therefore represented by the vector AB and when δt is very small this vector is perpendicular to the plane XOY. Just as vector AA′ (representing the change of momentum of the reaction wheel) is normal to the couple causing the change of momentum, so vector AB is normal to the couple required to cause this change of axis; thus the couple is in plane XOY, and application of the right-handed screw rule tells us that the sense of the couple is clockwise when looking in the direction OZ. The rate of change of angular momentum in this case can be shown to be:

$$T = I\omega_p\omega \tag{2.10}$$

where ω = angular velocity about the axis and
ω_p = angular velocity of the spin axis in the plane XOY

A couple that causes the spin axis of a wheel to change direction while the angular velocity remains constant is known as a gyroscopic couple and the movement of the spin axis is known as precession. The plane of spin, the plane of precession and the plane of the couple are mutually perpendicular.

From expression 2.10 it can be seen that, for a given torque, the rate of precession can be made smaller (i.e. the stability of the satellite can be improved) by making ωI larger; large ωI is said to give the satellite 'gyroscopic stiffness', 'inertial stiffness' or 'momentum bias'. This is the principle of the momentum wheel and spin stabilisation.

The high inertia of a spin-stabilised satellite about its spin axis gives it gyroscopic stiffness even though the satellite is only spinning at one or two revolutions per second. Momentum wheels have smaller mass and must therefore be spun relatively fast; INTELSAT V satellites include a momentum wheel which operates at about 60 revolutions per second (and has a nominal momentum of 35 Nms). By speeding up or slowing down a momentum wheel it can be used in the same way as a reaction wheel, to apply a torque to the satellite about its axis of spin. Limits must be set to the variation in speed of the wheel in order to avoid large variations in the momentum bias of the satellite; the control system must therefore arrange for some of the energy in the wheel to be dumped (or for energy to be added) when the speed of the wheel varies from nominal by more than a specified amount (say, about 10%).

Although spin-stabilisation greatly reduces the effects of couples acting about the roll and yaw axes there will still be a tendency to slow precession of these axes. This precession is limited by means of a counteracting torque applied by means of a pair of thrusters. The operation of these thrusters is synchronised with the rotation of the satellite; by choosing the appropriate instant during rotation a couple can be applied in any plane through the axis of rotation. The thrusters are operated by means of solenoid valves. Tangential thrusters are also required in order to maintain the rotation speed of the satellite.

If the solar panels on opposite sides of a satellite are set at slightly different angles relative to the sun the force exerted on the panels by the sun's radiation will be in slightly different directions and this can be used to exert a small torque on the satellite (this is sometimes called 'solar sailing'). Electromagnetic coils interacting with the magnetic field of the earth may also be used to exert torques.

If a spin-stabilised satellite is spun about its axis of maximum inertia then it is inherently stable. If not, then there will be a tendency for nutation (i.e. oscillation of the spin axis); this must be controlled either by a system which detects the nutation and applies countervailing torque or by means of an internal damper which dissipates the kinetic energy.

The attitude of body-stabilised satellites is usually controlled in a similar way to that of spin-stabilised satellites, i.e. by means of a momentum wheel spinning about the pitch axis and by thrusters which control precession of the roll and yaw axes. Reaction wheels may be used, instead of thrusters, to apply torques about the roll and yaw axes, and in some satellites reaction wheels are used to control movement about all three axes (i.e. there is no momentum bias).

Attitude control has been discussed in the context of a satellite already on station but the attitude-control mechanisms are also required during the launch phase, for example to put the satellite into the right attitude before the apogee motor is fired.

For INTELSAT V the mass of the fuel provided for station keeping and attitude control is nearly 10% of the mass of the satellite at launch. The attitude determination and control system accounts for about 5% of the total power consumption.

2.4.2.6 *Telemetry, tracking and command (TT&C)*

No complex system can be efficiently operated and maintained without the use of supervisory and control equipment which continually monitors the state of the system and takes action to change that state when this is necessary. Earth-station managers may choose to supervise and control their stations either locally or from a remote site; satellite managers can only exercise supervision and control from a distance (although they may be aided by on-board control systems). Reliable means must therefore be provided on the satellite of:

(i) Monitoring the state of equipment, and encoding and transmitting the resulting (telemetry) data to a control centre
(ii) Receiving, decoding and executing commands from a control centre

Control centres are also responsible for tracking the satellite during the launch phase (the satellite must therefore start transmitting a beacon signal as soon as it is separated from its launch vehicle) and for correcting the orbit and position of the satellite once it is on station. Telemetry, tracking and command (TT&C) are usually considered as one package of related functions. The control centre may, in addition, be responsible for the general management of the satellite system and for the monitoring of satellite transmissions to ensure that earth stations are not adversely affecting operation of the system, for example by radiating on the wrong frequency or at too high a power. (When monitoring is included with the tracking, telemetry and command functions, then TT&C becomes TTC&M.)

During launch the control centre exercises the TT&C functions via a worldwide network of tracking stations. When a geostationary satellite is in its operational position the control centre can exercise its functions via a single earth station.

The INTELSAT control complex, which is located in Washington, has a Spacecraft Control Centre (SCC), an INTELSAT Operations Centre (IOC) and a Technical and Operational Control Centre (TOCC) for each Ocean Region.

Control of the satellite is divided between on-board systems and the control centre. The functions of the latter may include, for example:

- Firing the apogee motor
- Running up momentum wheels
- Using thrusters to correct satellite attitude and position
- Steering antennas
- Switching transponders on and off
- Operating switching matrices (for example to provide the required connections between input and output beams)
- Charging, conditioning and operating batteries
- Switching standby equipment to the operational state and removing malfunctioning equipment from service.

Telemetry data should include all the information necessary for the efficient operation and maintenance of the satellite, for example:

- State of switches
- Values of important voltages and currents
- Temperatures
- Amount of propulsion fuel left
- Environmental information such as radiation levels and meteor impacts.

The telemetry data is not only necessary for the correct operation and maintenance of the satellite but can be of great use in identifying the cause of malfunctions and indicating how satellite and launcher design may be improved.

Errors in telemetry information and (particularly) in command messages could have dire consequences. Fortunately, the bit rates required are usually modest (of the order of 100 bit/s for commands and up to about 1000 bit/s for telemetry) and this makes it practicable to use complex modulation and encoding methods which minimise the number of errors and ensure that any errors that do occur are detected. For critical operations, such as firing the apogee motor or deployment of antennas, the commands may be retransmitted to the ground for confirmation that they have been received correctly before they are acted on.

Telemetry transducers and sensors may produce digital or analogue outputs; analogue outputs are usually digitised and multiplexed with the other (digital) telemetry signals.

Satellites are usually provided with an omnidirectional antenna for the transmission of telemetry and beacon signals and reception of command signals at those times when the communications antennas are either not deployed or are not pointing at the earth (e.g. during launch). However, the communications antennas may also be used for command and telemetry purposes when the satellite is on station and functioning normally. Specific frequencies are allocated to telemetry and command; allocations may be made both in the bands used for satellite communications (e.g. 6/4 GHz or 14/11 GHz) and in S-Band (i.e. around 2 GHz) or at VHF.

2.4.3 The communications module

2.4.3.1 General

The communications modules of all the civil communications satellites now in service comprise a number of transponders which select one or more received signals (by means of a bandpass filter), translate these signals to a new frequency band and retransmit them. The satellites do not perform any other processing of the signals except that it is usually possible to switch the output of a transponder to one of a number of down beams; this switching is necessary in order to provide as good a match as is practicable between the capacities of the paths through the satellite and the pattern of traffic flow between the earth stations which the satellite serves (a poor match results in inefficient use of the satellite). Communications modules of this type are said to be 'transparent' (or comprised of 'bent-pipe' transponders). Transparent modules have the great advantage that they impose minimal constraint on the characteristics of the satellite system; this is of particular importance when the system forms part of a larger communications network.

However, it is now possible to develop and manufacture satellites which process signals in more complex ways; for example, the communications module can be made to regenerate digital signals, or to combine a number of signals received (at low bit rates) from very small earth stations into a single time-division-multiplex (TDM) signal for transmission (at a high bit rate) to a large central earth station, or to adapt to changing traffic patterns and propagation conditions. Such on-board processing adds to the complexity and cost of the satellite but it can lead to more efficient use of system bandwidth and power, can facilitate the introduction of new services and is already in use on some military satellites. In commercial satellite-communications systems the transfer of complexity from earth-stations to satellites could bring about savings in overall cost for systems with many earth stations (for example, mobile communications systems or systems where a central antenna distributes data to, or collects it from, many small peripheral stations). On the other hand, on-board processing makes it more difficult to achieve high reliability in the hostile environment of space. On-board processing also results in loss of flexibility since the satellite is no longer able to work with as wide a range of signal formats and modulation methods.

Although the communications modules of satellites use the same basic techniques and (to some extent) the same types of circuits and components as terrestrial systems, the space environment imposes additional constraints. All equipment must be designed to have a very high probability of survival for the life of the satellite because it is not practicable to service satellites once they are in the geostationary orbit. On the other hand, designing for survival (and particularly the survival of solid-state equipment) is made more difficult because equipment above the earth's atmosphere is exposed to ionising radiation (e.g. high-energy protons generated by solar flares) and to extremes of heat and cold; moreover the mass of any protective shielding or devices must be minimised. It is thus necessary to select materials and technologies with great care and to provide standby equipment for all units which incorporate active devices.

2.4.3.2 *Transparent communications modules*

(a) INTELSAT V

An INTELSAT V satellite includes most of the features typical of large modern satellites intended for point-to-point communication between fixed earth stations in international, regional or large national systems. Such satellites generally comprise many (typically twenty-four) transponders, a moderate number of antenna beams (some at least of these beams being tailored to give high gain in the direction of geographical areas generating a lot of traffic) and complex switching matrices which enable interconnection between the antenna beams to be adapted to match the pattern of traffic flow. Satellites intended for point-to-multipoint relay of a few signals (e.g. direct-broadcast satellites) or satellites for mobile communications are usually more simple in concept (e.g. they have only a few transponders and simple, if any, switching matrices).

An INTELSAT V satellite has global-beam, hemi-beam, zone-beam and spot-beam antennas; Fig. 1.9 shows examples of zone-, hemi- and spot-beam coverage areas. The spot beams use the 14/11 GHz frequency bands; all other beams use the 6/4 GHz bands. The coverage areas of the zone and hemi beams overlap and they use the same frequency bands; the transmissions using these beams must therefore be orthogonally polarised.

Fig. 2.9 is a block diagram of the communications module of INTELSAT V with standby equipment omitted. There are seven working receivers, one each for transmissions received via the global beam, the east and west hemi beams, the east and west zone beams and the east and west spot beams. The first stage of each receiver is an LNA with a noise figure of about 6 dB; tunnel diodes are used for the 14 GHz LNAs and bipolar transistors for the 6 GHz LNAs (there is no point in using LNAs with lower noise figures because the satellite antennas are looking at the warm earth, which is at a temperature of about 290 K). After amplification, both the 6 and 14 GHz signals are downconverted to 4 GHz. The downconverters are followed by further transistor amplifiers.

The transmission in each beam (whether receive or transmit) comprises a number of channels (see Fig. 2.10). For example, the spot-beam transmissions comprise channel 1–2 with a bandwidth of 77 MHz, channel 5–6 with a bandwidth of 72 MHz and channel 7–12 with a bandwidth of 241 MHz. (The term 'channel' is used in this context to mean part of the total frequency spectrum which can be routed independently through the satellite and the

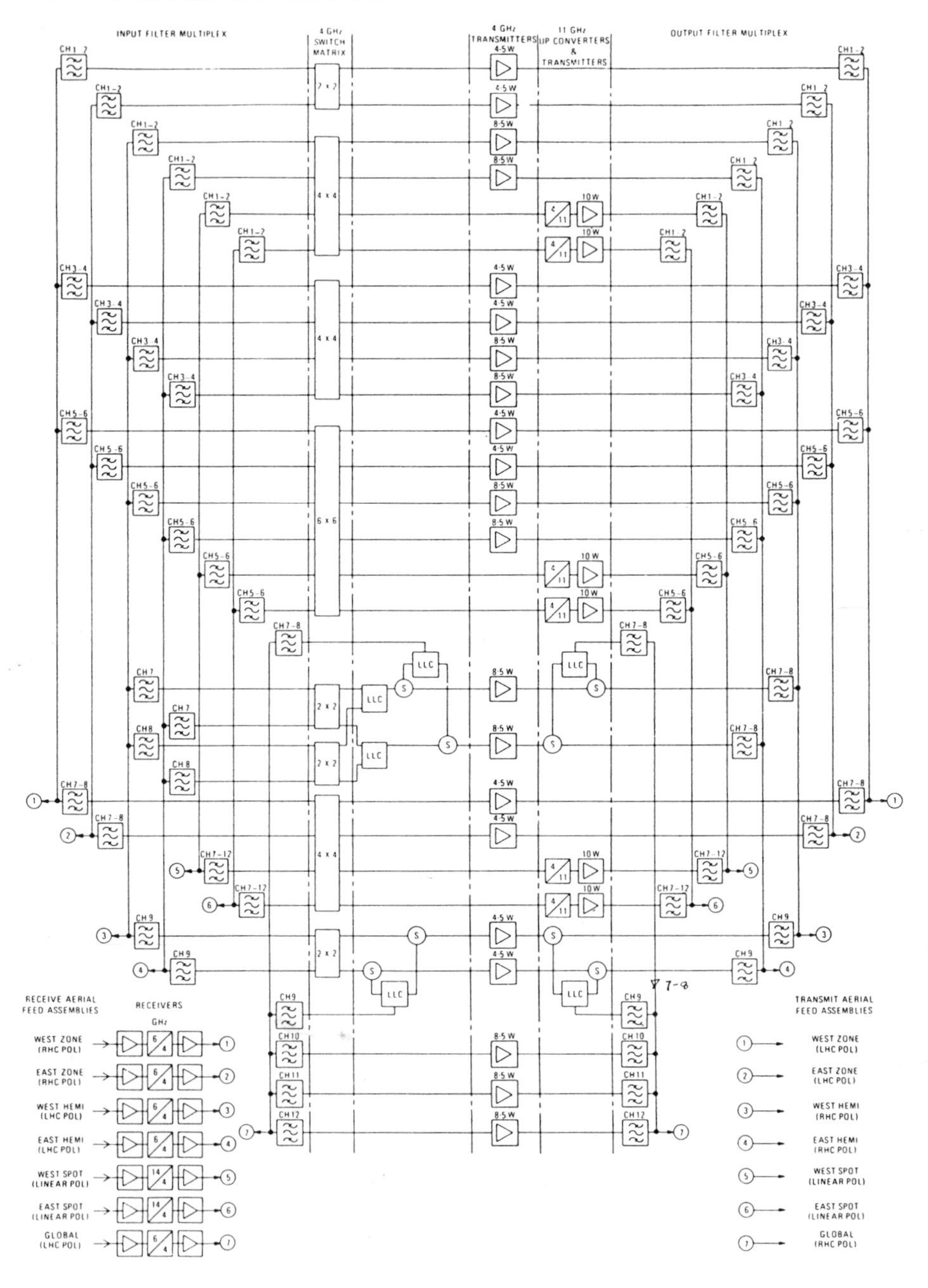

Fig. 2.9 *Communications module (INTELSAT V) (Reproduced from IPOEE Journal,* 1977, **70***)*

POL = polarisation
RHC = right-hand circular
LHC = left-hand circular
LLC = low-loss coupler
CH = channel
S = switch

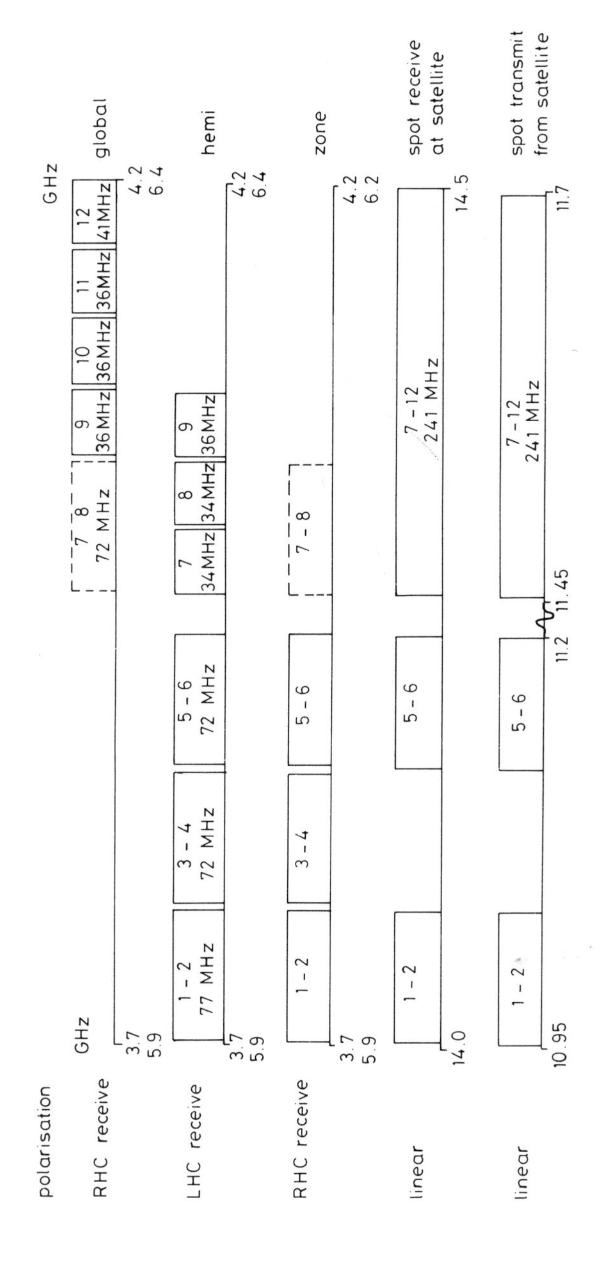

Fig. 2.10 *INTELSAT V channel plan*

Channels 7—8 and channel 9 cannot be used simultaneously for global and hemi transmissions as they use the same polarisations

channel numbering is a hangover from the days when all INTELSAT satellites had 12 contiguous channels, spaced at 40 MHz.) Each independent route through the satellite constitutes a transponder, but it is not so easy to identify the transponders in INTELSAT V as it is in a satellite which does not have switching between the input and output beams.

The traffic received by the satellite in a single beam may be intended for stations distributed between a number of the transmit coverage areas and it is therefore necessary to have some means of switching individual channels in a receive beam to any of a number of outgoing beams. The method of interconnection in INTELSAT V is as follows (see Fig. 2.9):

(i) The output from each receiver is passed to an array of bandpass filters (collectively known as the input multiplexer) which separate out the individual channels
(ii) The channels then go to a series of 4 GHz switching matrices
(iii) After amplification in travelling-wave-tube amplifiers (TWTAs) the channels are reassembled by a further array of bandpass filters (the output multiplexer) for retransmission.

The routings through the switching matrices can be altered by command from the ground to accommodate changing traffic patterns or the requirements of different ocean areas. It is not possible to interconnect every channel with the equivalent channel in every other beam but a great deal of flexibility is available. For example, channel 1–2 received via the east or west hemi beams; or the east or west spot beams, can be retransmitted via the east or west hemi, or east or west spot beams; on the other hand channel 1–2 received via a zone beam can only be retransmitted via a zone beam.

The ability to receive signals at 14 and 6 GHz and retransmit them at 4 and 11 GHz, respectively, is known as cross-strapping and is of particular importance because there are comparatively few 14/11 GHz stations in the INTELSAT system. Channels to be assembled for transmission via a global, hemi or zone beam can be passed directly to the output TWTAs and multiplexer filters, but amplifiers feeding filters which select channels for assembly into spot-beam transmissions must be preceded by a 4/11 GHz upconverter.

An attenuator which can be varied by command from the ground is provided before each TWTA. These attenuators (not shown in Fig. 2.9) give the satellite the ability to handle a wide range of signals. For example, if one of the TWTA's is to be used to amplify a few low-level transmissions received from, say, very-small-aperture earth stations (VSATs) then most of the attenuation in the appropriate path will be removed; on the other hand if a TWTA is to be used to amplify a single high-level transmission (e.g. a television carrier) then most, or all, of the attenuation will be inserted. Where a TWTA is being used to amplify a number of carriers the attenuator must be set so as to give the back-off required to keep intermodulation within acceptable limits (see Section 3.4).

The global-beam antennas are conventional horns which transmit and receive only one polarisation. The hemi and zone beams are formed by offset paraboloids fed by arrays of over 80 horns, each horn being able simultaneously to support both orthogonal polarisations; the offset geometry is essential in order to avoid blockage of the beams by the large feed arrays. The beams are shaped to the required coverage areas by activating the appropriate combinations of horns via networks which control the relative amplitude and phase of the signals applied to each of the horns. In the case of the zone beams alternative beam shapes can be selected by command from the ground, one for use if the satellite is serving the Atlantic Ocean or Pacific Ocean Region and the other for the Indian Ocean Region. Separate reflectors and feed arrays are used for the 6 GHz and 4 GHz antennas. The isolation between east and west beams is greater than 27 dB and the isolation between the orthogonally-polarised hemi and zone beams is about the same. (These figures take no account of the imperfections of the earth-station antennas.)

The spot-beam antennas are offset paraboloids fed by a single linearly-polarised feed; the reflectors are mounted on gimbal mechanisms so that they can be steered (by control from the ground) over a limited range.

The antenna reflectors, feed arrays and parts of the support structure are fabricated from graphite-fibre-reinforced plastic (GFRP). This is strong and light, and has a very low coefficient of expansion which minimises distortion of the antennas by the large temperature gradients which can occur on structures in space. GFRP is also used for the band-pass filters in the multiplexers; there are 60 of these and it is essential to reduce their mass as far as possible.

The Space Shuttle has recovered some geostationary satellites which failed to leave the parking orbit but it is not practicable at present to repair or recover satellites which have reached the geostationary orbit. Standby equipment is therefore provided for all units which include active components. Two standby receivers are provided for each pair of working receivers serving the hemi, zone and spot beams; two standby receivers are also provided for the single working receiver serving the global beam; there is thus a total of eight standby receivers for seven working receivers. One standby TWTA is provided for each working TWTA serving the global beam or amplifying the 241 MHz channels (channels 7–12), and one standby TWTA is provided for two working TWTAs in all other cases.

(b) *INTELSAT VI*

The communications module of INTELSAT VI is, like its predecessors, transparent. It carries further (see Section 1.5.8) the method of frequency reuse exemplified in INTELSAT V and it also employs satellite-switched TDMA (SS–TDMA).

SS–TDMA allows the interconnections between the receive and transmit zone beams and hemi beams, via channels 1–2 and 3–4, to be changed up to 64 times in each 2 ms TDMA frame period. This provides greater flexibility in matching satellite capability to the pattern of traffic flow; it also reduces the number of upconverters and downconverters which must be provided at earth stations. Although the interconnection between receive and transmit beams is varied during the course of a frame, the same pattern is repeated from frame to frame and is only altered (under command from the earth) at infrequent intervals when a new traffic plan is required.

2.4.3.3 On-board processing

It is not possible to draw a clear dividing line between transparent communications subsystems and those using on-board processing. Even the switching matrices in the INTELSAT V communications module have been called on-board processors but they hardly seem to merit this description. On the other hand, SS–TDMA can certainly be regarded as on-board processing of a relatively simple kind.

Some of the earlier communications satellites were described, in newspapers, as 'telephone exchanges in the sky' (which they were not); with the advent of on-board processing, satellites could really become exchanges in the sky (i.e. they could route connections between individual terminals on demand) but it is doubtful whether there would be any advantage in this. It seems more likely that on-board processing will find use in reducing the cost and increasing the flexibility of business communications-satellite systems using many small earth stations.

There are many possible forms of on-board processing, for example:

(i) *On-board regeneration:* Terrestrial digital systems are able to keep the number of errors arising from Gaussian noise to an arbitrarily low figure by reducing the distance between regenerators. The only point in a satellite channel where a regenerator could be located is the satellite itself, but this single stage of regeneration could be worthwhile. Consider, for example, a system for which the maximum allowable bit error rate (BER) is 1 in 10^6 and suppose that this corresponds, when using QPSK modulation, to a carrier-to-noise power ratio C/N at the demodulator of 14 dB. The noise power on the uplink and downlink will add so that, for equal C/N on each link and in the absence of regeneration, the ratio on each link must be 17 dB to give an overall ratio of 14 dB. However, with regeneration at the satellite it is the error rates on the two links which add (not the noise powers). Thus if the error rate on both links is the same, the allowable BER per link is 5 in 10^7 which would correspond to a C/N of about 14·3 dB; the use of a regenerator would thus reduce the required EIRP by nearly 3 dB on each link. There may be much less advantage in regeneration at the satellite if the *C/N* ratios available on the uplink and downlink are significantly different. For example, suppose that the allowable BER is 1 in 10^6, as in the previous case, but that in this instance a *C/N* of about 21 dB is available on the uplink. When using a transparent satellite, the required downlink *C/N* (to give an overall *C/N* of 14 dB) is 15 dB. Now suppose that a regenerator is introduced at the satellite. The BER on the uplink will be many orders below 1 in 10^6 and it would therefore be possible to reduce the power on the downlink by 1 dB (to give a *C/N* of 14 dB); alternatively, by making the error rates on the two paths equal, it would be possible (as before) to reduce the *C/N* on both paths to 14·3 dB which corresponds to a reduction of 6·7 dB in the up-path power but only 0·7 dB in the down-path power. Unfortunately, in most satellite systems it is satellite (i.e. down-path) power which is in short supply.

(ii) *Conversion and retiming of data:* It would be possible for the satellite to accept a range of lower-rate data streams from small earth stations and combine them into higher-rate streams for transmission to large stations and vice versa; this would enable earth stations to be dimensioned according to their traffic needs. It would also be possible for the satellite to receive bursts of data, store them and retransmit them as a TDM stream (using an on-board clock for timing); this would avoid the necessity of having burst demodulators at earth stations.

(iii) *Adaptive processing:* A range of bit rates and coding methods could be made available. The on-board processor would monitor the BER on each uplink and downlink and determine, for each link, which of the bit rates and coding methods were required to maintain the desired performance with the conditions existing on that link. The processor would then allocate the appropriate satellite channels and issue instructions to the earth stations. In this way the system could adapt to changes in propagation conditions, the advent of interference etc.

(iv) *Assignment of channels:* In some systems (particularly mobile systems) channels are allocated by processors at earth stations and system management data has to be transmitted (and possibly retransmitted) via the satellite. For example, in the INMARSAT Standard-A system a request from a ship earth station (SES) for a telephone circuit is received by a coast earth station (CES), the CES then passes a message via the satellite to a network co-ordination station (NCS) and the NCS allocates channels and transmits an assignment message giving details of the channels to be used; thus the satellite is used three times to set up one call. If an on-board processor were used to allocate channels the time required to set up a connection and the system overheads would be significantly reduced.

On-board processing must inevitably add to the complexity, power consumption, mass and cost of communications satellites, and increase the difficulties of achieving high reliability, long life and compatibility with other systems. The testing and debugging of software, which is a major problem with terrestrial equipment, will be even more difficult for processors on satellites and, once debugged, the software must be protected from damage by radiation and electrostatic discharges. Complex processors have, of course, been flown on military and scientific satellites but the cost of these is not judged by the same criteria as the cost of communications satellites. A good example of the problems that can arise was met in the development of INTELSAT VI; equipment for protecting the software for the relatively simple switching matrix takes up an undesirably large percentage of the payload and a proposed on-board monitor for the matrix had to be abandoned because the cost would have been prohibitive.

2.5 Reliability

Reliability is the probability that a unit will function to a specified standard for a given period of time in a prescribed environment.

The probability that a communications satellite will perform the functions required of it depends on the probability of a successful launch as well as on the reliability of the satellite. The reliability of both satellite and launcher must therefore be considered together and some of the biggest risks in the life of a satellite occur during launch.

Both satellites and launchers are complex systems which have to work in a very difficult environment. Achieving good reliability therefore requires extremely careful design and quality assurance, together with an extensive programme of testing.

It is often possible to make quite accurate forecasts of the reliability of communications equipment if it is made from mass-produced components; this is because sufficient data has been amassed for the probability of failure of the components (when operated in normal conditions) to be known with a high degree of confidence. However, satellites and launchers often use newly-developed components and processes for the production of modules which must survive and function under very difficult conditions. This can make the forecasting of reliability very difficult; for example, predictions by different forecasters of the probability of successful operation of the solid rocket boosters used with the US Space Shuttle ranged from 99·999% (i.e. a 1 in 500 chance of a failure in the first 200 flights) to about 99% (i.e. a 1 in 5 chance of failure in the first twenty flights). The disastrous failure of *Challenger* came after just over 20 flights.

Even the most careful testing on the ground cannot exactly simulate conditions in space or explore performance under every possible combination of circumstances and it usually takes many successful missions before confidence in a new design is firmly established. Furthermore, even apparently trivial changes to a successful design have been known to cause catastrophic failure. Once a design has been proved in space there is thus very good reason to use it with little, or preferably no, alteration for as long as possible. On the other hand, the rapid rate of development of satellite communications makes it very difficult to use only well-tried designs and components.

A successful launch of a communications satellite depends on the correct operation of the launch vehicle, proper separation of the satellite from the vehicle (a satellite and its launch vehicle have been known to collide after separation), and correct operation of the apogee stage. Of the large number of launch vehicles fired by the USA about 10% are known to have failed. However, the apparent reliability of launchers and satellites has varied quite widely over the years and predictions of the reliability of particular spacecraft cannot usually be made with a high degree of confidence. Three out of the eight INTELSAT III satellites (which were launched in 1969 and the early 1970s) failed to reach geostationary orbit, and two of the remaining five developed problems before the end of their design life. On the other hand, experience with INTELSAT IV and IVA (launched by the well-tried Atlas–Centaur) in the late 1970s was reassuring. In the early 1980s, launch cover could usually be obtained for 10% (or less) of the sum insured and cover against failure in orbit cost about 1% per annum (for the first three years of life).

Unfortunately, there were rather a large number of failures of launch vehicles and satellites in 1984, 1985 and the first half of 1986. The cost of a single failure can be as much as $100M and it has been claimed that the net loss to the underwriters in this period was over $500M. As a result premiums for launch shot up to as much as 25–30% and underwriters have at times been unwilling to give a firm quote more than three months ahead of launch. In consequence some communications satellites are now being launched without insurance and some purchasers are stipulating that they will only take delivery of their satellites when they are safely in orbit.

There is no doubt that the present problems with launchers and satellites will be overcome and confidence will be restored. In the meantime it seems possible that, until insurance rates come down, investment in communications-satellite systems may be reduced. This stresses the particular importance of quality assurance in the manufacture and launch of communications satellites.

The design life of a satellite is determined largely by the amount of fuel it can carry for correction of position and attitude, but deterioration or failure of active components such as TWTs is another important factor. As an example of the reliability which ought to be achievable, it is estimated:

(i) There is a high probability that all of the 20 TWT amplifiers on board an INTELSAT V will be operational two years after launch

(ii) On average, 18 TWTs will still be operational after three years
(iii) Sixteen TWTs will survive for over seven years.

It has also been estimated that there is a probability of at least 0·76 that 65% of the channels in each of the antenna beams will be operational at the end of seven years.

The most common method of improving reliability is to provide standby units which can be switched in either manually or automatically in case of failure of a working unit (this is known as providing 'redundancy'). If the reliability of a system subject to random failures over a given period is, say, 0·99 then the probability of failure during that period is 0·01. The probability of failure of both the working unit and an identical standby unit in the same period of time is approximately $(0{\cdot}01)^2 = 0{\cdot}0001$. Thus the reliability of a unit can be improved from 0·99 to about 0·9999 by the provision of a standby unit.

Chapter 3

The RF transmission path

3.1 The link equations

The signal-to-noise power ratio in a baseband channel of a radio transmission depends mainly on the coding and modulation methods used, and the radio-frequency (RF) carrier-to-noise power ratio at the input to the receiver (i.e. before demodulation). In this Chapter we consider the factors affecting the RF carrier-to-noise power ratio. The effects of modulation and coding are dealt with in Chapters 4 and 5.

The RF carrier-to-noise power ration (C/N) at the receiving end of a radio link depends on

- Power delivered to the transmitting antenna
- Gains of the transmitting and receiving antennas
- Propagation loss between the antennas
- Effective noise temperature of the receiving system.

The RF signal may also be degraded by interference (which is becoming an increasingly important factor in some well-used frequency bands).

3.1.1 *Path loss*

Consider an isotropic antenna, i.e. one which radiates power equally in all directions. If the total power radiated by the antenna is P_t watts then the power flux density (PFD) at a distance s metres from the antenna (in free space) is

$$\text{PFD} = P_t/(4\pi s^2) \text{ watts/m}^2 \tag{3.1}$$

$4\pi s^2$, the area of the surface of the sphere or radius s, is called the 'spreading area'.

If the effective area of the receiving antenna is A square metres, and it is at distance s metres from the transmitting antenna, the power received P_r is

$$P_r = A\ (\text{PFD}) = A\ P_t/(4\pi s^2) \text{ watts} \tag{3.2}$$

The relation between the gain G and the effective area of an antenna is

$$A = G\lambda^2/4\pi \text{ metres}^2 \qquad (3.3a)$$

where λ is the wavelength in metres

Thus the effective area of an isotropic antenna (which, by definition, has unit gain) is:

$$A_i = \lambda^2/4\pi \text{ metres}^2 \qquad (3.3b)$$

The power received by an isotropic receiving antenna at a distance s from an isotropic transmitting antenna (radiating P watts) is therefore

$$P_r = A_i \text{ (PFD)} = A_i\ P_t/(4\pi s^2) = P/L \text{ watts} \qquad (3.4)$$

where $L = (4\pi s^2)/A_i = (4\pi s/\lambda)^2$. L (which is the ratio of the spreading area to the effective area of an isotropic antenna) is called the free-space attenuation between isotropic antennas or the path loss.

Now if the transmitting and receiving antennas have gains (relative to isotropic) of G_t and G, respectively, then the power received C is

$$C = P_t\ G_t\ G/L \text{ watts} \qquad (3.5a)$$

$P_t\ G_t$ is called the equivalent isotropically radiated power (EIRP); thus

$$C = \text{EIRP} \times G/L \text{ watts} \qquad (3.5b)$$

Part of any transmission path between an earth station and a satellite traverses the atmosphere and additional losses occur on this section of the path because of the effects of water vapour, rain, cloud etc. (See Section 3.5.2.)

3.1.2 *Noise power*

The thermal noise power N at the input to a receiving terminal is

$$N = kTB \text{ watts} \qquad (3.6)$$

where k = Boltzmann's constant ($1{\cdot}38 \times 10^{-23}$ J/K)
T = effective noise temperature of the terminal, K
B = noise bandwidth of the receiver, Hz

The effective noise temperature of the terminal depends on a number of factors of which the most important are the angle of elevation of the antenna, the weather conditions on the transmission path between the satellite and the terminal, the type of antenna and the noise temperature of the first stage of the receiver.

3.1.3 *C/N and G/T*

From equations 3.5 and 3.6 the carrier-to-noise power ratio C/N is

$$C/N = \mathrm{EIRP}(1/L)(G/T)(1/kB) \tag{3.7}$$

Note that C and N must be referred to, or measured at, the same point; thus if N is the noise referred to the input of the receiver then C must be the RF power at the input to the receiver (i.e. any attenuation of the signal between the antenna output and the receiver input must be taken into account).

It will be seen that C/N is proportional to the ratio G/T (which is sometimes called the 'figure of merit'); G/T is usually the most important characteristic of an earth station or a satellite transponder operating in the receive mode. Both the gain of an antenna and the system noise temperature depend on the point at which they are measured or to which they are referred, but the ratio G/T is independent of the point of reference (see Section 7.4.2).

3.1.4 *The link equations*

Relation 3.7 can be rewritten as

$$[C/N] = [\mathrm{EIRP}] - [L] + [G/T] - [k] - [B] \text{ decibels} \tag{3.8}$$

where $[x]$ denotes $10\log x$.

Note: At one time the decibel was used only in connection with power ratios but it is now common practice to use it to denote the logarithmic ratio of other quantities; thus, for example, a bandwidth of 1 MHz is written as 60 dB(Hz), 200 K is written as 23 dB(K) and Boltzmann's constant is written as −228·6 dB(J/K).

In a satellite-communications system the overall RF link is made up of an uplink (i.e. the link from a transmitting earth station to the satellite) and a downlink (i.e. the link from the satellite to the receiving earth station). There are thus two link equations, namely

$$[C/N]_u = [\mathrm{EIRP}]_e - [L]_u + [G/T]_s - [k] - [B] \text{ decibels} \tag{3.9a}$$

$$[C/N]_d = [\mathrm{EIRP}]_s - [L]_d + [G/T]_e - [k] - [B] \text{ decibels} \tag{3.9b}$$

where the suffixes u, d, s, and e refer to the uplink, downlink, satellite and earth terminal, respectively.

The free-space path loss L is, by definition, $4\pi s^2 \times (1/A_i)$ and thus:

$$[L] = 10\log(4\pi s^2) + 10\log(1/A_i) \text{ decibels} \tag{3.10}$$

For links with satellites in the geostationary orbit, the first term of 3·10 (the spreading area) ranges in value from 162 dB(m^2) at 90° elevation to 163·3 dB(m^2) at 5° elevation; $10\log(1/A_i)$ is 33·5, 37·0, 42·3 and 44·4 dB($1/m^2$) at frequencies of 4, 6, 11 and 14 GHz, respectively. Thus $[L_u]$ and $[L_d]$ for 6/4 GHz links at an elevation angle of 5° are (163·3 + 37) = 200·3 dB and (163·3 + 33·5) = 196·8 dB, respectively.

Atmospheric losses (see Section 3.5.2) are significant, except in clear weather and at low frequencies, and must be added to the free-space path losses.

The civil communications satellites at present in use are all 'transparent', that is they amplify the received signals, change the frequencies of the RF carriers and then retransmit them but do not change the characteristics of the signals in any other way. The noise powers on the uplink and downlink add arithmetically when transparent satellites are used. Thus C/N at the receiving earth station is given by the expression

$$1/(C/N) = 1/(C/N)_u + 1/(C/N)_d \tag{3.11}$$

where the carrier-to-noise ratios (C/N) etc. are straightforward numerical ratios (i.e. they are not in decibels). For example, if $[C/N]_u$ is 20 dB and $[C/N]_d$ is 14 dB then $(C/N)_u = 100$ and $(C/N)_d = 25$.

Thus

$$1/(C/N) = 1/100 + 1/25 = 1/20$$

and $[C/N]$ is therefore 13 dB

When, in future years, satellites with on-board processing are used (see Section 2.4.3.3) the noise on the uplink and downlink will not necessarily add in this way.

(Note: the square brackets for quantities expressed in decibels will be omitted in future when there is no risk of confusion).

Only the degradation of the signal caused by the addition of thermal noise has been considered so far. There are other sources of degradation such as intermodulation in the satellite transponder (see Section 3.4), variation of group delay with frequency (see Section 4.3.2) and interference (see Section 3.6). The effects of these other sources may often be treated as noise and added arithmetically to the thermal noise without introducing significant errors. Thus, for example, if the carrier-to-thermal-noise ratio $[(C/N)_{th}]$ is 20 dB and the carrier-to-intermodulation-noise ratio $[(C/N)_i]$ is 15 dB then the resultant carrier-to-noise ratio (C/N) is usually derived as follows:

$$\begin{aligned} 1/(C/N) &= 1/(C/N)_{th} + 1/(C/N)_i \\ &= 1/100 \quad + 1/31{\cdot}6 \\ &= 1/24 \end{aligned}$$

that is $C/N = 13{\cdot}8$ dB

Satellite-communications systems using frequency modulation and conventional demodulators are usually designed so that the overall C/N ratio is more than 10 dB for all but a small percentage of the time.

3.1.5 C/N_0, C/T, E_b/N_0

Other ratios related to C/N are C/N_0, C/T and E_b/N_0 (the latter ratio only has a meaning in relation to digital systems). All three of these ratios are, unlike C/N, independent of bandwidth.

N_0 = noise power density, i.e. $N_0 = N/B = kT$ watts/Hz

T = effective system noise temperature, i.e. N/kB

E_b = energy per bit, i.e. C/R (where R = bit rate in bit/s)

Thus

$$C/N_0 = C/kT = (C/N)B$$

$$= \mathrm{EIRP} - L + G/T - k \text{ dB(Hz)} \quad (3.12)$$

$$C/T = \mathrm{EIRP} - L + G/T \text{ dB(W/K)} \quad (3.13)$$

$$E_b/N_0 = \mathrm{EIRP} - L + G/T - k - R \text{ dB} \quad (3.14)$$

Working values of C/T for point-to-point links range from about −130 dB(W/K) for carriers with large traffic capacities to about −174 dB(W/K) for single-channel-per-carrier (SCPC) or spread-spectrum (SS) systems.

The theoretical value of E_b/N_0 required to give an error ratio of 10^{-6} when using QPSK modulation is 10·5 dB; the practical value required is usually 2 or 3 dB greater than this to allow for the imperfections of equipment and the effects of interference.

3.1.6 Link budgets

The two link equations 3.9*a* and 3.9*b* are used to prepare the link budget which plays an essential part in the planning of every communications-satellite system. For example, when it is proposed to set up a new communications link via an existing satellite, a link budget is used to check that the required carrier-to-noise power ratio (C/N) at the receiver can be achieved using the bandwidth and satellite power allocated to the service, and to establish what EIRP and G/T are necessary at the transmit and receive earth stations, respectively.

The rest of the Chapter is devoted to further discussion of factors which may affect link budgets and some examples of budgets are presented in Appendix 3.7.

3.2 Earth terminal characteristics

3.2.1 Antenna gain

The gain of an earth terminal antenna is an important factor in achieving the required EIRP on the up path and G/T on the down path.

The gain can be derived from expression 3.3*a* which can be rewritten as:

$$G(\mathrm{dB}) = 10 \log E(\pi D f/0{\cdot}3)^2 \qquad (3.14)$$

where G = gain in dB relative to isotropic (the unit is, strictly speaking, dBi but the 'i' is usually omitted)
f = frequency, GHz
D = diameter of the aperture, m
E = efficiency of the antenna

The effective area of an antenna is less than the physical area of its aperture (i.e. E is less than 1) because of factors such as errors in the reflector profile, the spillover of energy past the reflector and blockage of the radiating aperture (see Section 7.3.3). The apparent efficiency also depends on the point at which the gain is measured. For example, the gain of an antenna may be given in manufacturers' literature as that which would be measured at the port of the orthomode transducer (OMT), which is the waveguide device used to combine or separate two orthogonally-polarised waves; however, there are usually additional losses between the OMT and the input to the earth-station receiver. For an antenna with a complex feed (e.g. one incorporating monopulse tracking components) these losses may amount to as much as 0·25 dB, which is equivalent to reducing the efficiency of the antenna by more than 5%, but for a small antenna with a simple feed the losses will probably be negligible, provided that the first stage of the receiver is mounted close to the feed.

The diameters of earth-station antennas range from less than 1 m to about 32 m. There are quite a lot of 32 m antennas in existence at earth stations carrying large amounts of traffic via 6/4 GHz links but antennas as large as this are not usually necessary with modern satellites and they are very expensive; the largest antennas built in future are therefore likely to have a diameter of 20 m or less. Efficiencies in the receive frequency band range from about 0·55 to 0·8 (the smaller antennas usually have efficiencies towards the lower end of this range). Because of the relatively low EIRP of satellites it is more difficult to achieve a satisfactory C/N on the down path than on the up path and earth-terminal antennas are therefore usually optimised for the receive frequency band; as a consequence their efficiency in the transmit frequency band is sometimes rather low.

The gains and efficiencies of some large and medium-sized antennas manufactured by some well-known companies are given in Table 3.1.

Table 3.1 *Antenna gains and efficiencies*

Diameter (m)	Frequency (GHz)	Gain (dB)	Efficiency	Notes
3	3·95	39·5	0·58	*a, b, c*
5	3·95	44·5	0·66	*c, d*
	6·175	47·3	0·51	
11	3·95	52·0	0·77	*c, d*
	6·175	54·8	0·60	
32	3·95	61·0	0·72	*d, e, f, h*
	6·175	64·2	0.61	
5·5	11·95	55·0	0·67	*c, d*
	14·25	56·3	0·63	
7·7	11·95	58·0	0·67	*c, d*
	14·25	59·3	0·63	
11·5	11·95	60·8	0·58	*e, f, g*
	14·25	61·9	0·53	

Notes
(*a*) Front-fed antenna
(*b*) Receive-only antenna
(*c*) Gain at OMT port
(*d*) Cassegrain antenna
(*e*) Dual-polarised operation in both receive and transmit bands
(*f*) Gain referred to interfaces between feed and earth station equipment
(*g*) Offset Cassegrain

3.2.2 *Noise temperature*

The noise temperature T_e of an earth terminal is the sum of the noise temperature of the antenna T_a and the noise temperature of the rest of the receiving system T_r referred to some specified point, which is usually the input to the low-noise amplifier (LNA). T_a depends mainly on the elevation angle, the frequency and the weather, and also (to a lesser extent) on the design of the antenna. As in the case of antenna gain, it is important to know to what point the quoted noise temperature refers; if, for example, the noise temperature quoted is 50 K at the OMT and there is a further loss of 0·25 dB between the OMT and an output port of the feed then the noise temperature at the output port will be 66 K. Typical values of antenna noise temperature (at the OMT flange) at a frequency of 4 GHz in clear weather are 20–25 K at an elevation angle of 40° and 45–50 K at an elevation angle of 5°; at 11 GHz and 5° elevation a noise temperature of 70–75 K is typical. T_r is usually dependent mainly on the noise temperature of the LNA but it is also affected by the attenuation of the cable or waveguide connecting the antenna to the LNA. The noise temperatures of the best thermo-electrically cooled LNAs are about 30 K at 4 GHz and 80 K at 11 GHz. Further discussion of antenna and system noise temperatures will be found in Sections 7.3 and 7.4.

The G/T of earth stations ranges from -4 dB(1/K) for an INMARSAT Standard-A ship earth station (SES), which uses an antenna with a diameter of about 1 m to receive transmissions at 1·6 GHz, to over 40 dB(1/K) for earth stations using antennas with a diameter of about 30 m to receive transmissions at 4 GHz. Mobile terminals with lower values of G/T will soon be coming into use; for example the INMARSAT Standard-C SES, which is intended for the transmission and reception of telex and low-speed data, will use an antenna with a G/T of about -23 dB(1/K).

3.2.3 Solar interference

Fig. 3.1*a* shows an earth station E and a geostationary satelite S. For simplicity, the station and the satellite are assumed to be at the same longitude and they are shown at the moment when they and the sun are in the same plane. The line of sight ES from the earth station to the satellite meets the equatorial plane at an angle ψ. At the equinoxes the sun is in the equatorial plane and an earth-station antenna looking at a geostationary satellite does not see the sun (unless the station is on the equator). During the period between successive equinoxes the declination (i.e. angular distance from the equatorial plane) of the sun rises from zero to a maximum of 23° at the solstice and then diminishes to zero again (see Fig. 3.1*b*). The declination ϕ in degrees at a time t days after the autumnal equinox in the northern hemisphere is given approximately by

$$\phi = 23 \sin(2\pi t/365) \tag{3.15}$$

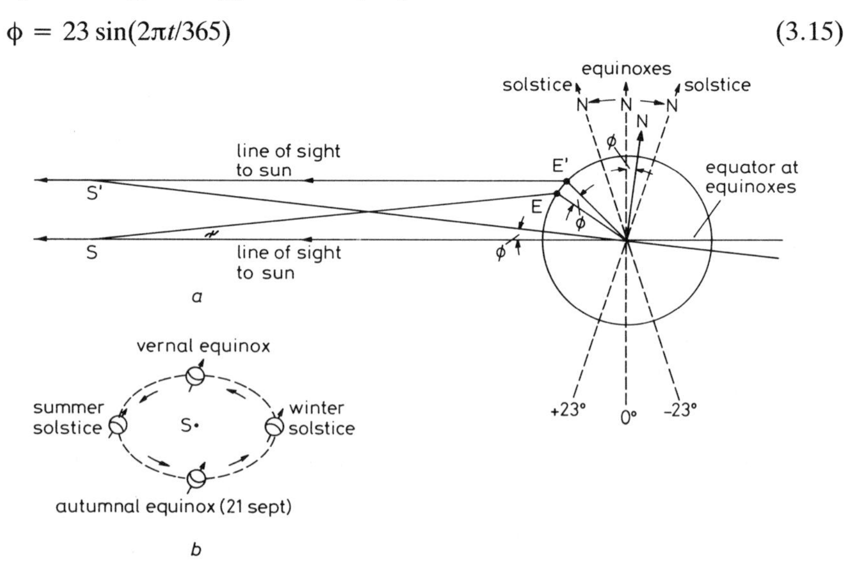

Fig. 3.1 *Solar interference*

a Geometry of solar interference
b Equinoxes and solstices (northern hemisphere)

(The convention used here is that declination S and declination N are positive and negative angles, respectively.)

When ϕ is approximately equal to ψ then the earth station (at position E′), the satellite (at position S′) and the sun will all be approximately in line at some time during the day and the sun will cross the beam of the earth-station antenna. The earth subtends an angle of about 17·4° from a geostationary satellite and the corresponding maximum value of ψ is therefore 8·7°. The values of t corresponding to a declination of ±8·7° are 22 days and 160 days. For a station in the northern hemisphere solar interference therefore occurs on a few successive days during the 22 days after the autumnal equinox and on a few successive days during the 22 days before the vernal equinox. Similarly, for a station in the southern hemisphere periods of solar interference occur during the 22 days after the vernal equinox and the 22 days before the autumnal equinox.

The noise temperature of the quiet sun decreases with rising frequency from about 20 000 K at 4 GHz to about 10 000 K at frequencies of 12 GHz and above (but can be significantly higher when there is sunspot activity). The sun subtends an angle of about 0·5° at the surface of the earth, and antennas with gains of greater than about 50 dB have 3 dB beamwidths of less than 0·5°. The sun can therefore fill the beam of a high-gain antenna; when this happens the antenna noise temperature rises to that of the sun and an outage occurs. For an earth-station antenna with a 3 dB beamwidth (θ_3) greater than 0·5° the maximum noise temperature caused by solar interference is reduced by a factor of about $(\theta_3/0{\cdot}5)^2$; thus terminals using relatively low-gain antennas may suffer periods of degraded service (rather than outages) and with VSATs (very-small-aperture terminals) the solar interference may go unnoticed. Solar interference is not usually very bothersome, even for terminals using large antennas, because it occurs at predictable times and only for short periods; however there is no way of avoiding it (except by the use of alternative means of transmission such as a second earth station at a different geographical location).

The maximum time per day for which interference is experienced depends on the angular speed of rotation of the earth (approximately 0·25° per minute) and the angular zone of interference (which is determined by the angle subtended by the sun at the earth, the beamwidth of the antenna and the level of solar interference which can be tolerated). Similarly, the number of days on which interference occurs depends on the rate of change of declination (about 0·4° per day during periods near the equinoxes) and the angular zone of interference. With high-gain antennas outages are experienced for about five days a year with the maximum outage being about 5–7 mins. Simple empirical expressions for forecasting the duration of solar interference corresponding to any size of antenna have been propounded (see, for example, p.212 of the CCIR Handbook on Satellite Communications, 1985) but these need to be used with caution.

3.2.4 EIRP

The EIRPs of transmissions from existing earth stations range from about 20 dBW (for a data transmission at about 20 kbit/s via a very small antenna) to nearly 90 dBW (for some television and multichannel telephony transmissions in the INTELSAT system). These EIRP's are achieved using:

(i) Antennas with gains ranging from about 20 dB (e.g. the gain of an antenna with a diameter of about 1 m transmitting at a frequency of around 1·5 GHz, as used in an INMARSAT Standard-A ship earth station) to about 66 dB (for an antenna with a diameter of 19 m transmitting at a frequency of 14 GHz)
(ii) Transmitter powers of 0–25 dBW (delivered to the antenna).

EIRPs even lower than 20 dBW will be transmitted by INMARSAT Standard-C terminals when these come into use in 1989.

It is necessary to limit the EIRP radiated by an earth station in directions away from the main beam in order to avoid unacceptable interference to other radio stations. In particular it is necessary to limit

(i) The EIRP of transmissions at low angles of elevation (which might interfere with terrestrial stations sharing the frequency band)
(ii) Off-axis radiation towards the geostationary orbit (which might interfere with nearby satellites).

The effect of an interfering transmission of wide bandwidth is usually much less than that of an interfering transmission with the same power but narrower bandwidth; the limits on EIRP are therefore usually expressed in terms of maximum allowable EIRP in a given bandwidth. Thus, for example, the ITU Radio Regulations state that the EIRP transmitted towards the horizon by an earth station shall not exceed $(40 + 3\theta)$ dBW in any 4 kHz band (for frequencies between 1 and 15 GHz) or $(64 + 3\theta)$ dBW in any 1 MHz band (for frequencies above 15 GHz), θ being the angle of elevation (in degrees) of the horizon as viewed from the centre of radiation of the antenna.

3.2.5 Antenna gain off axis

The gain of an antenna in directions more than a few degrees off its axis usually depends more on the physical and electrical characteristics of the antenna than on its size (see Section 7.3.3). Any limitation on off-axis EIRP therefore imposes a limit on the transmitter power which may be fed to the antenna and this can be a particular problem for systems using small antennas because the power required to give a satisfactory carrier-to-noise ratio may make it difficult to avoid interference to neighbouring systems.

3.2.6 *Antenna beamwidth and tracking*

An approximate expression for the half-power (3 dB) beamwidth θ_3 of an antenna is:

$$\theta_3^\circ = 65\lambda/D \qquad (3.16)$$

where λ = wavelength of the transmission
D = diameter of the antenna (in the same units as the wavelength)

The reduction in gain ΔG, relative to the on-axis value for a small angle offset $\Delta\theta$, is given by

$$\Delta G = 12(\Delta\theta/\theta_3)^2 \text{ decibels} \qquad (3.17)$$

For small antennas the beamwidth is sufficiently great for the variation in gain caused by the movement (say ±0·1° N–S and E–W) of a typical geostationary satellite to be acceptable; however, for antennas with a gain greater than about 50 dB a tracking system (see Section 7.3.5) is usually considered necessary. When tracking is used the loss of gain from pointing errors is normally less than 0·5 dB.

3.2.7 *Polarisation*

(Note: the use of orthogonally-polarised antennas is discussed in detail in Section 7.3.4)

Faraday rotation (see Section 3.5.1) is significant in the 6/4 GHz (and lower) frequency bands; circularly-polarised transmissions, which are unaffected by the rotation, are therefore normally used for satellite communications at these frequencies.

Linearly-polarised antennas are often used at the higher frequencies (where Faraday rotation is negligible) because the cross-polar discrimination between a pair of orthogonal linearly-polarised transmissions during rain is rather better than that between a pair of circularly-polarised transmissions; however, a linearly-polarised earth-station antenna has the disadvantage that its direction of polarisation must be lined up with that of the corresponding satellite antenna.

The voltage axial ratio (VAR) of an antenna is a major factor in determining what cross-polar discrimination (XPD) can be achieved; it must therefore always be specified for antennas working in a system using orthogonally-polarised transmissions. For example, INTELSAT specifies that:

(i) The VAR of a large circularly-polarised antenna for a Standard-A earth station (operating at 6/4 GHz) shall be not greater than 1·06 which corresponds to an XPD of about 30 dB.
(ii) The VAR of a small circularly-polarised antenna for a Standard D–1

(6/4 GHz) earth station shall be not greater than 1·3 which corresponds to an XPD of about 18 dB.

(iii) The VAR of a large linearly-polarised antenna for a Standard-C (14/11 GHz) earth station shall be not less than 31·6 which corresponds to an XPD of about 30 dB.

3.3 Satellite characteristics

The characteristics of a satellite transponder which are of importance in the preparation of an RF budget are:

(i) Figure of merit (G/T)

(ii) EIRP of the transponder (normally specified as the EIRP which would be developed if the transponder output amplifier were working at saturation, i.e. at maximum output power)

(iii) Power flux density (PFD) required at the satellite receive antenna to saturate the transponder output amplifiers.

If a transponder can be used in a number of different modes (e.g. with global, hemi, zone and spot beams) then the characteristics must be specified for each of these modes. Allowance must be made for the variation in gain of satellite antennas over their coverage areas because the gain in the direction of an earth station at the edge of the area may be 3 or 4 dB less than the maximum (on-beam) gain. The satellite specification usually states the minimum required EIRP and G/T in the direction of the edge of the coverage area.

3.3.1 Antennas

Satellite antennas, unlike earth-station antennas, are rarely pointed at a single station but must give service to any station in a designated coverage area; this imposes a constraint on antenna gain. For example, the INTELSAT V (west) spot-beam antenna has a beamwidth of 1·6° which corresponds to a gain of around 40 dB, while a global-coverage antenna has a beamwidth of about 18° which corresponds to a gain of only about 20 dB.

When the system uses orthogonally-polarised transmissions the VAR of the satellite antennas must be specified; the XPD required is typically about 30 dB. This XPD must be maintained for transmissions received from or directed to any part of the antenna coverage area (e.g. over the 3 dB beamwidth); this is generally a much more stringent requirement than that for earth-station antennas which are usually directed so that the satellite appears close to the centre of the antenna beam.

3.3.2 G/T

The noise temperature at the input to a satellite receiver is high compared with that at the input to most earth-station receivers because satellite antennas look at the earth, which has a physical temperature of about 290 K; the first stages of satellite receivers are usually therefore fairly simple transistor or tunnel-diode amplifiers.

The *G/T* of satellite transponders is limited by the relatively low antenna gains and the high input noise temperatures. Typical values for current satellites are +3 dB(1/K) for an INTELSAT V spot-beam transponder (receiving transmissions at 14 GHz) and −19 db(1/K) for a global-beam transponder (receiving transmissions at 6 GHz). (The specification for the global beam of an INTELSAT VII satellite is −12 dB(1/K) but the first of these satellites will not be launched before 1993.)

3.3.3 EIRP

The saturated output power of satellite output amplifiers for telephony and data transmissions, and for business services, ranges from about 2 W to about 30 W and EIRPs at the edge of the coverage area range from about 24 dBW to 46 dBW. However, the direct broadcast of TV to domestic earth stations and the point-to-multipoint relay of television programmes require higher output powers and EIRPs. For example:

(i) The French and German direct-broadcast satellites (TDF–1 and TV–Sat) have output amplifiers with powers of over 200 W and transponder EIRPs of 65 dBW.

(ii) EUTELSAT II satellites (the first of which should be launched in 1989) will include a transponder with an EIRP of about 50 dBW for the distribution of programmes to earth stations feeding cable-television systems.

The output stage of a transponder may be backed off (i.e. worked at less than the maximum power of which it is capable) in order to limit the effects of non-linearity. The optimum backoff depends on the multiplexing, modulation and multiple-access methods used and on the type of output stage; output backoffs for TWT amplifiers range from as little as 0·2 dB for an amplifier carrying a single time-division-multiple-access (TDMA) transmission to 10 dB for a transponder carrying many carriers in frequency-division multiple access (FDMA). The input backoff is greater than the corresponding output backoff because of the non-linear characteristic of the amplifier; for example, an output backoff of 5 or 6 dB might correspond to an input backoff of 10 dB for a TWT or klystron amplifier. Backoff is discussed further in Section 3.4.

The PFDs required to saturate the output stage of a satellite transponder vary with satellite design and method of use and depend in particular on the effective area of the receiving antenna and the overall gain of the satellite. The PFD required for the saturation of most transponders falls in the range −90 to −70 dB(W/m^2).

3.3.4 *Transponder gain*

The overall gain of a satellite transponder is, typically, about 100–120 dB. Thus, for example, the PFD required at the satellite to saturate a transponder might be $-67\,\text{dB(W/m}^2)$. If the receiving antenna is a global-beam antenna with a gain of 19 dB, then its effective area (at 6 GHz) is $-18\,\text{dB(m}^2)$ and the corresponding power collected by the antenna is -85 dBW. If the satellite EIRP at saturation is $+30$ dBW then the transponder gain is $(30-(-85)) = 115$ dB. If the transponder is worked with an output backoff of 5 dB and this corresponds to an input backoff of 10 dB then the transponder gain is increased by 5 dB to 120 dB.

3.4 Intermodulation resulting from non-linearity of power transfer characteristics

3.4.1 *Harmonics and intermodulation products*

The voltage transfer characteristics of an amplifier such as a travelling-wave tube (TWT) may be represented by a series of the form:

$$V_o = aV_i^2 + bV_i^2 + cV_i^3 + \ldots\ldots \tag{3.18}$$

where V_o = output voltage
V_i = input voltage and
a, b, c, etc. = constants

It can be shown that:

(i) When the input is a single signal at frequency X, the terms bV_i^2, cV_i^3 etc. generate components in the output of the amplifier at the harmonic frequencies $2X$, $3X$ etc.
(ii) When the input comprises two or more signals then the terms bV_i^2, cV_i^3 etc. generate intermodulation products at frequencies which are sums and differences of the input frequencies (X, Y etc.) and their harmonics ($2X$, $3X$ etc. . . . $2Y$, $3Y$ etc.)

Fig. 3.2 shows typical power transfer characteristics of a TWT amplifier. If the amplifier were distortionless (i.e. it had the voltage transfer characteristic $V_o = aV_i$) the power transfer characteristic would be the straight line in the Figure. For any given input power, the difference between the output power corresponding to this straight line and that corresponding to one of the characteristics is the total output power of the harmonics and intermodulation (IM) products. It will be seen that the power in the harmonics and intermodulation products:

(i) is negligible at output powers well below the maximum which the amplifier can deliver but increases rapidly as that maximum is approached

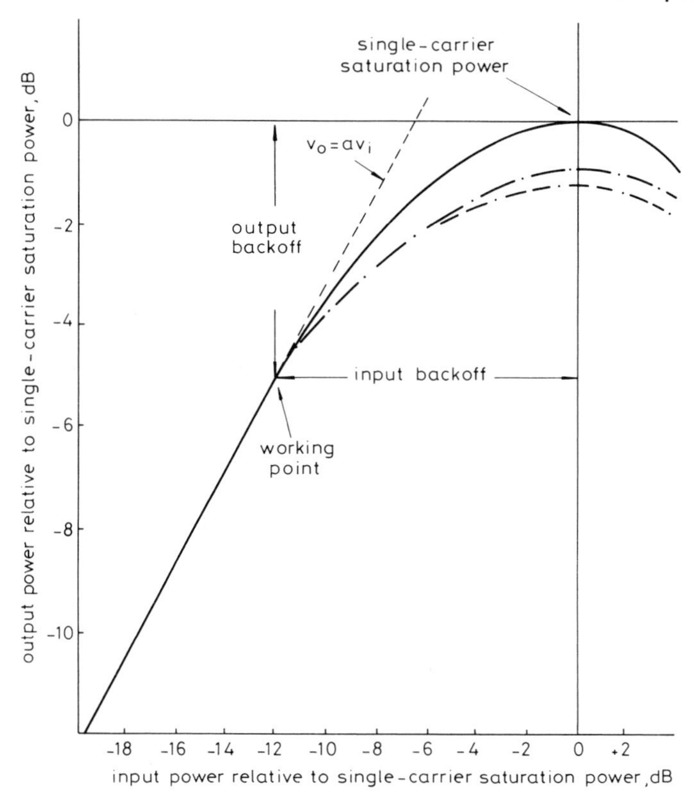

Fig. 3.2 *Power transfer characteristic of a TWT*

—— single carrier
—·— two carriers
·—— eight carriers

(ii) increases with the number of input signals contributing to a given total input power; this is because the number of intermodulation products increases rapidly with the number of input signals.

One of the ways of specifying the non-linearity of an amplifier is by means of the 1 dB compression point; this is the output power at which the power transfer characteristic for two input signals falls 1 dB below the ideal straight-line characteristic.

3.4.2 Third-order IM products

Only the odd-order terms (V_i^3, V_i^5 etc.) of the voltage transfer characteristic generate intermodulation products which fall back in the frequency band occupied by the wanted signals and thus may cause degradation of those signals. Moreover, under normal working conditions, it is usually only the products generated by the third-order term of the characteristic which need to be considered, the power of the products generated by the higher-order terms being small by comparison.

There are two types of third-order intermodulation product which can fall back in the frequency band of the generating signals; these products have frequencies $(2X - Y)$, $(2Y - Z)$ etc. and $(X + Y - Z)$, $(Y + Z - X)$ etc., where X, Y and Z are the frequencies of the signals.

For example, let X, Y and Z be 3740, 3750 and 3760 MHz, respectively (Fig. 3.3*a*). The third-order intermodulation products generated when these signals are amplified in the same TWT are given in Table 3.2 (and see Fig. 3.3*b*).

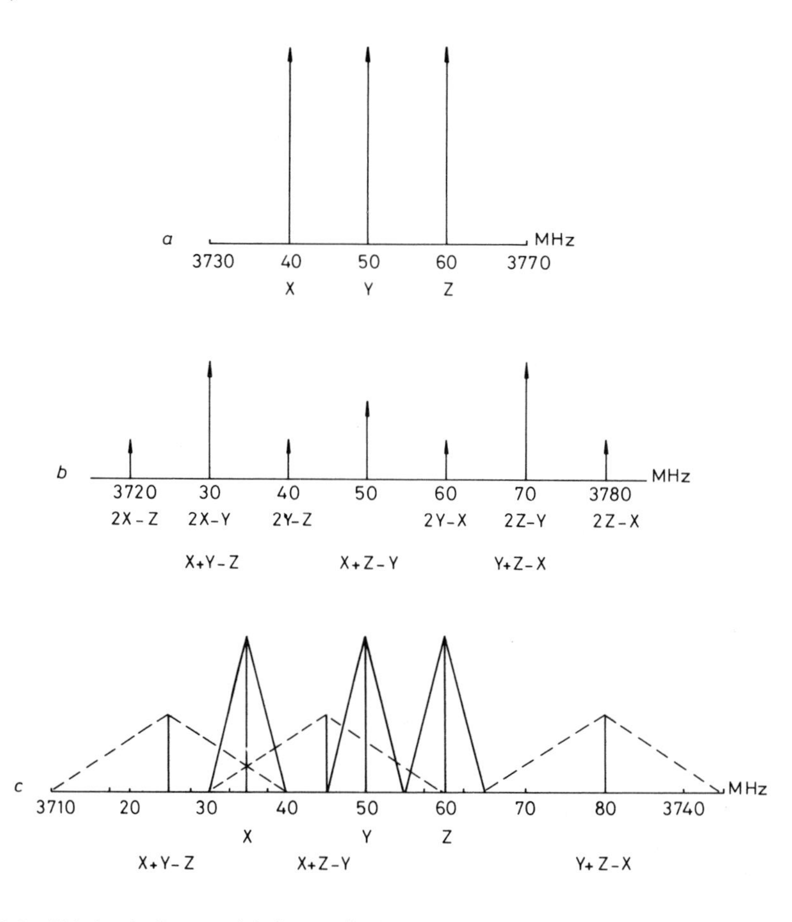

Fig. 3.3 *Third-order intermodulation products*

△ signals
△ intermodulation products

Table 3.2 *Third-order intermodulation products*

Product	Frequency (MHz)
$2X - Z$	3720
$X + Y - Z$	3730
$2X - Y$	3730
$2Y - Z$	3740
$X + Z - Y$	3750
$2Y - Z$	3760
$Y + Z - X$	3770
$2Z - Y$	3770
$2Z - X$	3780

It will be seen that some of the intermodulation products are identical in frequency with the wanted signals; this always happens when, as in this case, the signals have a regular pattern in frequency. The effects of intermodulation can therefore be reduced by irregular spacing of carrier frequencies. However, with some systems (e.g. single-channel-per-carrier systems) the use of irregular frequency patterns would lead to undesirable complication in the earth-station equipment.

It will be seen that the distribution of the intermodulation power varies with frequency: for example, there are two intermodulation products centred on 3730 MHz, one of type $(X + Y - Z)$ and one of type $(2X - Y)$. In this case the maximum intermodulation power is not centred on one of the signal frequencies; however, when there are more than three input signals and they are evenly spaced in frequency the maximum number of intermodulation products always falls on the central signal (or pair of signals).

In practice the signals are usually modulated carriers which occupy a band of frequencies. The bandwidth of an intermodulation product formed from a number of modulated carriers is a function of the bandwidths of the signals. Thus, for example, the bandwidth of a third-order product of form $(X + Y - Z)$ is the sum of the bandwidths of the individual carriers X, Y and Z. Fig. 3.3*c* shows three transmissions whose carrier frequencies have been chosen so that the centre frequencies of the $(X + Y - Z)$ products are not the same as the carrier frequencies; it will be seen that the spectra of two of the intermodulation products overlap the spectra of the signals and interference to the signals may therefore result.

It is often possible to reduce the level of interference from intermodulation products by a suitable choice of carrier frequencies. For example, a satellite transponder transmitting to small earth-station antennas may be power limited and unable to use all of the available bandwidth. There is not usually enough spare bandwidth for it to be practicable to avoid all overlap between intermodulation products and signals but it may be practicable to assign frequencies in such a way that much of the intermodulation power falls in the unused parts of the spectrum. Even when virtually the whole of the allocated bandwidth is occupied by the wanted transmissions the effects of intermodulation can still be reduced by good frequency planning. Thus, where there is a requirement to pass a number of carriers of widely differing power through a single transponder, the effects of intermodulation can be reduced by putting the largest carriers towards the edges of the transponder frequency band; this causes many of the largest intermodulation products to fall outside the band (these products are prevented from interfering with other transponders by means of the band-limiting filters associated with the transponders).

If the number of input signals is N then the number of products of type $(2X - Y)$ is $N(N - 1)$ and the number of products of type $(X + Y - Z)$ is $(1/2)\ N(N - 1)(N - 2)$. This means that:

(i) The total number of products increases rapidly with the number of input signals.
(ii) If N is large there are many more products of type $(X + Y - Z)$ than of type $(2X - Y)$.

For example, for $N = 5$ and $N = 10$ the total number of products are 50 and 450, respectively, and the number of products of type $(X + Y - Z)$ are 30 and 360, respectively. This, together with the fact that products of type $(X + Y - Z)$ have powers about 6 dB greater than products of type $(2X - Y)$ for generating carriers of equal power, means that the $(2X - Y)$ type of product can generally be disregarded except when there are only two or three generating carriers.

The total power of the third-order products is proportional to the cube of the input power and the power of any specific product of type $(X + Y - Z)$ is roughly proportional to P_x, P_y, P_z where P_x, P_y and P_z are the powers of the signals at frequency X, Y and Z, respectively.

3.4.3 Backoff

When overlap between intermodulation products and wanted signals cannot be avoided (as is the case in most FDMA systems) it is usually necessary to use amplifiers at power levels well below the maximum which they are capable of handling.

It will be seen from Fig. 3.2 that as the input power to a TWT amplifier is increased the output power of the wanted signal (or signals) reaches a maximum and then decreases. The maximum signal power available from an amplifier when it is transmitting a single carrier is known as the single-carrier saturated output power. The backoff is defined as the ratio of this single-carrier saturated output power to the total output power at the working level; input backoff (which because of the non-linearity of the power transfer characteristic is always greater than output back-off) is, of course, the ratio of the input power required for saturation to the input power at the working point.

Fig. 3.4 gives an example, for a typical TWT, of the variation of carrier-to-intermodulation power ratio with changing output backoff (approximate input backoff is also given); in this case the input carriers are equally spaced in frequency and of equal power and the ratio (C/I_m) is that of the power of one of the central carriers to the sum of the powers of the intermodulation products falling on it (i.e. the worst case). The figure shows that, for output backoffs of greater than a few decibels, the ratio of the carrier power to the third-order intermodulation power increases by 2 dB for every dB reduction in the input power. This is because a reduction of 1 dB in the carrier power results in a reduction of 3 dB in the power of the third-order intermodulation products.

Although an increase in backoff results in an improvement in the ratio of the total carrier power to the total intermodulation power $(C/N)_{im}$, the resulting decrease in carrier power on both the up path and the down path degrades the corresponding carrier-to-thermal noise ratios, $(C/N)_u$ and

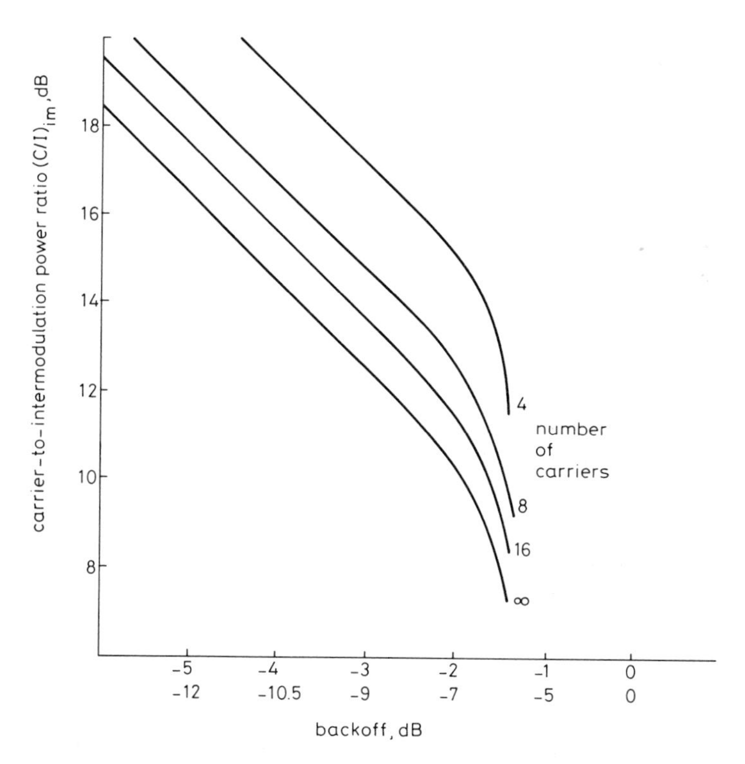

Fig. 3.4 *Carrier-to-intermodulation power ratio as a function of backoff*

$(C/N)_d$. The optimum backoff (see Fig. 3.5) is that which results in the maximum overall ratio of carrier power to intermodulation-plus-thermal noise power (C/N) at the input to the receiver. (In this example it is assumed that the gain of the transponder is fixed. A better overall C/N ratio may be achieved if the gain is variable.)

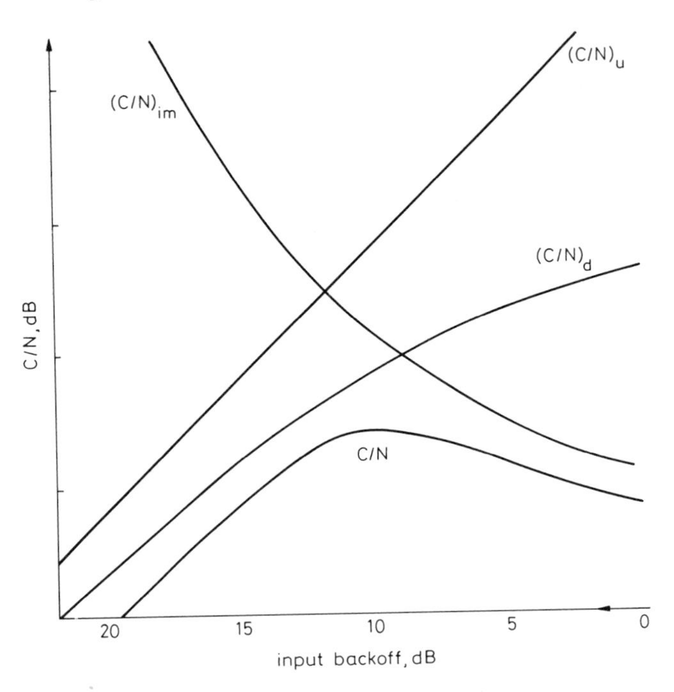

Fig. 3.5 *C/N versus backoff*

Output backoffs of up to about 6 dB are used with FDMA systems using FM and PSK, and up to 10 dB with systems using SSB (the latter being less resistant to interference than FM or PSK).

The spectrum resulting from intermodulation of a large number of carriers can usually be treated as noise. However, it is difficult to calculate the effects of intermodulation between a large number of carriers and, in this case, computer simulation is often used to find the optimum backoff point and frequency plan.

Backed-off amplifiers are inefficient but the avoidance of intermodulation does not usually impose a limit on the EIRP of an earth terminal; any backoff can be offset by using a TWT with a sufficiently high saturated output power or a separate amplifier can be used for each carrier. By contrast there are strict limits on the mass which a satellite can carry and the power which it can generate, and therefore on the number of its output amplifiers and their total saturated output power. Because of this the reduction of efficiency caused by backoff can have serious effects on transponder capacity, and the total traffic capacity of a satellite transponder when it is used to carry a number of small carriers may be 40% or less of the capacity it would have if it were used to carry a single large carrier.

3.5 Propagation

3.5.1 Ionospheric effects

The ionosphere comprises several layers of free electrons and ions which mostly occur at heights between about 50 and 500 km. The ionisation is caused mainly by high-energy radiation from the sun, and the state of the ionosphere varies with sunspot activity, time of day, the season of the year and geographical location.

Transmissions at frequencies below about 30 MHz may be completely reflected or absorbed by the ionosphere. Transmissions at higher frequencies pass through the ionised layers but are attenuated, and with linearly-polarised transmissions the plane of polarisation is rotated (the latter effect being known as Faraday rotation). The magnitudes of both the attenuation and the Faraday rotation vary with the state of the ionosphere and decrease with increasing frequency. The attenuation is negligible at frequencies above a few hundred megahertz. The upper limit of the Faraday rotation is about 100° at 1 GHz, about 5° at 4 GHz and less than 1° at frequencies above 10 GHz.

The variability of Faraday rotation may make it difficult to keep linearly-polarised satellite and earth-station antennas in alignment with the received transmissions; systems working in the 6/4 GHz (or lower) frequency bands therefore often use circular polarisation to avoid this problem.

The attenuation of the ionosphere sets a lower limit to the frequencies which can be used for satellite communications. Frequencies as low as 135 MHz have been used experimentally but the lowest frequency band in use operationally for civil satellite communications is 1·5–1·6 GHz.

3.5.2 Atmospheric effects

3.5.2.1 Clear-sky attenuation

Radio signals travelling through the atmosphere suffer attenuation even during fine weather; this clear-sky attenuation is mainly the result of absorption of energy from the transmissions by water vapour and oxygen molecules. The magnitude of the attenuation rises to very high levels over certain bands of frequencies (the absorption bands). Fig. 3.6 shows the specific attenuation (i.e. attenuation per unit distance) of oxygen and water vapour at frequencies up to 100 GHz and it will be seen that the two lowest-frequency absorption bands are centred at about 22 and 60 GHz. The frequencies used for communication between earth stations and satellites must be in 'windows' well clear of the absorption bands. The frequencies used so far have all been above those where ionospheric attenuation is significant but well below the absorption band at 22 GHz. The ITU has, however, already allocated frequencies in the window between the first two absorption bands for future use by satellite-communications systems; the band 27·5–31·3 MHz has, for example, been allocated to earth-to-space links. Frequencies within the absorption bands could, of course, be used for transmissions between space vehicles.

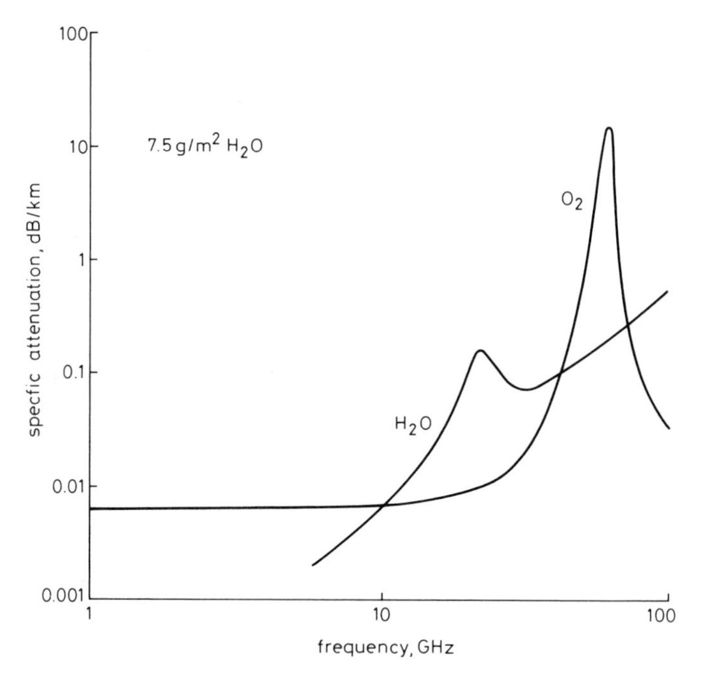

Fig. 3.6 *Atmospheric attenuation (clear sky)*

Table 3.3 gives examples of the loss resulting from water vapour in the atmosphere (when skies are clear) and shows that, in general, the loss increases:

(i) With increasing frequency
(ii) With decreasing elevation (which causes the path length through the atmosphere to become longer).

The loss is also dependent, to a lesser extent, on humidity and temperature.

Table 3.3 *Attenuation (dB) caused by water vapour (clear sky)*

Frequency (GHz)	Angle of elevation 5°	10°	20°	90°
4	0·40	0·21	0·11	0·04
6	0·45	0·24	0·12	0·04
11	0·60	0·30	0·16	0·06
14	0·85	0·45	0·21	0·08

3.5.2.2 Rain attenuation

With the frequencies at present in use for satellite communications the clear-sky attenuation causes no problems but the losses resulting from rain can be serious. It is therefore essential to allow for rain attenuation when designing communications-satellite systems (especially systems working at 11 GHz and higher frequencies). Rain attenuation is, like clear-sky attenuation, a function of frequency and elevation angle; it is also a function of the characteristics of the rain (of which rain rate is by far the most important).

Weather is very variable in most parts of the world and it is necessary to use statistical methods to describe its effects. It is usual to present data on rain attenuation in the form of cumulative distributions, the variable on one axis being attenuation and that on the other axis being the percentage of time (in the long term) for which that attenuation will or will not be exceeded. The percentage of time is usually the percentage of an average year but separate distributions are sometimes prepared for each month of the year. Fig. 3.7 is an example of a cumulative distribution of rain attenuation.

The acquisition of meaningful statistics of rain attenuation for small percentages of the time requires more or less continuous measurements extending over many years; this is because the statistics may show large variations from year to year and also because, for example, the events providing the data for 0·001% of a year occupy only about five of the half a million minutes in a year. A large amount of data have been accumulated for a few sites; however, rain attenuation is very much dependent on the location of the site and it is clearly impracticable to make measurements over many years at every potential earth-station site. Some method is therefore needed for deriving, from information that already exists, an approximate distribution of attenuation with time for any site at any frequency and any elevation angle.

A large number of models have been developed for the prediction of rain attenuation; virtually all of them are based on a relationship between the attenuation at a particular site and the rain-rate statistics at some point nearby. No method has been universally accepted but some have been shown to give very good predictions, at least for particular regions of the world and

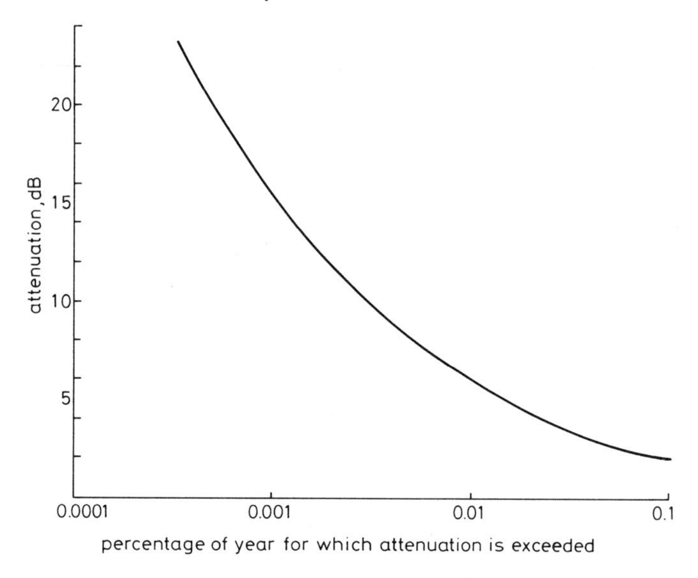

Fig. 3.7 *Cumulative distribution of rain attenuation*

Site : central UK
Elevation 5°
Frequency 11 GHz

frequencies below about 20 GHz. The quality of the predictions is of course dependent on the quality of the rain-rate statistics used.

To understand some of the difficulties of predicting the cumulative distribution of rainfall attenuation it is necessary to consider rainfall, and the relation between attenuation and rainfall, in a little more detail (but the following discussion only sketches in a few aspects of the complete picture).

Rain varies from prolonged fine drizzle which may extend over hundreds of square kilometres to exceptionally heavy cloudbursts which may last only a few minutes and affect areas of only a few square kilometres. Heavy rain tends to occur as cells. The intensity of the rain within a cell may vary rapidly with time while the cell stays more or less stationary, or the cell may move while the rainfall rate remains more or less constant, or the cell may move and the rainfall rate may vary at the same time. The variation of rainfall statistics with location is influenced both by large-scale weather patterns and by local features such as hills.

If the characteristics of the rain along a transmission path were the same at all points on the path then the attenuation in decibels would be proportional to the length of the path, the constant of proportionality (the specific attenuation) being dependent mainly on the number of raindrops per unit volume and the frequency of the transmission. It has been shown that the specific attentuation α can be approximated by the expression:

$$\alpha = aR^b \text{ decibels/km} \qquad (3.19)$$

where R = rain rate at the surface of the earth
a and b = constants for a given temperature, frequency and polarisation.

This relation forms part of most methods of predicting attenuation and the constants a and b have been tabulated for a wide range of frequencies and temperatures.

In practice, the conditions along the transmission path through the atmosphere are not constant but all we usually know, for any specific event, is the rainfall rate at a point on or near the earth station (in particular, we do not know over what length of the path it is raining or how the rainfall rate varies with distance along the path). However, although the instantaneous correlation between attenuation over the atmospheric path and rainfall rate at a nearby point on the surface of the earth may be poor, the correlation between the two quantities over the long term is quite good. Thus, despite the apparent complexity and variability of rainfall characteristics, it has been possible to develop a number of useful models for predicting the cumulative distribution with time of the total attenuation along a slant path from rainfall data gathered at a nearby point.

The CCIR prediction method, which is widely accepted, starts from eqn. 3.19 and calculates the specific attenuation corresponding to the rainfall rate for 0·01% of the year. (If the rainfall rate for 0·01% of the year is not known then an approximate value may be obtained from CCIR rain-climatic-zone maps.) Once the specific attenuation has been calculated, the path attenuation for 0·01% of the time A is calculated from the expression

$$A = \alpha L r \text{ decibels} \qquad (3.20)$$

Where L = path length through rain
r = a factor, dependent on path length, which allows for the variation of rainfall along the path.

L is a function of the elevation angle and the height to which the rain extends; CCIR Report 563 explains how the latter, which is a function of latitude, may be derived.

When the attenuation for 0·01% of the time has been found, a simple procedure allows extrapolation to other percentages of the time from 0·001 to 1%.

Typical losses at frequencies from 4 to 30 GHz and an elevation angle of

30°, for a temperate climate with an annual rainfall of 500 mm, are shown in Table 3.4

Table 3.4 *Attenuation due to rain*

Percentage of time	Attenuation (dB) not exceeded for the stated frequency and percentage of the time					
	4	6	11	14	20	30 (GHz)
1	<0·1	<0·1	0·6	0·9	2·3	2·8
0·5	<0·1	0·1	0·7	1·1	2·6	3·5
0·2	0·1	0·2	1·0	1·5	3·3	4·8
0·1	0·1	0·2	1·4	2·0	3·8	6·4
0·05	0·2	0·3	2·0	2·8	4·7	8·3
0·02	0·3	0·4	2·8	4·2	7·2	13·0
0·01	0·4	0·6	3·6	5·2	8·7	18·0

In determining the effect of rainfall on a satellite link it must be remembered that the introduction of an attenuating device or medium into a communications system increases the noise temperature of the system and rain is no exception. The increase in sky noise temperature resulting from rainfall may be calculated on the basis given in Section 7.4.2, but the effective temperature of the rain should be taken as 270 K.

3.5.2.3 *Site diversity*

In a few areas of the world (e.g. parts of the eastern seaboard of continental USA) the rain attenuation for small percentages of the time is so high that, when working at frequencies of 14/11 GHz and above, it may be cheaper to use site diversity (i.e. two earth stations at different locations) than to provide a large rain margin by using higher satellite power per channel and a larger earth-station antenna.

Site diversity works because the cell structure of heavy rain means that the probability of such rain occurring simultaneously at two sites separated by, say, 10 km is very small compared with the probability of the same rainfall rate occurring at one of the sites. Large improvements in the performance of a system may thus be obtained by providing two separate earth stations and using the transmission path which has the lowest instantaneous attenuation. The improvement in performance resulting from site diversity is quantified either as diversity advantage or diversity gain. Diversity advantage is the ratio of the percentage of time that a given attenuation is exceeded at a single site to the percentage of time that the same attenuation is exceeded at both sites simultaneously; diversity gain is the difference (in dB) between the rain attenuation exceeded for a given percentage of the time at a single site and the attenuation exceeded for the same percentage of the time taking account of diversity. The optimum distance between diversity sites and the best orientation of the line between them depends on the local weather pattern.

3.5.2.4 Hydrometeors

Water can appear in the atmosphere as snow, hail, cloud, fog and ice particles as well as rain. All these forms of atmospheric water are called 'hydrometeors' and rain, hail and snow are 'precipitation'. Cloud may cause attenuation but this attenuation is much less than that caused by relatively light rain rates of, say, 10 mm/h. The attenuation caused by fog and dry snow is usually negligible. However, snow accumulating in the reflectors of antennas can cause severe losses and the reflectors of large antennas situated in snowy areas are therefore usually fitted with heaters.

3.5.2.5 Depolarisation

(Note: If the concepts of polarisation state and cross-polar discrimination used in this Section are unfamiliar then Section 7.3.4 may prove helpful).

When frequencies are reused by means of orthogonally-polarised transmissions, the isolation between the transmissions is limited by imperfections in the satellite and earth-station antennas, and by the effects of the atmosphere.

Rain drops start their life as spheres but become non-spherical as a result of air resistance as they fall. In theoretical studies a raindrop is generally assumed to be an oblate spheroid (i.e. the shape generated by rotating an ellipse about its minor axis, the minor axis in this case being approximately vertical). The axes of symmetry of the raindrops are tilted by the effects of wind, and the resulting angle between the minor axis and the vertical is known as the canting angle.

The polarisation of a linearly-polarised wave remains unchanged when the wave passes through a raindrop if the plane of polarisation is parallel to an axis of symmetry of the raindrop. However, when, as is usually the case, the polarisation is not parallel to an axis of symmetry then the components of the wave parallel to the two axes of the raindrop suffer different attenuations and different phase shifts; this changes the polarisation state of the wave (i.e. the wave is depolarised). Elliptically-polarised waves can be resolved into two linearly-polarised waves and thus also suffer depolarisation when they pass through rain.

Experiments have shown that there is correlation between the attenuation resulting from rain and the maximum cross-polar discrimination (XPD) which can be achieved. An empirical relationship (which can be shown to have some basis in theory) is:

$$\text{XPD} = X - Y\log A \text{ decibels} \tag{3.21}$$

where A = (copolar) attenuation, dB
Y = a constant with a value of about 20
X depends on (i) frequency, (ii) the angle between the plane of polarisation and the horizontal at the earth station (the tilt angle), (iii) the canting angles of the raindrops and (iv) the elevation angle

For low angles of elevation (less than 45°) and canting angles of about 45°, X is a minimum and approximately equal to $30\log f$ (where f is the frequency of the transmission in GHz).

Early attempts to correlate XPD and rain attenuation showed that significant depolarisation can occur when it is not raining; this is called anomalous depolarisation and it has been shown to be caused by ice crystals at high altitudes. Anomalous depolarisation can cause XPD to fall to about 27 dB and 20 dB for 0·1% and 0·01% of the time, respectively, and, in some climates, it can be the dominant cause of depolarisation.

3.6 Interference

3.6.1 Sources of interference

Possible sources of interference to satellite transmissions are:

(i) Interference from ignition systems, industrial equipment etc.; this is usually only of importance when the sources are very close to an earth station.
(ii) Radiation at harmonic or spurious frequencies from equipment operating in other frequency bands; well-designed and properly-operated equipment should not cause interference to other systems but problems do sometimes occur.
(iii) Interference from satellite systems in adjacent frequency bands. This is not a widespread problem but there is a possibility that high-power transmissions from satellites used for direct broadcasting to homes (operating in the 12 GHz band) may interfere with earth stations of the fixed satellite service receiving transmissions in the neighbouring frequency bands. This possibility is discussed in CCIR Report 712–1.
(iv) Interference arising from terrestrial and satellite systems sharing the same frequency bands, and interference resulting from frequency reuse within the system. The ever-increasing traffic carried by satellites and terrestrial systems, the consequent rapid increase in the number of satellites, earth stations and terrestrial stations, and the trend towards smaller earth-station antennas continually exacerbate the problem of keeping this type of interference within acceptable limits.

The general question of interference between systems sharing frequency bands is considered in the following Section. Interference to earth stations is considered in Section 7.2.2.

3.6.2 Interference between systems sharing the same frequency bands

3.6.2.1 Allocation of frequencies

The radio-frequency spectrum (and particularly that part of it which it is economical to exploit with present techniques) is a limited resource whereas the demand for radio channels continues to grow and is apparently without limit. It is thus often necessary for a transmission to share part or the whole of its frequency band and, as a result, many transmissions must suffer a limited amount of interference. For the continued satisfactory operation of radio systems it is necessary that there shall be agreement, for each type of system, on the maximum amount of interference that is permissible and on means of regulating the use of the radio spectrum so that this maximum is not exceeded. In the international context it is the ITU that provides the forum in which nations reach agreement on permissible interference levels and the means by which interference shall be limited to these levels.

For example, for a telephone channel of a communications-satellite system using FM–FDM transmissions CCIR specifies the permissible total interference from all terrestrial stations as 1000 pW0p and the permissible total level of interference from all other satellite systems as 2000 pW0p.

The ITU Radio Regulations (RRs) comprise a code of conduct for planning and regulating the use of the radio-frequency spectrum. The Radio Regulations are divided into Articles (which correspond, more or less, to chapters in a book).

Article 8 comprises a series of tables of frequency allocations which divide the radio-frequency spectrum into a large number of bands and specify the uses to which each band may be put. Examples of these uses (or services in ITU terminology) are Fixed (Terrestrial) Service, Fixed Satellite Service (FSS), Mobile (Terrestrial) Service, Mobile Satellite Service (MSS), the Broadcasting Satellite Service (BSS), Radiolocation, Radionavigation and Radioastronomy.

There are three types of frequency allocation — primary, permitted and secondary. Stations using registered frequencies in the primary or permitted categories can claim protection from 'harmful' interference (i.e. interference exceeding the permitted maximum) but stations in the primary service have first choice of frequency when plans are being prepared. Stations in a secondary service can only claim protection from interference from other stations in the same or other secondary services.

It is easier to get agreement on the allocation of frequencies on a regional basis than on a worldwide basis. The ITU therefore divides the world into three regions. Region 1 includes Europe, the USSR, Africa and Asia west of Iran; Region 2 includes North and South America; Region 3 includes Australasia, China, India, Japan and most of the rest of Asia (RR8–2 gives definitions and a map of the Regions). Thus, for example, the table of frequency allocations for the band 2500–2655 MHz (RR8–111) shows that the fixed satellite service (FSS) may use the whole of the frequency band in Region 2, none of it in Region 1, and only that part of the band between 2500 amd 2535 MHz in Region 3. In addition to the regional variations, many other variations in allocations to particular countries and groups of countries are specified in notes to the Table (and to virtually every other table in Article 8). As a result, Article 8 needs careful reading.

3.6.2.2 *Shared frequency bands*

When satellite communications first became practicable, all of the radio-frequency spectrum (up to 40 GHz) had already been allocated to other services. It was therefore necessary either:

(i) to reallocate existing frequencies to the satellite service or
(ii) to allow the satellite service to use (simultaneously) frequencies which were already in use by another service.

An Extraordinary Administrative Radio Conference (EARC) of the ITU decided, in 1963, that the best solution was to let the fixed satellite service (FSS) share frequency bands with the terrestrial services. For example, two frequency bands included in the allocations by EARC for shared use by the FSS and the terrestrial service were 3700–4200 MHz and 5925–6425 MHz; these 6/4 GHz bands have been extensively used by satellite systems ever since.

The original allocations have been modified and added to by a series of ITU meetings in subsequent years, for example WARC/ST (World Administrative Radio Conference on Space Telecommunications) in 1971. WARC/ST allocated further shared bands, notably at 10 950–11 200 MHz, 11 450–11 700 MHz and 14 000–14 500 MHz, and these 14/11 GHz bands are now also extensively used by the FSS and the terrestrial service.

Other allocations have of course been made to both the FSS and other satellite services. However, few of these allocations lead to such difficult problems in the avoidance of interference as do the allocations in the 6/4 and 14/11 GHz bands, and some bands at relatively high frequencies (e.g. 29·5–30 GHz) have now been reallocated to the exclusive use of satellite-communications services.

3.6.2.3 *Factors affecting interference*

Interference to a satellite link may arise on both the up-path (i.e. interference to satellites) and on the down-path (i.e. interference to earth stations). Interference on the up-path may arise from terrestrial stations, from earth stations of other systems or from earth stations of the same system (in the latter case the interfering signals will be either nominally orthogonally-polarised transmissions or will be radiation from the sidelobes of another earth station). Interference on the down-path may arise from adjacent satellites or the wanted satellite (in the latter case the interfering signals will be either nominally orthogonally-polarised transmissions or radiation from the sidelobes of antenna beams directed at other areas of the earth).

Interference from satellite systems to terrestrial systems may be caused by both earth stations and satellites.

The major factors affecting interference between systems sharing the same frequency bands are:

(i) *Modulation and coding:* A system may be designed to have good resistance to interference through, for example, the use of methods such as wide-band FM, carrier-energy dispersal, and error detection and correction.

(ii) *Antenna characteristics:* For directions well off antenna boresight, sidelobe levels are (to a first approximation) independent of antenna size; antenna radiation patterns can, however, be improved by suitable design of the antenna. Systems using smaller antennas may be less susceptible to some types of interference because they work with higher wanted signal levels (but, for the same reason, they are likely to cause greater interference to other systems).

(iii) *EIRP:* The EIRP of an interfering carrier in the direction of the interfered-with station depends on the antenna radiation pattern and the transmitter power. Interference from terrestrial stations to earth stations and satellites is limited by placing restrictions on the EIRP radiated by terrestrial stations (particularly in the direction of the geostationary orbit). Interference from satellites to terrestrial stations is limited by placing restrictions on the power flux density of satellite transmissions arriving at the surface of the earth at low angles of elevation and interference from earth stations to terrestrial stations is limited by preventing earth stations from radiating at elevation angles of less than 3°, and by placing restrictions on the EIRP radiated at elevation angles below 5°.

(iv) *Satellite spacing:* Decreasing the spacing between satellites increases interference between satellite systems on both up-paths and down-paths.

(v) *Choice of sites for earth stations and terrestrial stations:* Earth-station sites should be chosen so that, as far as possible, they are screened from terrestrial stations and other earth stations by obstacles such as hills or buildings (see Section 7.2.2.4). The sites of terrestrial stations should be chosen, if possible, so that the line of shoot of the antennas does not pass very close to an earth station.

(vi) *Frequency planning:* It may be practicable to reduce interference between two systems by careful choice of carrier frequencies.

Systems with similar characteristics can, in general, coexist more easily than systems with dissimilar characteristics.

Most ways of increasing the traffic capacity of an arc of the geostationary orbit also result in increased interference between systems. For example, using more satellites reduces the angular spacing between them, and reducing RF bandwidth per channel usually increases the susceptibility of a signal to interference and increases its potential for causing interference to other systems.

3.6.3 Procedures of the ITU

The restrictions imposed by the RRs on the EIRPs radiated by earth and terrestrial stations, and on the PFD at the surface of the earth produced by satellites, help to limit interference but cannot (by themselves) ensure that unacceptable interference between systems does not occur. The additional elements which ensure that systems do not cause unacceptable interference to one another are co-ordination and the registration of frequencies by the International Frequency Registration Board (IFRB).

3.6.3.1 The frequency register

The IFRB maintains a master register of all frequency assignments, each assignment being a record of all the information required to determine that the use of the frequency in any system to which it is assigned will not cause unacceptable interference to other systems. The process of determining that there will be no unacceptable interference is called co-ordination. The administrative and technical procedures required before frequency assignments are formally recorded by the IFRB are very complex and it is practicable only to give the barest outline in this book. The full details are to be found in the RRs themselves.

3.6.3.2 Advance notification

An administration that intends to establish a satellite network should notify the IFRB and send it certain basic information such as the proposed location of the satellite, the gain of the satellite and earth-station antennas and the maximum spectral power density which will be fed to the antennas. This information (which is specified in Appendix 4 of the RRs) should be sent not more than five years, but preferably at least two years, before the network is to be brought into service.

The IFRB includes the information in its Weekly Circular. If any administration thinks the proposed new system could cause unacceptable interference to one or more of its existing or notified communications-satellite links then it must respond by informing the administration, making the advance notification and sending a copy of its comments to the IFRB within four months.

For advance notification:

(i) The characteristics of the new system need not be completely defined; in particular, it is not necessary to specify the frequencies to be used and the positions of the earth-station sites (but the frequency band to be used and the area in which the earth stations will be situated must be stated)

(ii) There is no requirement to consider whether there might be interference between the earth stations of the system and any terrestrial stations (this is left until the positions of the earth stations have been decided)

3.6.3.3 The Delta–T method

An accurate determination of the effect of interference between systems is a complex matter. It is therefore desirable to have some relatively simple procedure to establish whether the probability of interference is negligible, in which case there is no need to make an accurate evaluation.

The effect of interference on a baseband signal is directly related to the radio frequency (RF) interference power at the input to the interfered-with receiver but the relationship between the two is usually complex and dependent on the characteristics of both the interfering and interfered-with transmissions. In order to simplify the problem the ITU has agreed that, in determining whether the probability of interference is negligible, the interference may be treated as if it were thermal noise.

The RF interference power in a reference bandwidth at the input to the interfered-with receiver is thus added to the thermal noise power in the same bandwidth at the same point and the effect on the apparent noise temperature (NT) is calculated; if the rise in NT is less than 4% then the risk of interference is considered negligible. The relatively low value of 4% is taken as the criterion for acceptable interference from a single source to try to ensure that the cumulative effects of interference are acceptable even if there is more than one source. This method of estimating interference is often referred to as the 'Delta–T' method; it is fully described in appendix 29 of the RRs.

3.6.3.4 Co-ordination

If the Delta–T method suggests that there is a possibility of interference then the notifying administration must consult with the responding administration as soon as possible and try to resolve the problem; the IFRB will help if it is asked.

Even if the administrations concerned do not, by means of this informal co-ordination, succeed in solving the problems they must (six months after publication of the advance notification) move on to formal co-ordination. This second phase of the regulatory procedures takes account of the specific frequencies to be used, the precise positions of any fixed earth stations, and (as far as practicable) the detailed characteristics of the systems. The notifying administration must send the additional information required (which is specified in Appendix 3 of the RRs) to any administration which may be concerned (and copy it to the IFRB who will publish it in the Weekly Circular).

Allowable interference levels are given in CCIR Recommendations and methods of predicting actual interference levels are given in the RRs and a series of CCIR Reports.

Methods of reducing interference so that satellite systems can coexist include:

(i) Careful selection of carrier frequencies to reduce overlapping of the spectra of transmissions
(ii) Use of antennas with better discrimination between on-beam and sidelobe gain (i.e. antennas with higher on-beam gain or lower sidelobe levels)
(iii) Changing the characteristics of the modulation or coding methods to increase protection against interference
(iv) Increasing the separation between satellites (this is not usually easy as interference is most likely to occur between satellites which are in an overcrowded sector of the geostationary orbit).

Formal co-ordination includes co-ordination between earth stations and terrestrial stations. The technical aspects of this can be even more difficult than the co-ordination between satellite systems because of the complexities associated with propagation paths lying entirely within the troposphere.

Formal co-ordination is governed by a set of rules and by a timetable laid down in the RRs and it can be a lengthy business. The IFRB must be told, every six months, what progress is being made but the negotiations are a matter for the administrations concerned (although the IFRB will help if it is asked). With good will, problems can usually be solved by means of relatively minor changes in frequency plans or the characteristics of systems. If all else fails it may be necessary to accept interference levels higher than the limits recommended by the CCIR.

The rules and procedures devised by the ITU must cover every conceivable situation and must therefore be conservative. Moreover, the international regulations can be revised (in the light of technical progress and other changes) only infrequently. National or regional regulations can usually be updated more often and (where there is no risk of interference outside the country or region) this may allow more efficient use to be made of the spectrum than is possible using the ITU procedures. Relaxations of the ITU procedures may also be used by agreement between two or more nations (provided that these relaxations do not adversely affect other countries).

3.7 Appendix: Examples of Link Budgets

3.7.1 Introduction

Examples of link budgets for a variety of communications-satellite systems are given in this Appendix.

The most important factors influencing the choice of modulation and multiple-access method are:

(i) Nature of the traffic and its pattern of flow (e.g. whether the information to be communicated is in the form of speech or data, whether the traffic loads are heavy or light and whether the links are point-to-point or point-to-multipoint)
(ii) Any requirement to integrate the system with another network such as the public switched telephone network (PSTN).

On trunk routes carrying a large number of telephone channels the channels are usually grouped as FDM (frequency-division-multiplex) assemblies if they are in analogue form or as TDM (time-division-multiplex) assemblies if they are in digital form. FDM and TDM assemblies are usually modulated onto carriers by means of FM (frequency modulation) and PSK (phase-shift keying), respectively. The carriers, whether FDM–FM or TDM–PSK arrive at the satellite in FDMA (frequency-division multiple access), i.e. each carrier occupies a different frequency band. Intermodulation products will be generated if (as is usually the case) several carriers are amplified by the same transponder. Link Budget 1 is a budget for a FDM–FM–FDMA system.

Television signals are usually transmitted by means of a frequency-modulated carrier and require relatively wide bandwidths; thus there is usually only room for one or two transmissions in a 36 MHz transponder. Link Budget 2 is a budget for a TV relay link using a single carrier per transponder.

The generation of intermodulation products resulting from the simultaneous amplification of a number of carriers in one satellite transponder may be avoided by allocating a separate time slot to each earth-station transmission, i.e. by using time-division multiple access (TDMA). TDMA requires each earth station to store information arriving via the terrestrial interface and transmit it as a burst of information in a time slot allocated to the station; the information carried by TDMA systems is therefore virtually always in digital (e.g. PCM) form because this simplifies the problems of storing the information, and assembling and transmitting the bursts. Link Budget 3 is a budget for a TDMA system.

Single-channel-per-carrier (SCPC) systems using analogue channels with FM or digital channels with PSK are often favoured for links carrying small numbers of speech or medium-speed data channels. Most SCPC systems use voice-switched carriers and companding; syllabic companders are used with analogue channels and companding is generally included as a matter of course in digital codecs. Link Budget 4 is a budget for an SCPC system.

In systems carrying relatively heavy traffic between relatively few stations it is usual to employ earth stations with antennas with a diameter of at least 11 m. On the other hand, if a system includes a large number of earth stations it is essential that the stations shall be small and simple in order to minimise total system costs. For example, in a data-distribution system a central hub station may broadcast only a single data channel, at a bit rate of (say) 64 kbit/s, to hundreds or even thousands of receive-only earth stations. This type of system uses very-small-aperture terminals (VSATs), which have antennas with a diameter of one or two metres and cost only a few thousand dollars. In order to achieve an acceptable carrier-to-noise ratio with such small antennas it is necessary for the transmission from the satellite to have a high EIRP (and this usually requires a high EIRP from the hub station); this is economic because the cost of the space sector and the central station is divided between a very large number of users. Link Budget 5 is a budget for a VSAT system.

Optimising the characteristics of a communications-satellite link is usually an iterative proceiss which is best carried out with the aid of a computer program. The budgets in this Appendix are, however, intended solely to illustrate the principles and no attempt has been made to optimise the performance of the links. The values used are nevertheless broadly in line with those of practical systems.

The examples given are all for clear-sky conditions. Rain, cloud and other forms of precipitation result, at times, in significant additional attenuation which degrades channel quality. This attenuation varies with angle of elevation, frequency and climate (see Section 3.5) as well as with time. Performance objectives for radio systems usually comprise standards that should be met for a large percentage of the time (say 80%) together with a lower standard which should be met for all but a very small percentage (say 0·3%) of the time.

For systems working at frequencies of 6 and 4 GHz the attenuation caused by precipitation is usually quite small; thus if the system is designed to meet the performance standards specified for the 80% of the time when the sky is clear (or there is only light rain) then the degradation of the channel caused by the heavy rain occuring for a small percentage of the time will usually be within the allowable limits. Heavy rain may cause serious attenuation of transmissions at higher frequencies and the characteristics (such as the EIRP of the transmission from the satellite and the size of the earth station) may be determined by the objectives for short percentages of the time. Whenever there is any doubt, the *C/N* or *C/T* corresponding to each objective must be determined and a separate budget must be prepared for each case.

In presenting a Link Budget it is important to state explicitly all the assumptions that were made in preparing it (e.g. the angle of elevation and the frequency, both of which will affect the path attenuation). It is best to keep the initial budget simple and to avoid, in the initial stages, including margins against contingencies which may not arise.

Budget 1 is developed in more detail than the subsequent budgets but they all follow similar methods. Budget 1 also includes explicit allowances for intermodulation and interference; in the other budgets the *C/N* or *C/T* ratios include any margins required to allow for these sources of degradation (except where otherwise stated).

3.7.2 Link Budget 1: Multiple FDM–FM–FDMA transmissions at 6/4 GHz via a 72 MHz transponder

The budget which follows is an overall budget using the total power and (usable) bandwidth of a typical 72 MHz transponder; thus the EIRPs given for the satellite and the earth station are the sums of the EIRPs of a set of transmissions which would make maximum use of the transponder, and the *C/N* ratios derived are mean values. In practice the transmissions would usually be received from (and relayed to) a number of earth stations with differing characteristics (e.g. they would see the satellite at different angles of elevation). Moreover, practical systems using FDM/FM carriers usually employ a range of transmission characteristics corresponding to a range of values of *C/N* at the earth stations (see note iv after the Budget); in this case a separate budget must be prepared for each type of carrier and this is usually the responsibility of the manager of the space sector, who tells the earth-station operators what earth-station EIRP's are required and what *C/N* ratios can be expected. An overall budget of the type given here ignores these practical complications but the example illustrates the principles of Link Budgets, and such an overall budget can be of help in determining the approximate traffic capacity of a transponder when it is used for FDM–FM–FDMA transmissions.

Up path

(a)	Power flux density (PFD) to saturate transponder: see note (i)	$-67{\cdot}1$ dB(W/m^2)
(b)	Input backoff: see note (ii)	10·0 dB
(c)	Working PFD	$-77{\cdot}1$ dB(W/m^2)

(d)	$10\log 4\pi s^2$ at an elevation of 5°	163·3 dB(m^2)
(e)	Clear-sky atmospheric attenuation (5°)	0·5 db
(f)	Required earth-station EIRP = $(c + d + e)$	86·7 dBW
(g)	Area of isotropic antenna at 6 GHz	−37·0 dB(m^2)
(h)	Power which would be collected by an isotropic antenna = $(c + g)$	−114·1 dBW
(k)	G/T	−9·2 dB(1/K)
(l)	C/T (up) = $(h + k)$	−123·3 dB(W/K)
(m)	k	−228·6 dB(W/(Hz K))
(n)	B (usable bandwidth: see note (iii)	78·1 dB(Hz)
	$(C/N)u = C/kTB = (l - m - n)$	27·2 dB

Down path

(p)	Satellite EIRP at saturation: see note (i)	31·0 dBW
(r)	Output backoff: see note (ii)	5·6 dB
(s)	Working EIRP	25·4 dBW
(t)	Path loss at 4 GHz and 5° elevation = $(4\pi s/\lambda)^2$)	196·8 dB
(u)	Atmospheric attenuation	0·4 dB
(v)	Earth-station G/T: see note (i)	40·7 dB(1/K)
	$(C/N)_d = (s - t - u + v - m - n)$	19·4 dB

Intermodulation

$(C/N)_{im}$ 22·5 dB

Interference from other earth-stations

$(C/N)_{iu}$ (uplink) 30·0 dB

Interference from other satellites

$(C/N)_{id}$ (downlink) 28·0 dB

Interference from frequency reuse

$(C/I)_{ru}$ (uplink) 22·0 dB
$(C/I)_{rd}$ (downlink) 22·0 dB

Overall (C/N)

$(C/N) = 1/((N/C)_u + (N/C)_d + (N/C)_{im} + (N/C)_{iu} + (N/C)_{id} + (N/C)_{ru} + (N/C)_{rd}) = 14{\cdot}9$ dB: see note (iv)

Notes

(i) The characteristics assumed for the satellite correspond approximately to those of a transponder of an INTELSAT satellite working via hemi or zone beams. The G/T of 40·7 dB(1/K) corresponds to that of an original INTELSAT Standard-A earth-station, i.e. an earth-station with an antenna of diameter of 30 m or more. From March 1986 the minimum G/T for an INTELSAT Standard-A station was reduced to 35 dB(1/K); this G/T can be achieved with an antenna of about 18 m diameter. The reduction in the required G/T was made possible by the increased power available from modern satellites and by a reduction in the margins allowed (e.g. for deterioration of performance with time).

(ii) For an explanation of backoff for FDMA working see Section 3.4.3.

(iii) 10% of the transponder bandwidth (i.e. approximately 7 MHz) is assumed to be required to provide guardbands between the carriers, leaving a usable bandwidth of 65 MHz or 78·1 dB(Hz).

(iv) The value of C/N (in clear weather) specified by INTELSAT for its regular FDM/FM carriers ranges from approximately 13 dB to over 20 dB. This range of values is achieved by exchanging power for bandwidth (see Section 4.2.2). For example, one of three combinations of C/N and bandwidth may be allocated to a 132-channel carrier, viz. 20·7 dB and 5 MHz, 14·4 dB and 7·5 MHz and 12·7 dB and 10 MHz. These alternative sets of characteristics give the system planners much-needed flexibility in meeting the demands of the earth stations and at the same time developing an efficient plan for fitting the carriers into the transponders. It will be seen from the Link Budget that the transponder considered would support nine of the 7·5 MHz 132-channel carriers (i.e. a total of over a thousand channels). In practice the transponders are usually required to support a mix of carriers carrying different numbers of channels at different C/N ratios; this makes the calculations rather tedious unless a suitable computer program is available.

3.7.3 Link Budget 2: 14/11 GHz TV distribution service using a single carrier per transponder

Up path

PFD to saturate transponder	−78·9 dB(W/m^2)
Input backoff	0 dB
Spreading area (20° elevation)	162·9 dB(m^2)
Atmospheric attenuation (clear sky)	0·2 dB
Required earth-station EIRP	84·2 dBW
Area of an isotropic antenna at 14 GHz	−44·4 dB(m^2)
Satellite G/T	−5·1 dB(1/K)
Bandwidth (27 MHz)	74·3 dB(Hz)
$(C/N)_u$	25·9 dB

Down path

Satellite EIRP at saturation	44·6 dBW
Output backoff	0 dB
Path loss (11 GHz and 20° elevation)	205·2 dB
Atmospheric attenuation (clear sky)	0·2 dB
Earth-station *G/T* (see note)	22·0 dB(1/K)
$(C/N)_d$	15·5 dB

Overall C/N	15·1 dB

Note: Using, for example, an antenna with a diameter of 3 m and a *NT* of 400 K.

3.7.4 Link Budget 3: TDMA–PSK system at 14/11 GHz

Up path

PFD to saturate transponder	−74·5 dB(W/m²)
Input backoff (see note)	2·0 dB
Spreading area (20° elevation)	162·9 dB
Atmospheric loss (clear sky and 20° elevation)	0·2 dB
Required earth-station EIRP	86·6 dBW
Area of isotropic antenna at 14·25 GHz	−44·5 dB(m²)
Satellite *G/T*	−5·2 dB(1/K)
$(C/T)_u$	−126·2 dB(W/K)

Down path

Transponder EIRP at saturation	43·6 dBW
Output backoff (see note)	0·2 dB
Path loss (11·5 GHz and 20° elevation)	205·6 dB
Atmospheric attenuation	0·2 dB
Earth-station G/T	39·0 dB(1/K)
$(C/T)_d$	−123·4 dB(W/K)

Overall C/T	−128·0 dB(W/K)

Note: Section 5.6.2 includes an explanation of why backoff may be necessary when using PSK–TDMA.

3.7.5 Link Budget 4: SCPC system

Suppose that:

(i) We wish to establish an SCPC system carrying a mixture of speech and data channels (with 16% of the total number of channels being data channels) and we want to know how many channels can be supported using a 6/4 GHz satellite transponder and existing earth-stations.

(ii) The earth-stations have a G/T of 31·7 dB(1/K)

(iii) We can purchase channel equipment with the following characteristics:

Carrier spacing	45 kHz
Channel (IF) bandwidth	38 kHz
Required C/T at input to receiver	−169·3 dB(W/K)

(iv) The characteristics of the satellite transponder are:

Bandwidth	36 MHz
PFD to saturate transponder	−67·1 dB(W/m^2)
EIRP at saturation	31·0 dBW
G/T	−18·6 dB(1/K)

Assume that the carriers for the speech channels are voice activated, i.e. no carrier is transmitted when there is no speech on the channel (thus the carrier will be absent when the other party to the conversation is talking and during pauses between words and syllables). This reduces the number of carriers required to about 40% of the total speech channels (possibly less but we will assume 40%). The data channels will, however, be active for 100% of the time.

There is enough bandwidth for $(36 \times 10^6)/(45 \times 10^3) = 800$ channels. If all 800 frequency slots are occupied there are $800 \times (16/100) = 133$ data channels and, on average, $(800 - 133) \times 0{\cdot}4 = 267$ active speech channels. The average total number of active channels is therefore $(267 + 133) = 400$.

We now need to determine whether there is enough transponder power to support this number of channels.

Up path

	PFD to saturate transponder	−67·1 dB(W/m^2)
	Input backoff	10·0 dB
(a)	Working PFD	−77·1 dB(W/m^2)
(b)	10log400	26·0 dB
	Working PFD per carrier = $(a - b)$	−103·1 dBW
	Spreading area (5° elevation)	163·3 dB(m^2)
	Atmospheric attenuation (clear sky)	0·5 dB
	Required earth-station EIRP	60·7 dBW
	Area of isotropic antenna at 6 GHz	−37·0 dB(m^2)
	Satellite G/T	−18·6 dB(1/K)
	$(C/T)_u$	−158·7 dB(W/K)

Down path

	Transponder EIRP at saturation	31·0 dBW
	Output backoff	5·5 dB
(c)	Total working EIRP	25·5 dBW
(d)	10log400	26·0 dB
	EIRP per carrier = $(c - d)$	−0·5 dBW
	Path loss at 4 GHz and 5° elevation	196·8 dB
	Atmospheric attenuation	0·4 dB
	Earth-station G/T	31·7 dB(1/K)
	$(C/T)_d$	−166·0 dB(W/K)

Overall (C/T)

$$1/(C/T) = 1/(C/T)_u + 1/(C/T)_d$$

$$(C/T) = -166{\cdot}7\ \text{dB(W/K)}$$

There is therefore a margin of 2·6 dB between the available C/T in clear weather and that required at the input to the receiver, i.e. −169·3 dB(W/K).

3.7.6 Link Budget 5: Data distribution to VSATs via 14/12 GHz links

Up path

PFD to saturate transponder	−80·2 dB(W/m^2)
Input backoff	10·0 dB
PFD for each carrier (of 15)	−102·0 dB(W/m^2)
Spreading area (at 20° elevation)	162·9 dB(m^2)
Atmospheric attenuation (clear sky)	0·2 dB
Required earth-station EIRP: see note (i)	61·1 dBW
Area of isotropic antenna at 14·25 GHz	−44·5 dB(m^2)
Satellite G/T	3·0 dB(1/K)
$(C/T)_u$	−143·5 dB(W/K)
Occupied bandwidth (154 kHz)	51·9 dB(Hz)
$(C/N)_u$	33·2 dB

Down path

Transponder EIRP at saturation	40·0 dBW
Output backoff	7·0 dB
Output EIRP for each carrier (of 15)	21·2 dBW
Path loss (12·5 GHz and 20° elevation)	206·3 dB
Atmospheric attenuation (clear sky)	0·2 dB
Earth-station G/T: see note (ii)	16·0 dB(1/K)
$(C/T)_d$	−169·3 dB(W/K)
$(C/N)_d$: see note (iii)	7·4 dB

Notes:

(i) This corresponds, for example, to an antenna with a diameter of 8 m (which would have a gain of about 61 dB at 14 GHz) and an input power to the antenna of about 1 W.

(ii) A VSAT with a 1·2 m-diameter antenna is assumed.

(iii) Note that the *C/N* on the uplink is so much higher than that on the downlink that the overall (earth-station to earth-station) *C/N* is virtually the same as that of the downlink.

Chapter 4

Frequency modulation

4.1 Introduction

4.1.1 Systems using FM

Frequency modulation (FM) was used for all the early satellite-communications transmissions, firstly because it made it possible to achieve acceptable performance with the very limited power available from the satellites of the 1960s and secondly because much of the technology required had already been developed for terrestrial radio-relay systems.

4.1.1.1 FDM–FM

Most early satellite-communications systems followed the practice of terrestrial microwave radio-relay systems by combining a number of 4 kHz telephony channels into a frequency-division-multiplex (FDM) baseband assembly which was frequency modulated onto a radio-frequency (RF) carrier. This method (FDM–FM) became very widely used and, although most new earth-station equipment is digital, a large amount of FDM–FM equipment will remain in use for many years to come.

Fig. 4.1 shows, as an example, how a 132-channel multiplex for the INTELSAT system is assembled. Note the channels below 12 kHz which comprise the dispersal channel (see Section 4.4) and the engineering service channels (which provide the communications necessary for the operation, management and maintenance of the system).

The combination of channels into FDM assemblies is of advantage because it reduces the number of RF carriers and therefore the amount of earth-station equipment required.

When a number of carriers at different frequencies are transmitted through a satellite this is called frequency-division multiple access (FDMA) and, if the carriers are FDM–FM carriers, the method of transmission is FDM–FM–FDMA.

FDM–FM–FDMA systems often combine a number of small streams of

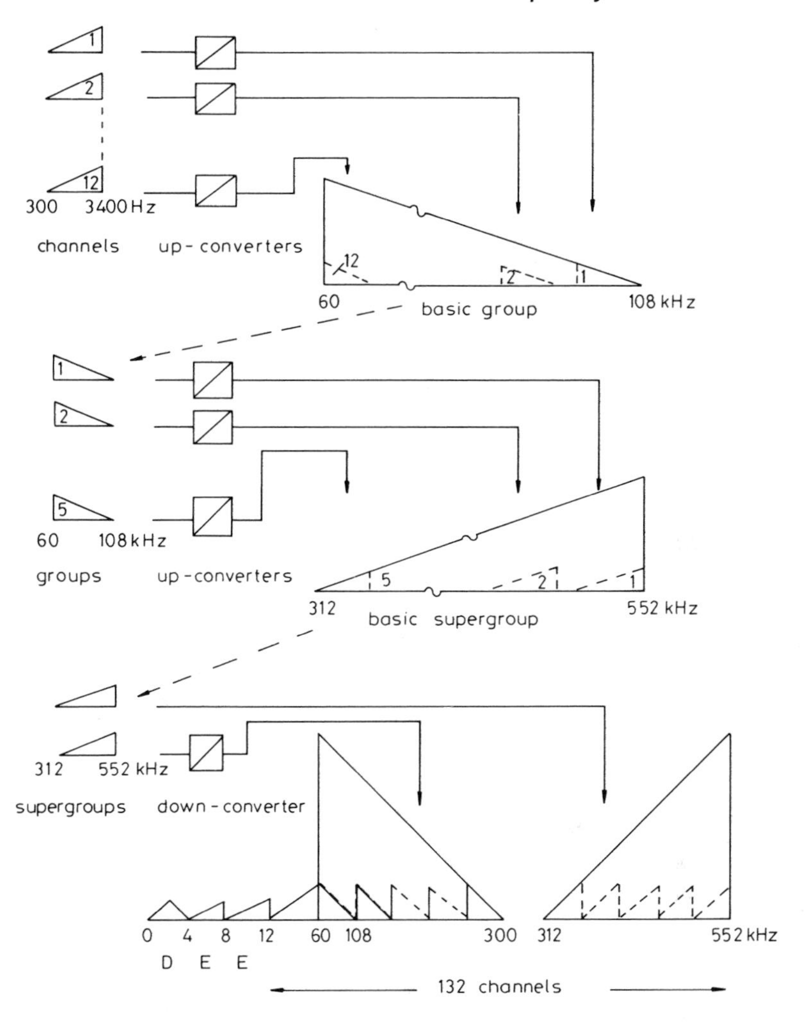

Fig. 4.1 *Assembly of 132-channel FDM baseband*

E = engineering service circuits
D = energy dispersal signal

traffic, each intended for a different earth station, into a single FDM assembly which is modulated onto a multidestination carrier. The use of multidestination carriers reduces the total number of carriers to be amplified and this helps to reduce intermodulation noise. Multidestination carriers also decrease the amount of equipment required on the transmit side of earth stations but have the disadvantage that each station receiving them must provide equipment to separate out the channels it wants from those intended for other stations.

4.1.1.2 SCPC–FM

FDM–FM–FDMA systems provide economical service on routes carrying medium or heavy traffic loads but lead to a high cost per channel on very lightly-loaded routes. In single-channel-per-carrier (SCPC) systems every channel is modulated onto an individual RF carrier and these systems are better suited to earth stations requiring only a few channels per route. One reason for this is that earth stations can start operation with the minimum number of channel units necessary to give the desired grade of service with their initial traffic and then add further units as traffic grows. The first SCPC systems used pulse code modulation (PCM) but there are now many systems using FM. An SCPC carrier is usually voice-activated (i.e. it is transmitted only when the user of the channel is speaking); this results in large savings in satellite power.

4.1.1.3 TV–FM

Virtually all transmissions via satellite of standard television programmes are made using frequency-modulated carriers and the use of FM for this purpose seems likely to continue for many years to come. This is because, at present, digital television transmissions by satellite require very large bandwidths. Digital transmissions are, on the other hand, very satisfactory for video conferencing because the bit rate (and therefore the bandwidth) of the signals can be greatly reduced by special coding methods.

4.1.2 Specifying system performance

4.1.2.1. Overall channel quality

Information passing through a telecommunications channel is degraded by attenuation of the signal and addition of thermal noise, and in other ways (e.g. by distortion of the signal and by interference). The degradation should, ideally, never be allowed to reach a level where it significantly affects the quality of the information being transmitted or causes difficulties in signalling. However, it is impracticable to provide telecommunications systems (especially radio systems) which give completely satisfactory service for 100% of the time. The usual way of specifying the performance of a radio system is by means of a completely acceptable standard which must be met for most of the time and a minimum acceptable standard which must be met for all but a very small percentage of the time. The standards required and the terms in which they are specified depend very much on the nature of the information carried by the system (e.g. whether it is speech, data or a television signal). In the case of signals such as speech and television the required performance is determined by what is heard or seen by the customer to be satisfactory. This acceptable subjective performance is translated into an objective requirement; for analogue channels this usually means specifying a minimum allowable ratio of signal power to noise power in the baseband channel or, what amounts to the same thing, a maximum noise power at some reference point where the signal power is known.

Once the required overall system performance has been determined it is necessary to apportion the total allowable degradation to the various links comprising a channel or circuit. Thus, for example, intercontinental circuits via the public switched telephone network (PSTN) usually include local lines (from users to the local exchanges), long-distance circuits (from local exchanges to international switching centres) and connections between international switching centres via submarine cables or satellites; on the other hand, a channel carrying specialised business services may comprise only very short links between the users' terminal equipment and small earth stations (sited at or close to the users' premises) together with a satellite link connecting the earth stations. The percentage of the total allowable degradation which can be allocated to the satellite link is clearly much greater in the latter case.

There are many factors which have to be considered in specifying the performance of telecommunications systems. For telephone circuits via the PSTN, these factors include the difference in level between loud and soft talkers, the background noise in the places where the users of the system may talk and listen (ranging from quiet homes to noisy industrial sites), the fact that some customers are close to their local exchange and some are relatively far away, and all the characteristics of the national and international links which may be interconnected.

It would be impossible to make satisfactory interconnections between national communications systems without the large measure of standardisation of characteristics and interfaces which has been achieved by the efforts of engineers of all nations working together in the two consultative committees of the International Telecommunications Union — the International Telegraph and Telephone Consultative Committee (CCITT) and the International Radio Consultative Committee (CCIR). However, it should be remembered that the recommendations of these committees may not be appropriate for all systems; for example, it may be practicable to adopt less stringent standards, with a consequent saving in cost, for private networks which are not to be interconnected with the public international network.

4.1.2.2 Hypothetical reference circuit (HRC)

In apportioning the total allowable noise between the links of a telecommunications channel or circuit it is desirable to specify carefully what is included in each link so that, for example, the designers of two contiguous links do not each think that the other is allowing for the noise from some piece of equipment on the boundary between the links. To assist in demarcation of the boundaries, CCIR and CCITT define a 'hypothetical reference circuit' (HRC) for each type of link.

For analogue transmissions in the fixed satellite service (FSS), i.e. transmissions not intended for use for mobile communications or broadcasting, CCIR defines the HRC as consisting of 'one earth–space–earth link' and says that the input to the reference circuit 'should correspond to the input of the modulator carrying out the translation from baseband to the radio-frequency carrier and the output should correspond to the output of the demodulator carrying out the reverse operation'.

4.1.2.3 Zero relative level

Because all communications systems comprise lossy sections (e.g. the transmission paths between earth stations and satellites) followed by amplifiers (e.g. satellite transponders or earth-station receivers) the levels of the signals and noise vary widely from point to point in the system. In specifying the power of signals or noise it is therefore also necessary to specify some location at which the power is to be measured or calculated. The latter is called a point of 'zero relative level' and for a telephone channel it is usually defined as a point where a specified test signal would have a power of 1 mW.

4.1.2.4 Allowable noise power

The noise in an analogue telephony channel is usually specified as the mean noise power over a period of 1 min or 5 ms; these periods were originally chosen as being of the same order as the durations (respectively) of a telephone call and an element of a signalling message or a telegraph message. CCIR states that, for an analogue channel of the HRC for the fixed satellite service, the noise power at a point of zero relative level should not exceed:

10 000 pW0P one-minute mean power for more than 20% of any month
50 000 pW0P one-minute mean power for more than 0·3% of any month
1 000 000 pW0 unweighted (with an integrating time of 5 ms) for more than 0·01% of any year

The abbreviation pW0P stands for picowatts, measured at a point of zero relative level and psophometrically weighted.

Psophometric weighting allows for the amplitude-frequency response of the ear. In order to measure psophometrically-weighted noise it is necessary to place a weighting network (i.e. a filter), with an amplitude/frequency characteristic similar to that of the ear, in front of the power meter. The combination of weighting network and power meter is called a psophometer.

Weighting networks with differing amplitude/frequency characteristics are in use in different countries (for example C Message Weighting is commonly used with the terrestrial systems of North America) but CCITT has specified a standard characteristic for use when measuring the noise in a 4 kHz telephone channel of an international link. The psophometrically-weighted power of white noise measured in such a channel, using the CCITT weighting network, is approximately 2·5 dB less than that measured without the network. Under normal circumstances the noise in a telephone channel of a satellite system using FM is white noise and it is therefore customary to add 2·5 dB to the calculated signal-to-noise power ratio to allow for the subjective effect of the ear.

4.2 Basic characteristics of FM

4.2.1 *Spectrum of an FM signal*

4.2.1.1 *Frequency deviation*

Consider a carrier modulated in frequency by a sinusoidal signal. The angular frequency of the modulated carrier at time t may be written

$$2\pi F_c + aV_m \sin 2\pi F_m t \tag{4.1}$$

where F_c = frequency of the carrier when unmodulated
V_m and F_m = amplitude and frequency, respectively, of the modulating signal
a = sensitivity of the modulator, rad s^{-1} V^{-1}

The modulation index m is defined as

$$m = F_p/F_m$$

where F_p is the peak deviation $= aV_m$

4.2.1.2 *Frequency spectrum*

The spectrum of this frequency-modulated signal comprises a component at the carrier frequency and an infinite number of pairs of components spaced at F_m, $2F_m$, $3F_m$ and so on above and below the carrier frequency (see Fig. 4.2 which shows only the first three pairs of side frequencies).

The amplitudes of the components vary with the modulation index m. If V_c is the amplitude of the unmodulated carrier then the amplitude of the component of the modulated signal at frequency F_c is $V_cJ_0(m)$ and the amplitudes of the first, second and nth side frequencies are $V_cJ_1(m)$, $V_cJ_2(m)$,

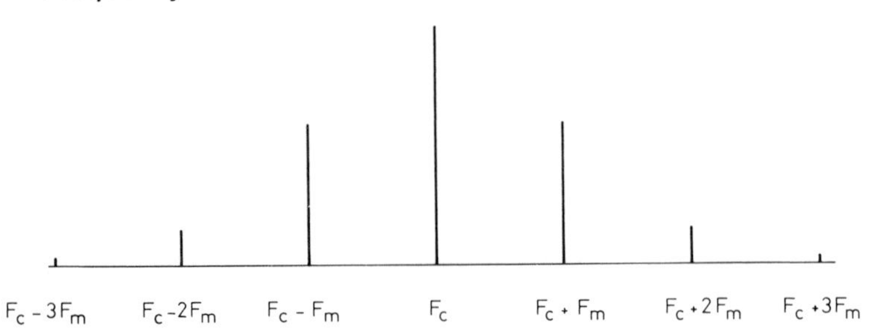

Fig. 4.2 *Spectrum of FM carrier*

and $V_c J_n(m)$, where $J_n(m)$ is the amplitude of the Bessel function (of m) of the first kind and order n.

From Fig. 4.3 it will be seen that the values of the functions decrease rapidly with increasing order as soon as n is greater than m; thus, for example, if $m = 6$ then J_7 (6), J_8 (6), J_9 (6) and J_{10} (6) are 0·13, 0·06, 0·02 and 0·007, respectively. Thus nearly all the power in the FM signal is contained in the carrier and the first n pairs of side frequencies, where n is the first integer greater than m. FM signals can therefore be transmitted in a finite bandwidth with negligible distortion. The minimum bandwidth required for practical transmissions is something which is considered further in Section 4.2.5.

4.2.2 Signal-to-noise power ratio of an FM transmission after demodulation

The unweighted signal-to-noise power ratio in the baseband of an FM transmission after demodulation (s/n) can be shown to be:

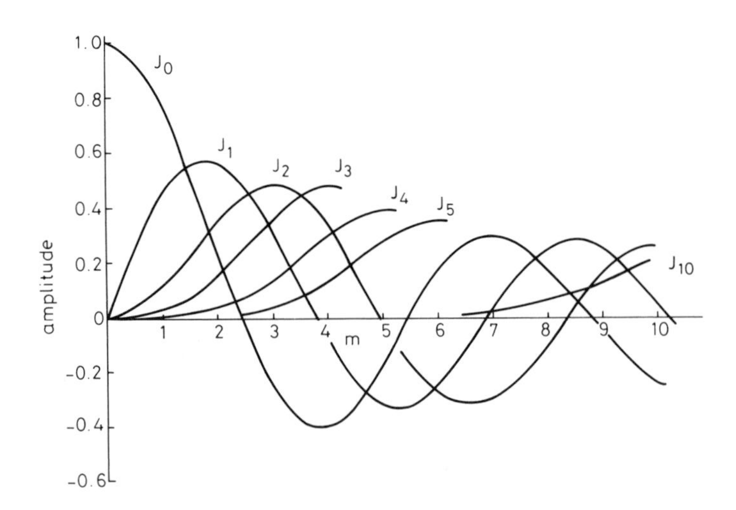

Fig. 4.3 *Bessel Functions*

$$s/n = (C/N)\,3B\,F_r^2/(F_u^3 - F_l^3) \quad (4.2a)$$
$$= (C/kT)\,3\,F_r^2/(F_u^3 - F_l^3) \quad (4.2b)$$

where C = carrier power at the input to the receiver, W
B = RF bandwidth of the receiver, Hz
N = noise power kTB in the bandwidth B, W
F_r = RMS frequency deviation, Hz
F_u and F_l = upper and lower frequencies of the baseband channel, Hz
k = Boltzmann's constant, J/K
T = effective noise temperature of the system, K

Relation 4.2*a* is known as the FM equation; it is the foundation on which the design of all FM systems is based.

The ratio s/n is usually significantly greater than C/N and the improvement in s/n relative to C/N is known as the FM advantage. Expressions 4.2*a* and 4.2*b* show that the FM advantage can be made larger by increasing the frequency deviation (which, in turn, increases the RF bandwidth); thus the C/N required for a given s/n can be reduced at the expense of using more bandwidth (this is known as exchanging power for bandwidth). However, the FM threshold (see Section 4.2.4) places a limit on the RF bandwidth which it is practicable to use, and therefore on the maximum available FM advantage.

The RMS frequency deviation F_r must correspond to a known signal power. In the case of a telephone channel this cannot be the speech power because we do not know what this is going to be (for example we do not know whether the subscriber using the channel will be a loud talker or a soft talker); F_r is therefore defined as the deviation corresponding to a test tone at the arbitrary level of 1 mW measured at a point of zero relative level (0 dBm0). Of course, if we define the deviation in terms of an arbitrary test-tone power then the signal-to-noise ratio which we derive by using expression 4.2*a* or 4.2*b* will be the ratio of test-tone power to noise power. Thus the criterion that the noise in a telephone circuit shall not exceed 10 000 pW corresponds to a requirement that the ratio of test-tone power to noise power shall be not less than:

$$(1 \times 10^{-3})/(10\,000 \times 10^{-12}) = 10^5 \text{ (i.e. 50 dB)}$$

4.2.3 Signal-to-noise power ratio in a FDM–FM channel

The bandwidth of a telephone channel (i.e. $F_u - F_l$) is 3·1 kHz. This is small compared with the mid-frequency of even the bottom channel of an FDM assembly. Because of this we can write:

$$F_u^3 - F_l^3 = (F_u - F_l)\,(F_u^2 + F_uF_l + F_l^2)$$
$$= b\,3F^2 \text{ (approximately)} \quad (4.3)$$

where $b = F_u - F_l = 3{\cdot}1$ kHz
$F = (F_u + F_l)/2$ (i.e. the mid-frequency of the telephone channel in the FDM assembly)

As an example, F_u and F_l for the bottom traffic channel of a 12-channel FDM assembly (as used by INTELSAT) are 12·3 and 15·4 kHz, respectively. With these values:

$$F_u^3 - F_l^3 = 1791 \times 10^9$$

$$b3F^2 = 1784 \times 10^9$$

The difference is less than 0·5%, and for any other channel (of any assembly) the error is even less.

Substituting $b3F^2$ *for* $(F_u^3 - F_l^3)$ in expressions 4.2*a* and 4.2*b* we get the relations:

$$s/n = (C/N)\,(B/b)\,(F_r/F)^2 \qquad (4.4a)$$

$$= (C/kT)\,(1/b)\,(F_r/F)^2 \qquad (4.4b)$$

Expressions 4.4*a* and 4.4*b* are the forms which are most often used in conjunction with FDM–FM transmissions.

Note that s/n is the signal-to-noise ratio in a single channel of the FDM assembly; F_r in expressions 4.4*a* and 4.4*b* is therefore the RMS deviation per channel (not the RMS deviation corresponding to the complete multiplex signal). As in the case of expressions 4.2*a* and 4.2*b*, the deviation is taken as that corresponding to a test tone at 0 dBm0.

In order to use expression 4.4*a* we need to know the RF bandwidth B. The calculation of B for an FDM–FM transmission is dealt with in Section 4.2.5 and this is followed, in Section 4.2.6, by an example of the calculation of s/n for a channel of an FDM–FM transmission.

4.2.3.1 Pre-emphasis

It will be seen from expression 4.4 that, if nothing is done to correct the situation, the noise in a telephone channel of an FDM–FM transmission is proportional to the square of the mid-frequency of the channel. It is obviously undesirable for the s/n ratio in the top channel of an assembly to be much worse than that in the bottom channel and it is therefore usual to insert a pre-emphasis network immediately before the modulator and a de-emphasis network immediately after the demodulator. The pre-emphasis network is designed to increase the amplitude of the signal in the higher baseband channels and decrease the amplitude in the lower baseband channels without

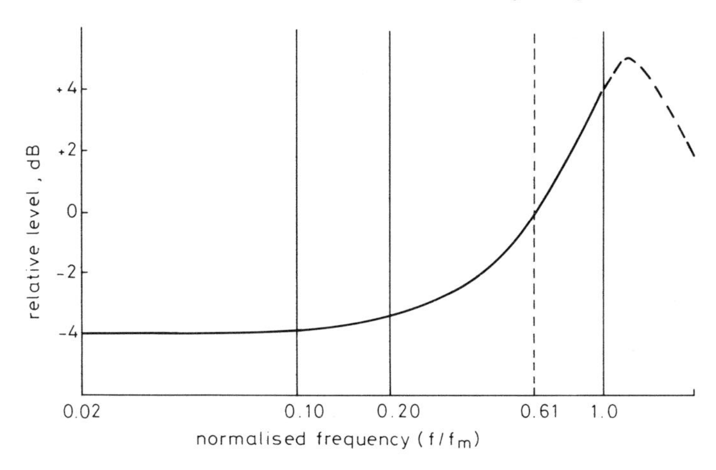

Fig. 4.4 *Pre-emphasis characteristic for FDM–FM telephony*

changing the total signal power (or frequency deviation). The de-emphasis network has the inverse characteristic to the pre-emphasis network and it therefore restores the relative amplitudes of the channels to what they were before pre-emphasis was applied. However, nearly all the noise is generated after the modulator and the noise in the higher-frequency channels is therefore reduced by the de-emphasis network (at the expense of an increase in the noise in the lower-frequency channels).

Because FM noise increases as the square of frequency, it seems that it should be possible to make the signal-to-noise ratio the same in every channel of an FDM assembly by using emphasis networks with a slope of 6 dB per octave; such networks would theoretically give an improvement of 4·8 dB in the signal-to-noise ratio of the top channel. However, when subscribers are remote from the earth stations, as may be the case when calls are made via the public switched international network, there can be quite a lot of noise on the lines incoming to the stations and the introduction of networks with a slope of 6 dB would make the bottom channels of the assemblies too noisy.

Better overall results are therefore usually achieved by using emphasis networks with 'roll-off' (i.e. progressive reduction in the slope of the characteristic) over the bottom half of the baseband. The roll-off characteristic sketched in Fig. 4.4 is the pre-emphasis characteristic recommended by CCIR; this gives an improvement of 4 dB in the top channel. Every emphasis network has a 'crossover' frequency at which the level of the signal is left unchanged; the crossover frequency for the CCIR network is approximately $0{\cdot}61F_m$ where F_m is the mid-frequency of the top telephone channel of the FDM assembly.

4.2.4 FM threshold

From expressions 4.2 and 4.4 it looks, at first sight, as if there is no limit to the improvement in *s/n* which can be obtained by increasing the deviation. However, increasing the deviation increases the RF bandwidth *B* required for satisfactory transmission (see Section 4.2.5) and this decreases the *C/N* ratio (because *N* is proportional to *B*). There is a lower limit to the *C/N* which can be used (as we shall see in the following paragraphs) and there is therefore a corresponding upper limit to the deviation.

Fig. 4.5 shows *s/n* plotted against *C/N* for a conventional FM demodulator. It will be seen that the characteristic can be divided into two regions. At high *C/N* ratios, *s/n* is proportional to *C/N* (as is predicted by expressions 4.2*a* and 4.4*a*) and the characteristic is a straight line. However for *C/N* values of less than about 10 dB, *s/n* falls off more and more rapidly as *C/N* decreases and the channel quickly becomes unusable. The region in which the *s/n* ratio begins to decrease more rapidly than the *C/N* ratio is known as the FM threshold and systems are usually designed so that *C/N* falls below threshold for only a small percentage of time.

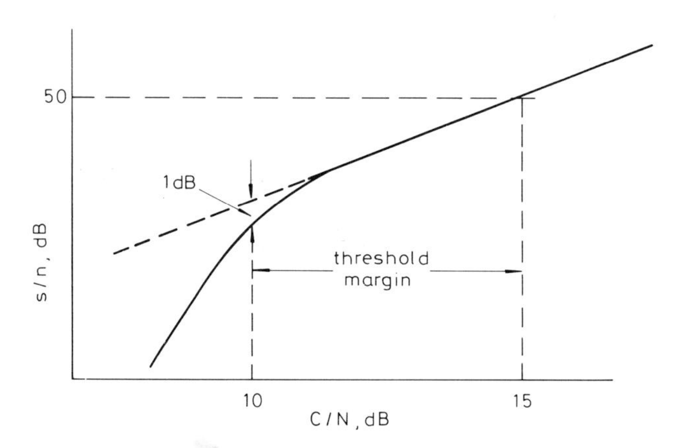

Fig. 4.5 *Characteristic of FM demodulator*

Because the departure from the straight-line relationship between *s/n* and *C/N* is (at first) very gradual, there is no obvious threshold point. Threshold is therefore usually defined arbitrarily as the point where *s/n* has fallen 1 dB below the straight-line characteristic (see Fig. 4.5). Conventional demodulators have thresholds of 10–12 dB but special threshold-extension demodulators (with thresholds down to about 6 dB) are available at extra cost. The difference between the normal working *C/N* and the *C/N* at threshold is known as the threshold margin.

The most simple way of understanding what causes FM threshold is first to consider what happens when an interfering signal is added to an FM signal.

Fig. 4.6*a* shows a wanted FM carrier of frequency F_c and amplitude V_w (represented as a stationary vector) and an interfering signal of frequency $F_c + F$ and constant amplitude V_i (represented as a vector rotating about the tip of the vector V_w). It can be seen that the effect of the interfering signal is to modulate the wanted signal in phase and amplitude. If the amplitude of the interfering signal is less than that of the wanted signal, the peak phase deviation (ϕ_p) is always less than 90° and the effect of the interfering signal is small (provided that the wanted signal has a high modulation index). However, if the amplitude of the interfering signal becomes V_i', where V_i' is greater than V_w, then the situation changes radically. The interfering signal is now modulated by the wanted signal (see Fig. 4.6*b*); this is known as the capture effect because the interfering signal is now dominant and suppresses the wanted signal. Thus when the RF power of the wanted carrier is only slightly greater than that of the interfering carrier, a small reduction in the power of the wanted carrier can result in a very large deterioration of the signal-to-interference power ratio in the baseband channel. In practice, therefore, it is essential to ensure that no interfering carrier exceeds the amplitude of a wanted FM transmission for a significant percentage of the time.

Now consider the effect of noise (rather than an interfering signal). The amplitude and phase of a noise voltage vary randomly. When the RMS power of the noise is very much less than the RMS power of the signal, the peaks of

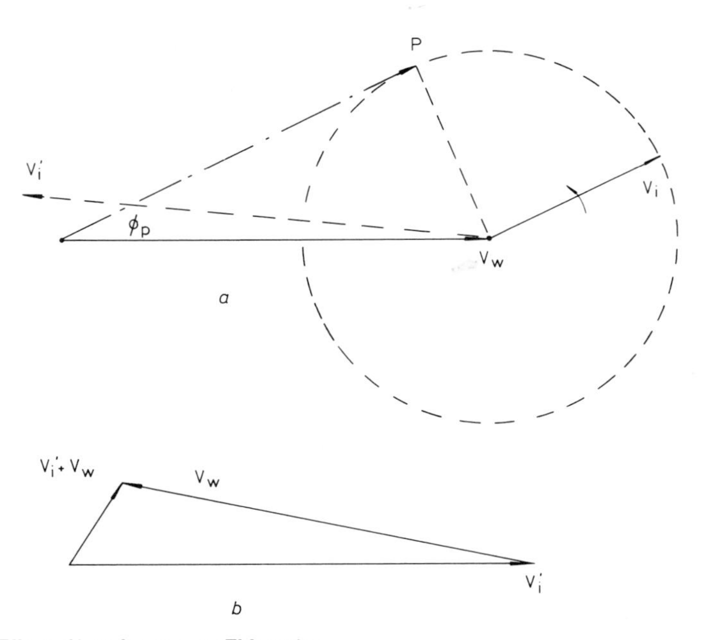

Fig. 4.6 *Effect of interference on FM carrier*

the noise voltage rarely exceed the amplitude of the wanted signal and the effect on the signal is negligible. However, when the noise power gets to about 10 dB below the signal power the peaks of the noise voltage begin to exceed the amplitude of the wanted signal for a significant percentage of the time. Each time the noise voltage exceeds the wanted signal, the noise will capture the channel for an instant and produce an audible click in the baseband channel; as the C/N ratio falls below about 10 dB there is a rapid increase in the number of clicks per second and a rapid fall in the s/n ratio. This is the threshold effect.

4.2.5 Bandwidth required for FDM–FM transmissions

4.2.5.1 Carson's rule

In Section 4.2.1 we considered the spectrum of a carrier modulated by a sinusoidal signal (of frequency F_m) and noted that nearly all the energy of the signal is contained in the carrier and the first n pairs of side frequencies, where n is the first integer greater than m (the modulation index). Thus, if we pass the signal through a bandpass filter with a bandwidth B where:

$$B = 2(m + 1)\,\mathrm{F}_m = 2(F_\mathrm{p} + F_m)\text{ hertz} \qquad (4.5a)$$

and F_p is the peak deviation, then the 'truncation' distortion caused by the loss of the rest of the side frequencies is negligible for most practical purposes.

An expression of the same form as eqn. 4.5*a* is frequently used to estimate the bandwidth required for the transmission of FDM–FM signals. We now write:

$$B = 2(F_p + F_m) \qquad (4.5b)$$

where F_m = top frequency of the FDM assembly
F_p = 'quasi-peak' deviation (the meaning of quasi-peak is explained in the next paragraph).

(Relation 4.5*b* is known as Carson's rule, after J.R. Carson who made important contributions to FM theory in the 1930s).

In the case of a sinusoidal signal there is no doubt what we mean by peak deviation F_p because there is no difficulty in defining or measuring the peak voltage of a sinusoid. However, when we try to determine the peak deviation resulting from frequency modulation by telephone channels we encounter two problems: first that the mean power in telephone channels may vary widely from channel to channel and from time to time (because, for example, the user of the channel may be a quiet talker or a loud talker); secondly that the voltage distribution with time of speech signals or FDM assemblies can only be described statistically (i.e. there is no well-defined peak voltage as there is in the case of a sinusoid). It is therefore necessary to use a quasi-peak deviation corresponding to a quasi-peak signal amplitude (i.e. an amplitude exceeded only for a very small percentage of the time).

The quasi-peak deviation is usually calculated in two stages:

(i) By determining a mean signal power relative to the test-tone power (and thus an RMS deviation relative to the test-tone deviation)
(ii) By determining the quasi-peak deviation by applying a quasi-peak factor to the RMS deviation.

In what follows the term 'peak' should be understood to mean quasi-peak where this is appropriate. Both the mean power and the peak factor are chosen in the light of measurements of the power distribution with time of FDM assemblies (which usually include a proportion of low-speed data channels as well as speech channels).

If the values chosen for the mean power or the peak factor are too high then bandwidth will be wasted and the RF carrier-to-noise ratio will be lower than it need be; if the values are too low then unacceptable truncation distortion will result.

4.2.5.2 *Mean power*

The mean power of an FDM assembly of n channels, at a point of zero relative level, is usually taken, by international agreement, as:

$$-15 + 10\log n \text{ dBm for } n \geqslant 240 \tag{4.6a}$$

and

$$-1 + 4\log n \text{ dBm for } n < 240 \tag{4.6b}$$

It may seem surprising that the mean power of assemblies of less than 240 channels is estimated on a different basis from that used for assemblies of over 240 channels since they are both made up from the same population of channels. The reason is that, although the long-term mean power is the same whatever the number of channels, the occasional very loud talkers have a greater effect on the short-term mean power as the number of channels becomes smaller.

The short-term mean powers per channel given by expressions 4.6*a* and 4.6*b* for multiplexes of 240 channels (or more) and a 12-channel multiplex are about –15 dBm and –7·5 dBm, respectively. Thus the RMS deviation per channel of a 12-channel FDM–FM transmission can be as much as antilog (7·5/20) = 2·4 times greater than that of a 240-channel transmission (all other factors being equal). Smaller multiplexes thus make less efficient use of bandwidth.

4.2.5.3 *Multi-channel peak factor*

The multi-channel peak factors commonly used are 3·16 for FDM assemblies of more than about 120 channels and 4·17 for assemblies of fewer channels. The choice of these factors (which correspond to power ratios of 10 dB and 13 dB) is based on a mixture of theory and practical experience.

4.2.5.4 *RF bandwidth*

At this point it may be helpful to give an example of the calculation of the bandwidth required for a typical FDM–FM transmission through a communications satellite:

(i) Assume a 252-channel carrier with RMS deviation per channel of 357 kHz (for a test tone of 0 dBm0)
(ii) The mean multiplex power is (–15 + 10log252) = 9·01 dBm and the ratio (multiplex RMS voltage)/(test-tone voltage) is therefore 2·82 (= antilog [9·01/20])
(iii) The multi-channel RMS deviation is thus 2·82 × 0·357 MHz = 1·007 MHz
(iv) The multi-channel peak factor is 3·16 and the peak deviation F_p is therefore 3·16 × 1·007 = 3·182 MHz
(v) The top frequency of the 252-channel multiplex F_m is 1·052 MHz
(vi) By Carson's rule the required bandwidth is:
$B = 2(F_p + F_m) = 2(3{\cdot}182 + 1{\cdot}052) = 8{\cdot}47$ MHz

The figures quoted in this example are those applicable to a 252-channel carrier transmitted via INTELSAT IVA, INTELSAT V or INTELSAT VI satellites. The earth-station filters used with such a carrier are designed to have a bandwidth equal to the occupied (or Carson's rule) bandwidth of 8·5 MHz but the bandwidth allocated in the satellite is 10 MHz. This is because, in the INTELSAT system, satellite bandwidth is almost always allocated to the various sizes of FDM–FM carrier in multiples of 2·5 MHz in order to simplify frequency planning. The guard band between carriers provided by the difference between the occupied and allocated bandwidths also:

(i) gives protection against interference between adjacent carriers (because practical band-pass filters do not have very large attenuations immediately outside the passband) and
(ii) helps to reduce the power of the intermodulation products falling in the signal bands.

4.2.6 Calculation of required carrier-to-noise power ratio for an FDM–FM transmission

Now that we have decided how to calculate the bandwidth of an FDM–FM transmission we can return to the problem (which we started to consider in Section 4.2.3) of the relation between the signal-to-noise power ration s/n in a baseband channel and the RF carrier-to-noise power ratio C/N before demodulation.

It will be recalled (from Section 4.1.2) that the total noise power in a telephone channel of the HRC should not exceed 10 000 pW0P for more than 20% of any month. The sources which may contribute to this noise are:

(i) Thermal noise on the uplink
(ii) Thermal noise on the downlink
(iii) Noise resulting from intermodulation between carriers amplified by the same transponder
(iv) Interference arising from frequency reuse (on orthogonal polarisations)
(v) Interference from other satellites and earth stations (using the same frequency band)
(vi) Noise arising from multi-carrier intermodulation in other earth stations of the system
(vii) Interference from terrestrial systems sharing the frequency band
(viii) Noise (other than thermal noise) arising in the transmitting and receiving earth-station equipment (for example, noise arising from variation of group delay with frequency)

Of the total allowable channel noise of 10 000 pW0P, INTELSAT allocates 1000 pW0P to item (vii), 1000 pW0P to item (viii) and 500 pW0P to item (vi); 7500 pW0P is thus left for items (i) to (v).

The owners of earth stations must ensure that interference from terrestrial systems is limited to 1000 pW0P and must procure earth-station equipment (in accordance with INTELSAT specifications) which does not introduce noise, other than thermal noise, in excess of 1000 pW0P. For its part INTELSAT must allocate sufficient satellite power to a transmission to ensure that the noise in the worst baseband channel resulting from items (i) to (v) is limited to 7500 pW0P (provided, of course, that the EIRP and deviation of the transmitting earth station and the G/T of the receiving earth station meet the INTELSAT specifications).

As an example, consider an INTELSAT 252-channel FDM–FM transmission with an allocated bandwidth of 10 MHz; this has a top baseband frequency of 1050 kHz and uses a test-tone deviation of 357 kHz.

The s/n ratio corresponding to 7500 pW0P is 51·2 dB (including 4 dB pre-emphasis and 2·5 dB psophometric weighting) or 44·7 dB without emphasis and weighting. Expression 4.4*a* can be used to find the required C/N. This expression can be rewritten as:

$$[s/n] = [C/N] + [B] - [b] + 2[F_r] - 2[F]$$

where $[x] = 10\log x$

Thus, substituting the values given in the previous paragraphs:

$$44{\cdot}7 = [C/N] + 69{\cdot}3 - 34{\cdot}9 + 2(55{\cdot}5) - 2(60{\cdot}2)$$

i.e. $[C/N] = 19{\cdot}7$ dB

C/N is the carrier-to-noise power ratio which INTELSAT must maintain in the presence of all the sources of noise listed in items (i) to (v) above. Thus, for example, if the C/N values corresponding to sources (iii), (iv) and (v) are $(C/N)_{im} = 27{\cdot}5$ dB, $(C/N)_r = 26$ dB and $(C/N)_i = 31$ dB then the required ratio of carrier power to the total thermal noise power arising on the up-and down-paths C/N_{th} is:

$$1/(C/N)_{th} = 1/(C/N) - 1/(C/N)_{im} - 1/(C/N)_r - 1/(C/N)_i$$

This gives $(C/N)_{th} = 22{\cdot}5$ dB

The carrier-to-thermal noise power on the uplink is usually made significantly greater than that on the downlink because it is more economic to provide high EIRPs from earth stations than from satellites; thus, for example, the uplink and downlink ratios corresponding to an overall C/N of 22·5 dB might be 29·5 and 23·5 dB, respectively.

4.2.7 *Noise power ratio (NPR)*

An FDM assembly has a frequency spectrum which is very similar to the continuous uniform spectrum of white noise (provided that the number of channels in the assembly is not too small). White noise is easily generated and is therefore often used as a signal for testing FDM transmission systems.

The sending equipment of a white-noise test set (see Fig. 4.7) comprises a white-noise generator, a bandpass filter F1 which limits the bandwidth of the output signal to that of the FDM assembly and a bandstop filter F2 with a bandwidth which is usually equal to or less than that of a single telephone channel (this is sometimes called a notch filter).

At the input to the system under test the power in the stopband is negligible. Intermodulation noise and thermal noise are however, generated as the test signal passes through the system and some of this noise falls within the stopband.

The receiving equipment includes a bandpass filter F3 with the same upper and lower frequency limits as the bandstop filter in the sending

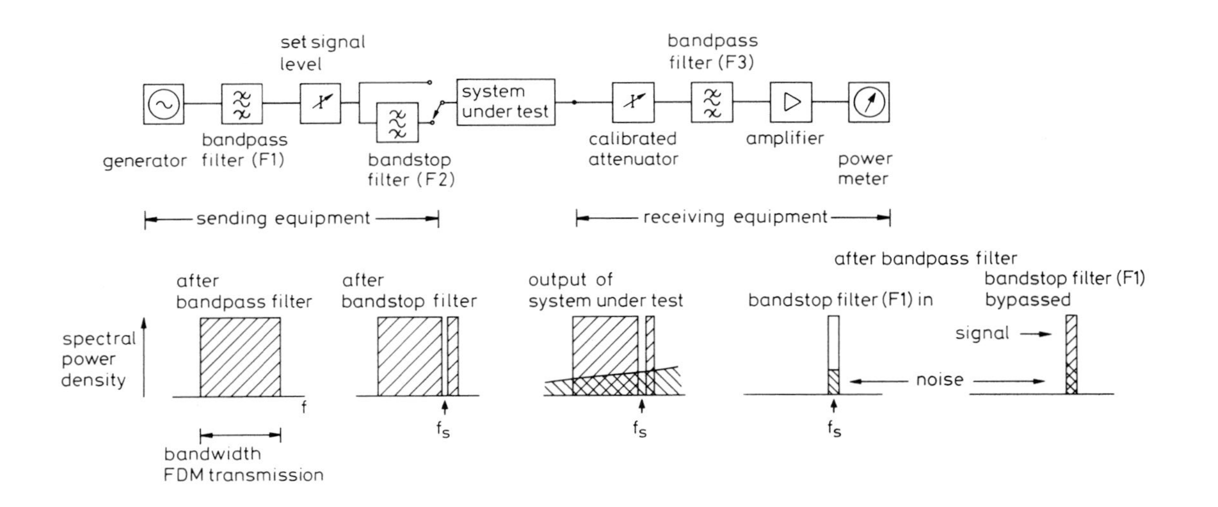

Fig. 4.7 *White-noise testing*

Filters F2 and F3 have the same centre frequency f_s and the same bandwidth

equipment. Thus the output from this bandpass filter is solely the noise power generated in the system and falling within the measurement channel.

The noise power is amplified and measured. The bandstop filter in the sending equipment is now bypassed and attenuation is inserted at the input to the receiving equipment until the power meter gives the same reading as before. The spectral power density of the white-noise signal is constant across the signal bandwidth and the spectral power density of the noise is virtually constant across the narrow measurement channel. The added attenuation is therefore equal to the ratio of the spectral power density of the signal (plus the noise which is negligible in comparison) to the spectral power density of the noise. This ratio is known as the noise power ratio (NPR).

Filters of type F1 with passbands corresponding to all the commonly used sizes of FDM assembly are available. A range of filters (of types F2 and F3) are also available to enable measurements to be made in the highest-frequency channel of any FDM assembly (and also in mid-frequency and low-frequency channels).

The total multichannel mean power used for NPR measurements is that which we have already used for calculating the occupied RF bandwidth namely:

$$-15 + 10\log n \text{ dBm0} \qquad n \geqslant 240$$

and $$-1 + 4\log n \text{ dBm0} \qquad n < 240$$

where n is the number of channels

The relationship between the signal-to-noise power ratio in a telephone channel (with the standard bandwidth of 3·1 kHz) and the NPR can be found as follows:

$$\text{NPR} = P_0/N_0 = (P/B_{fdm} \times 10^3)(1/N_0)$$

or $$1/N_0 = \text{NPR} \times B_{fdm} \times 10^3/P \tag{4.7a}$$

where P_0 = spectral power density of the signal, W/Hz
N_0 = spectral power density of the noise, W/Hz
P = total multichannel signal power, W
B_{fdm} = FDM base bandwidth, kHz

The signal-to-noise power ratio s/n is, by definition, the ratio of a test-tone signal power of 1 mW (at a point of zero reference level) to the total noise power n in the (3·1 kHz) channel, i.e.:

$$s/n = (1\text{ mW})/n = (1\text{ mW})/[N_0 \times 3{\cdot}1 \times 10^3] \tag{4.7b}$$

Substituting (from eqn. 4.7a) for $1/N_0$ we get

$$[s/n] = [\mathrm{NPR}] + [B_{fdm}/3{\cdot}1] + [(1\,\mathrm{mW})/P] \qquad (4.7c)$$

where $[x]$ means 10logx

As an example:

(i) Assume that the measured NPR in a channel of a 132-channel multiplex is 34 dB

(ii) The top and bottom frequencies of a 132-channel multiplex are 12 and 552 kHz, respectively, so that B_{fdm} is 540 and $[B_{fdm}/3{\cdot}1) = 22{\cdot}4$ dB

(iii) $[P] = -1 + 4\log 132$ dBm0
$= 7{\cdot}5$ dBm0

and $[1\,\mathrm{mW}/P] = (0 - 7{\cdot}5)$
$= -7{\cdot}5$ dB

(iv) $[s/n] = 34 + 22{\cdot}4 - 7{\cdot}5$
$= 48{\cdot}9$ dB (unweighted)

4.3 Distortion of FM transmissions

4.3.1 General

One of the main sources of distortion in FM–FDMA systems is non-linearity of the power-transfer characteristic; this results in the generation of intermodulation products. This type of distortion is common to all FDMA systems (whatever form of modulation is used) and it is dealt with in Section 3.4.

Other sources of significant distortion of FM signals are variation of group delay with frequency and AM/PM conversion.

4.3.2 Group-delay distortion

When a modulated carrier is passed through a network the modulating signal is transmitted without distortion only if the slope $d\phi/d\omega$ of the phase-shift versus frequency characteristic (see Fig. 4.8) is constant over the frequency band occupied by the signal. $d\phi/d\omega$ is known as the group delay.

For most practical networks group delay varies with frequency and for FDM–FM transmissions this variation causes intermodulation between the channels of the FDM assembly (which appears as additional noise in the baseband).

The chief sources of variation of group delay in satellite systems are bandpass filters in the satellites and earth stations. The calculation of the amount of noise which will result from variation of group delay is usually complex. Most of the methods used require the resolution of the group-delay characteristic into linear, parabolic and ripple components. The system

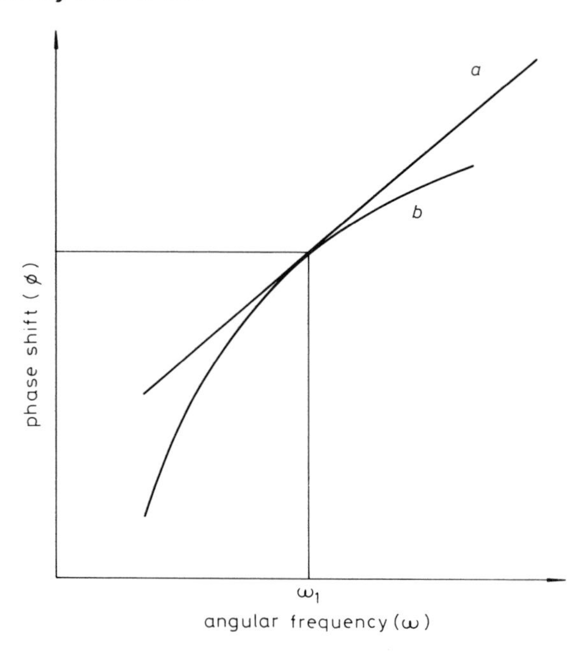

Fig. 4.8 *Phase-shift/frequency characteristic*

a Group delay ($d\phi/d\omega$) at frequency ω_1

b Phase-shift/frequency characteristic

designer may therefore specify the maximum variation of group delay directly in terms of these components or indirectly by means of a 'mask' within which the characteristic must fall; the latter is the method adopted by INTELSAT.

Figure 4.9 shows an example of a mask. A mask such as this may be used to specify the characteristics of a range of networks with different bandwidths; in this case it is necessary to provide a table giving the values of *A*, *B*, *C* and *a*, *b*, *c* for each bandwidth. Thus, for example, for an INTELSAT carrier occupying a bandwidth of 32·4 MHz:

(i) *A*, *B* and *C* are specified as 8·6 MHz, 25·9 MHz and 29·9 MHz, respectively
(ii) *a*, *b* and *c* are specified as 3 ns, 5 ns and 15 ns, respectively.

As will be seen, the allowable variation of group delay increases towards the edges of the passband (where there is relatively little power in the components of the signal and the effect of distortion is therefore comparatively small).

Fig. 4.9 also shows how the linear, parabolic and ripple components of the group-delay characteristic may be approximately determined.

The linear component is found by drawing a straight line between two points on the curve at the edges of the frequency band over which the group delay is to be specified. Thus if, in the Figure:

(i) The points are O and L, and
(ii) The scales of bandwidth and group delay are 1 MHz and 5 ns per division

then the bandwidth is 10 MHz, the difference in group delay between O and L is 10 ns and the coefficient of linear group delay is (10/10) = 1 ns/MHz.

If the linear element of group delay is removed from the characteristic then the parabolic component of group delay is given by

$$\triangle t = c(\triangle f)^2 \tag{4.8a}$$

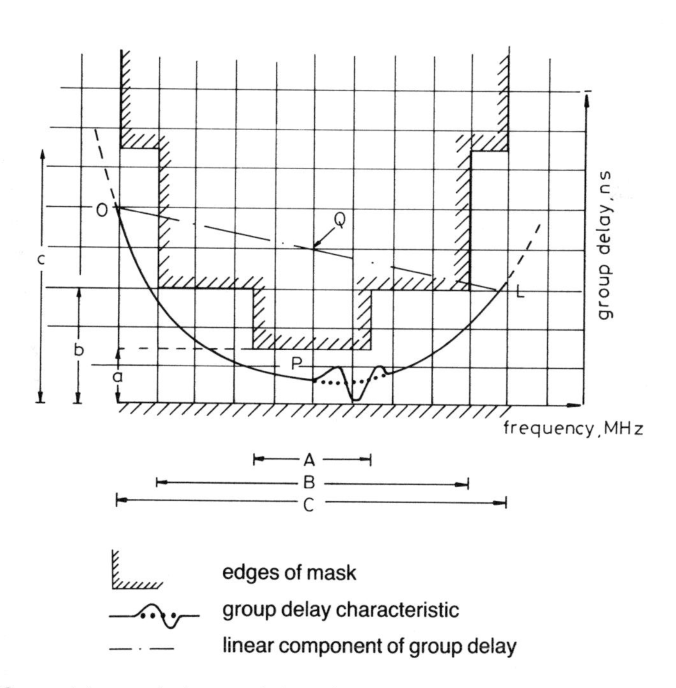

Fig. 4.9 *Group-delay mask characteristic and mask*

where $\triangle t$ = difference between the group delay at the centre of the passband and that at a frequency $\triangle f$ from the centre of the passband

c = constant (coefficient of parabolic group delay)

c may be determined approximately by measuring the difference in group delay $\triangle t$ represented by the distance QP between line OL and the group-delay characteristic at the mid-frequency of the passband. c is now given by

$$c = \triangle t/(C/2)^2 \tag{4.8b}$$

Using the same scales as before, QP in the Figure is 16·5 ns, the half-bandwidth $C/2$ is 5 MHz and the coefficient of parabolic group delay is 16·5 ns/(5 MHz)^2 = 0·66 ns/(MHz)^2.

If the characteristic includes a ripple component then, before determining the parabolic component, a smoothed characteristic is drawn as shown by the dotted line in the Figure. The ripple is specified as the peak-to-peak departure from the smoothed curve; this can be seen to be approximately 5 ns peak-to-peak in the example taken.

Although the group-delay characteristic may include cubic and higher-order terms the distortion caused by these is negligible, for most systems, compared with that caused by the linear and parabolic terms.

The necessity of making good use of the restricted bandwidth available for satellite communications means that the guard bands between channels must be as narrow as is practicable and thus the filters used to isolate the channels must be very selective. Filters with a sharp cut-off usually have large variations of group delay at the band edges. In order to achieve acceptable overall group-delay characteristics it is therefore often necessary to provide equalisers; these are networks whose group-delay characteristics can be adjusted to be approximately the inverse of the group-delay characteristics of part or all of the channel. Group-delay noise in the baseband channels of a satellite link can usually be limited to less than 500 pW by means of good filter and equaliser design. In the INTELSAT system, earth stations may be required to provide equalisation of the variation of group delay introduced by the satellite transponders (as well as equalisation of the group delay associated with the earth-terminal equipment); this reduces the mass of the satellite payload at the expense of the provision of additional equipment at every earth terminal in place of the one set of equalisers which would have been needed on the satellite.

4.3.3 AM/PM conversion and intelligible crosstalk

4.3.3.1 AM/PM conversion

The phase delay through a TWT (or any similar velocity-modulated device such as a klystron) is dependent on the amplitude of the input signal. The amplitude of a single FM transmission should be constant with time but the amplitude of the envelope of two or more FM carriers (of different frequencies) varies. Thus, when a number of carriers are amplified simultaneously in a TWT or klystron the amplitude variation of the envelope causes variation of delay which in turn results in phase modulation of the carriers. This is AM/PM conversion and the AM/PM conversion coefficient is specified in degrees/dB.

When a large number of carriers are amplified simultaneously the amplitude variations of the composite signal are random and the distortion can be treated as noise. It is usually the sum of the products resulting from amplitude non-linearity and AM/PM conversion which is measured in practice.

4.3.3.2 Intelligible crosstalk

When an FM carrier is passed through a circuit whose gain varies with frequency it will acquire amplitude modulation which is a function of the modulating signal. If this carrier now passes through a TWT (or similar device) in company with one or more other signals then the AM on the first carrier will be converted to PM of all the carriers. Thus the modulating signal impressed on the first carrier may become audible as an intelligible or partly-intelligible signal when the other carriers are demodulated in an FM or PM demodulator. This 'intelligible crosstalk' is much more annoying to a listener than noise of the same power and CCITT recommends that the ratio of signal power to intelligible-crosstalk power shall be not less than 58 dB. (Note that the signal level is assumed to be –15 dBm0, not the test-tone level of 0 dBm0, and thus the intelligible crosstalk is required to be at a level not exceeding –73 dBm0. This requires stringent control of the gain/frequency characteristics of the equipment in the earth stations and satellite transponders).

4.4 Energy dispersal

An FM transmission from a satellite or earth station may interfere with the transmissions of other systems sharing the same frequency band (or vice versa). Consider for example an FM transmission from an earth station interfering with an FM transmission of a terrestrial system. We have seen (Section 4.2.4) that an interfering carrier phase-modulates the wanted carrier at a frequency equal to the difference between the two carrier frequencies. Thus if the carriers are unmodulated and the frequency difference is F hertz and this frequency is within the baseband of the wanted transmission then all the interference power will fall within the baseband channel encompassing frequency F. On the other hand when either (or both) of the carriers is modulated then the frequency difference F varies over a range which is dependent on the power of the modulating signal(s), and the interference power is spread over the same range (centred on frequency F). If either (or both) of the transmissions occupies a large bandwidth then only a small fraction of the interference power will fall in any 4 kHz channel.

4.4.1 *Spectral power density*

The RF spectral power density (i.e. power per hertz) of FDM–FM transmissions with a high modulation index is given (to a good approximation) by the Gaussian distribution:

$$P/[\sqrt{(2\pi)}\, F_{ms}]\exp(-f^2/2F^2_{ms}) \text{ watts/Hz} \tag{4.9}$$

where P = total power of the transmission, W
F_{ms} = RMS multichannel frequency deviation, Hz and
f = frequency relative to the (unmodulated) carrier frequency

Fig. 4.10 shows a plot of this RF spectral distribution. (The Figure does not look like the usual Gaussian-distribution curve because the scale of power density is dB relative to the maximum value; i.e. the ordinate scale is logarithmic whereas a linear scale is more usual.) For FDM–FM transmissions with relatively low modulation indices, or for TV–FM transmissions, the spectral distribution will usually have peaks of significantly greater power density than the maximum of the corresponding Gaussian distribution.

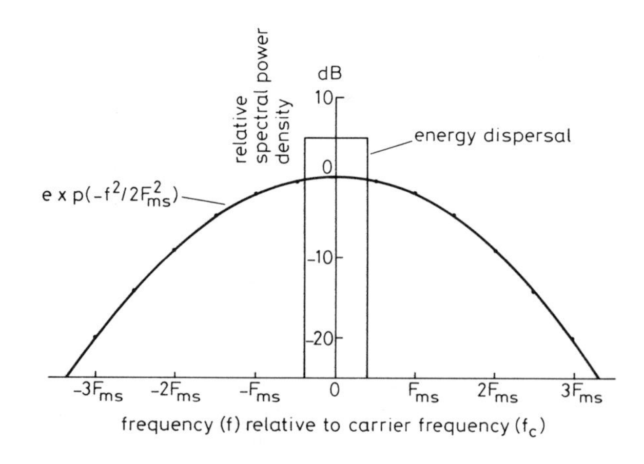

Fig. 4.10 *Spectral power density of FM transmission*

The ratio of the maximum power in a 4 kHz frequency slot of the RF carrier to the total power is (to a good approximation):

$4000/\sqrt{(2\pi)}F_{ms}$ decibels

For example, the ratio of the maximum power in a 4 kHz slot of the RF bandwidth of a fully-loaded 252-channel satellite-communications transmission with a multichannel deviation of 1627 kHz to the total signal power is:

$10\log[4/\sqrt{(2\pi)} \times 1627] = -30\,\text{dB}$

Thus the interference from this transmission into a 4 kHz channel of another transmission when both carriers are lightly loaded could be as much as 30 dB greater than the interference to the same channel when the satellite carrier is fully loaded. (The corresponding values for INTELSAT FDM–FM carriers with capacities of 12–972 channels are in the range 20–35 dB).

4.4.2 Energy dispersal

It would be difficult (if not impossible) to avoid interference between satellite systems and terrestrial systems sharing the same frequency bands if systems of both types could become lightly loaded at the same time. Artificial loading is therefore added to satellite-communications transmissions; this is known as energy dispersal. The aim is to keep the maximum spectral energy density of the transmissions during periods of light loading as close as possible to the density during periods of full loading without adding significantly to the RF bandwidth occupied by the signal.

One of the most effective methods of energy dispersal, for FDM–FM transmissions, is to add a low-frequency triangular waveform to the modulating signal. If the waveform were truly triangular then the deviated carrier would move through the frequency spectrum at constant speed and the spectral energy density across the RF bandwidth would be uniform (see Fig. 4.11). However, the dispersal signal is usually added beneath the FDM assembly and it may be necessary to pass it through a low-pass filter to avoid interference with the lowest-frequency channels of the baseband. The effect of removing the higher harmonics of a triangular waveform is to round off the peaks and this causes an increase in spectral density at the edges of the RF energy-dispersal bandwidth (see Fig. 4.11). To limit this effect to an acceptable level it is usual to choose a low fundamental frequency (less than about 150 Hz) for the energy-dispersal signal so that a large number of harmonics can be preserved.

The ratio between the total power and the power in any 4 kHz of a transmission modulated solely by a triangular dispersal waveform is:

$10\log(B/4000)$ decibels

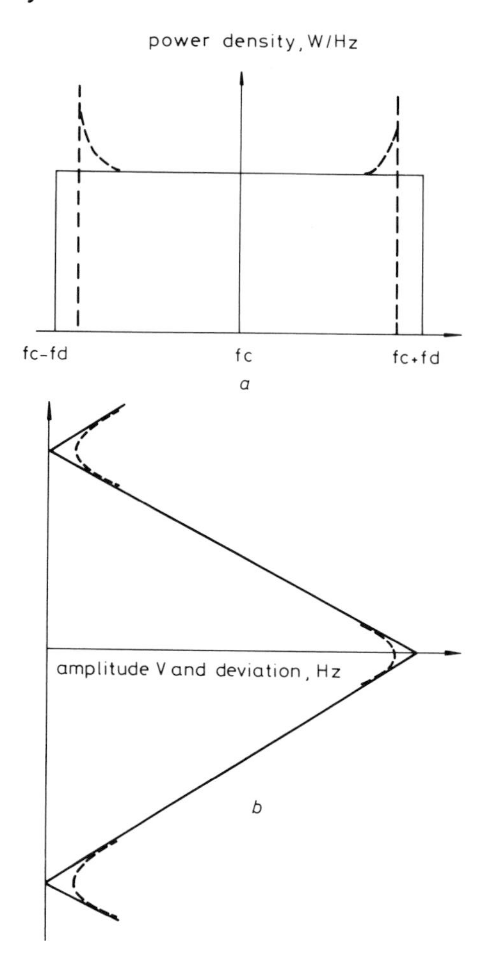

Fig. 4.11 *Energy dispersal*

a RF spectral energy density
b Energy dispersal waveform

where B is the peak-to-peak deviation corresponding to the dispersal signal.

Thus, for example, consider once again a 252-channel transmission for which the multichannel RMS deviation (F_{ms}) is 1·627 MHz. If, to the baseband signal of this transmission, we add a triangular waveform giving a peak-to-peak deviation of 1·26 MHz ($= 0{\cdot}77F_{ms}$) this will reduce the power per 4 kHz channel (in the absence of traffic loading) to $10\log[(1{\cdot}26 \times 10^6)/(4 \times 10^3)] = 25$ dB below the total transmission power, which is only 5 dB higher than the maximum power per channel which results when the carrier is fully loaded by the FDM assembly. The RF spectrum produced by a triangular waveform giving a peak-to-peak deviation of $0{\cdot}77F_{ms}$ is shown in Fig. 4.10 to the same scales as the spectrum of the Gaussian distribution. (Note that the total area under the Gaussian curve would be equal to the area of the rectangle if a linear scale of power density were used). The occupied bandwidth of the 252-channel transmission is 12·4 MHz, so that addition of this energy-dispersal signal increases the bandwidth required by about 10%. If a large amount of dispersal is needed and this would increase the bandwidth to an unacceptable extent then the dispersal signal must be switched out when the traffic loading rises to some predetermined level (alternatively the amplitude of the dispersal signal may be continuously decreased as the loading increases). The dispersal signal must, of course, be removed at the receiver.

A slightly different form of energy-dispersal signal is used in the case of TV signals (see Section 4.7.2.5).

Because the distribution of interfering energy over the baseband depends on the way in which the difference frequency between wanted and interfering signals varies (not on how the frequencies of the individual signals vary) energy dispersal will not only reduce the interference power which an FM communications-satellite transmission can inject into a baseband channel of a FM terrestrial system but will also provide the same protection to the channels of the satellite system against interference from a terrestrial system. Applying the same energy-dispersal deviation to both a satellite system and a terrestrial system increases the protection against interference by only 3 dB relative to that given by the application of dispersal to one of the systems; dispersal is therefore nomally applied only to transmissions via communications satellites.

In this Section we have been discussing the limitation of interference between FM transmissions which can be achieved by spreading the total energy of the transmissions over a wide RF bandwidth and thus limiting the spectral energy density. The principle is, however, not confined to interference between FM transmissions; in general, interference between transmissions using any type of modulation can be limited by using large RF bandwidths to reduce the maximum spectral energy density of the transmissions.

4.5 Companding

4.5.1 Syllabic compandors

Companding is a way of reducing the effect of noise on speech channels. A compandor comprises a compressor and an expander. In the type of compandor which is usually employed with analogue telephone channels, the gains of the compressor and expander are controlled by the speech power at syllabic rate; these compandors are therefore known as syllabic compandors.

Fig. 4.12*a* shows the characteristics of a typical compressor and expander. It will be seen that the gain of the compressor varies in such a way that an input level of x dB (relative to some reference level) results in an output level of $x/2$ dB. For example, input levels to the compressor of +10, 0 (the reference level) and −60 dB result in output levels from the compressor of +5, 0 and −30 dB, respectively. The characteristic of the expander is the inverse of that of the compressor; thus all signals that pass through both the compressor and the expander end up with their relative levels unchanged. It is not usually practicable to use compressors and expanders with a ratio of greater than 2 : 1 and the absolute power level corresponding to the unaffected level must be chosen with care in the light of the characteristics of the signals to be handled by the compandor.

Fig. 4.12*b* shows the effect of companding on the relative levels of a signal and the noise added during transmission (i.e. after the compressor and before the expander). The noise is assumed to have a relative level (without companding) of −40 dB and the figure shows that:

(i) The compressor raises the power of weak signals relative to the noise; thus for an input signal at a level of −40 dB (i.e. a signal which would be at the same level as the noise in the absence of companding) the effect of the compressor is to raise the signal-to-noise power ratio from 0 to 20 db.

(ii) The expander restores the original level of the speech signal and, in the absence of a signal, it reduces the level of the noise; in the example given the level of the signal is restored to −40dB and, in the absence of a signal, the noise is reduced from −40 to −80 dB.

The improvement in the quality of the channel is thus partly objective and partly subjective. The compressor raises the level of weak signals and thereby improves the signal-to-noise ratio by an amount which is measurable. The expander produces a large reduction in the level of the noise during pauses in the speech and this makes the channel sound much quieter even though the average noise power in the channel may not have been greatly reduced. The quality of a companded channel must therefore be judged by its effect on users (i.e. by subjective tests) and not by measurements of noise power.

Suppose that a group of telephone users are asked to use two different sets of telephone circuits; one set having no compandors and channel noise at a power level of (say) −50 dBm, and the other set having compandors and channel noise which would be at a level of −40 dBm in the absence of compandors. If the group of users cannot detect any significant difference between the two sets of circuits then the compandors are said to give a companding advantage of 10 dB.

The companding advantage obtained in practice is dependent on the compandor, the characteristics of the satellite system and on the character-

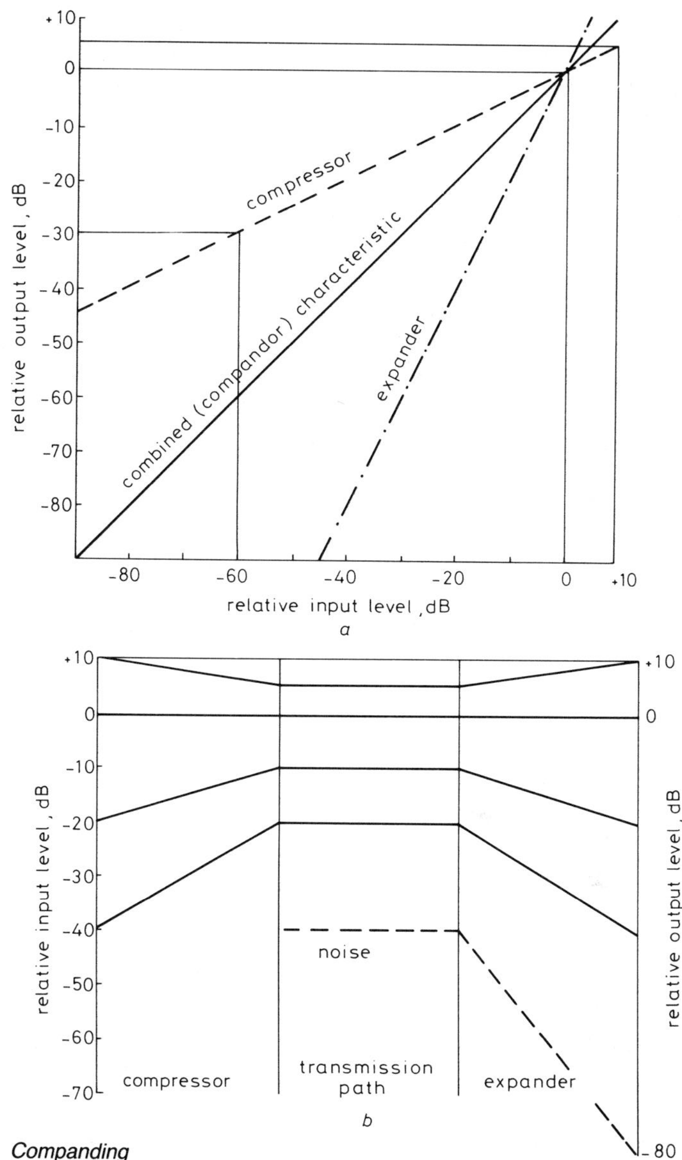

Fig. 4.12 *Companding*

a Compandor characteristics
b Principle of companding

istics of the speech signals (e.g. the advantage usually varies significantly with the speech power and can be dependent on language). It is thus better to carry out listening tests to establish the advantage for any new system rather than depending on a forecast of companding advantage. Listening tests on the

INTELSAT CFDM–FM system showed the companding advantage to be 9 dB; higher advantages, in the range 10–20 dB, are usually claimed when companders are used with SCPC–FM systems.

Companding has been used for many years, particularly in systems where it would otherwise have been difficult to achieve satisfactory quality at an acceptable cost. In the past, however, some administrations have objected to the use of compandors on circuits forming part of the public switched telephone network (except in special circumstances). Two of the reasons for this were:

(i) Some of the earlier types of compandors made it difficult to maintain stable signal levels.
(ii) It was feared that degradation of circuit quality would result from the connection of a number of compandors in tandem in long circuits via switched networks.

Improvements in design have, to a large extent, removed the dangers of instability and it is now generally accepted that the inclusion of syllabic companders in satellite circuits is unlikely to lead to problems.

4.5.2 CFDM

Companded channels may be combined into FDM assemblies in just the same way as uncompanded channels and the companding advantage can be particularly useful in offsetting the comparative inefficiency of FDM–FM transmission of small assemblies of channels. INTELSAT provides for the transmission of FDM assemblies of 24–252 companded channels on FM carriers (CFDM–FM transmission). Where the channels are part of the international public switched telephone network the compandors may be installed at international switching centres, thus extending the benefits of companding to the terrestrial links between switching centres and earth stations.

The benefit of CFDM may be illustrated by comparing the INTELSAT 24-channel FDM–FM and CFDM–FM transmissions (see Table 4.1).

Table 4.1 *Characteristics of INTELSAT 24-Channel FDM–FM and CFDM–FM transmissions*

	Allocated bandwidth (MHz)	Occupied bandwidth (MHz)	RMS deviation 0 dBm0 test tone (kHz)	C/T (dB(W/K))	C/N (dB)
FDM	2·5	2·0	164	−153·0	12·7
CFDM	1.25	1·1	81	−156·0	12·2

It will be seen that the companding advantage of 9 dB is used in two ways:

(i) To reduce the required C/T by 3 dB
(ii) To reduce the test-tone deviation, and therefore the required bandwidth, by a factor of 2 (i.e. 6 dB).

4.6 SCPC–FM

SCPC–FM channels are frequently used for the provision of service in systems which include many remote earth stations, each of which requires only a small number of telephony channels. The earth-station equipment is usually simple and easy to maintain, and start-up costs are relatively low since each station is able to commence operation with as few channels as is desired and to add more equipment as traffic increases or communication with additional stations is needed. SCPC–FM systems normally make use of syllabic companding and voice-activated carriers in order to reduce the power and bandwidth, and therefore the cost, per channel. The term narrow-band FM (NBFM) is often used in referring to SCPC–FM systems.

The INTELSAT VISTA service is a typical application of SCPC techniques. A global-beam transponder of an INTELSAT V satellite (with a saturated EIRP of 23·5 dBW) and Standard D–1 earth stations (with $G/T = 31{\cdot}7$ dB(1/K)) can provide 1200 VISTA channels using SCPC carriers spaced at intervals of 30 kHz across the usable bandwidth of 36 MHz. Frequency spacings of 22.5 and 45 kHz are used in some other systems and a transponder may be shared between SCPC and other types of transmission such as television transmissions.

The use of carriers spaced at regular frequency intervals simplifies the earth-station equipment but means that many of the large number of third-order intermodulation products will be coincident in frequency with the wanted transmissions; the power in these products must be limited by backing off the output amplifier of the transponder (a typical output back-off for a TWT amplifier is 5 dB). On the other hand, in systems which are severely power limited it is possible to use a frequency plan designed to ensure that much of the intermodulation power falls in the gaps between the wanted transmissions and this allows the output amplifiers to be used with lower back-off.

Expression 4.2*a* can be used to find the relation between s/n and C/N for a SCPC transmission; in this case F_u and F_l are the upper and lower frequency limits of a telephone channel, i.e. 3400 Hz and 300 Hz, and F_l^3 can be ignored compared with F_u^3. The relation therefore becomes:

$$s/n = (C/N) \times 3B \times F_r^2/F_u^3 \qquad (4.2c)$$

It will be recalled from Section 4.2.5 that the RF bandwidth required for a FDM–FM transmission was determined from Carson's rule, i.e.

$$B = 2(F_p + F_m)$$

where F_p is the quasi-peak deviation and F_m is the top baseband frequency. Carson's rule may also be used to estimate the bandwidth required for SCPC–FM transmissions but in this case there is no general agreement on a method of deriving the quasi-peak deviation F_p.

The mean power level of speech channels is around −16 dBm0 with a standard deviation of about 6 dB, i.e. about 95% of channels have a mean power level within the range −28 dBm0 to − 4 dBm0. The sensitivity of the modulator (i.e. the test-tone deviation) must be chosen to give acceptable quality in the channels used by soft talkers; on the other hand, the RF bandwidth per channel must be restricted to an economic value and this means that some clipping of the loud talkers must be accepted.

It is very difficult to estimate the amount of clipping that will occur (especially when the effects of pre-emphasis and companding are taken into account) and, as always, the final test of the system is whether it is acceptable to the users. It is therefore very desirable to conduct listening tests before the characteristics of a new system are finalised.

The ratio of quasi-peak deviation to test-tone (0 dBm0) deviation for current SCPC–FM systems is in the range 1 to 3 with most systems using a value of around 2. For example, INTELSAT specifies that the test-tone deviation for the VISTA system shall be 5·1 kHz and that the IF bandwidth shall be 25 kHz. A bandwidth of 25 kHz corresponds to a peak deviation of 9·1 kHz and the ratio of peak deviation to test-tone deviation is therefore approximately 1·8.

Most of the power of speech signals is contained in the lower-frequency components whereas the power density of the thermal noise in an SCPC–FM channel increases with the square of frequency. Significant improvement in the s/n ratio can usually therefore be achieved by the use of emphasis networks to reduce the power density of the noise at the higher frequencies. The characteristics of the networks are chosen to suit the particular system. Most networks have a crossover frequency of about 1000 Hz with a slope of 4–6 dB per octave above crossover and there may be marked roll-off at the lower frequencies. Improvements of up to 6 dB can be achieved.

The psophometric weighting factor is 2·5 dB and companding advantages of 10–20 dB may be obtained (the higher advantage being obtainable with lower speech levels).

Let us now calculate the signal-to-noise ratio corresponding to the system characteristics previously considered, i.e. a test-tone deviation F_r of 5·1 kHz, RF bandwidth B of 25 kHz and top baseband frequency F_u of 3·4 kHz. If C/N is 10 dB then (from eqn. 4.2c):

$$\begin{aligned}[s/n] &= [C/N] + [3B] + 2[F_r] - 3[F_u] \\ &= 10 + 48{\cdot}8 + 74{\cdot}2 - 105{\cdot}9 \\ &= 27{\cdot}1\ \text{dB}\end{aligned}$$

If we assume pre-emphasis improvement of 5·5 dB, companding advantage of 17 dB and add the usual psophometric weighting of 2·5 dB, the effective *s/n* becomes:

$$\begin{aligned} & 27{\cdot}1 + 2{\cdot}5 + 5{\cdot}5 + 17{\cdot}0 \\ = \; & 52{\cdot}1 \text{ dB} \end{aligned}$$

We have seen that the normal frequency spacing between adjacent SCPC transmissions in an INTELSAT VISTA network is 30 kHz and that the occupied bandwidth per transmission is about 25 kHz. Similarly, for systems with spacings between carriers of 22·5 and 45 kHz the occupied bandwidths are about 30 and 19 kHz, respectively. It is thus clear that carrier frequencies must be controlled within a few kilohertz in order to avoid interference between adjacent transmissions. Each earth station must therefore be provided with automatic frequency control (AFC) equipment and (at least) one of the stations in the system must transmit an AFC pilot (see, for example, Section 6.4.5).

4.7 TV–FM

4.7.1 Introduction

A colour-television signal comprises a luminance (brightness) signal, a chrominance (colour) signal and timing information (e.g. line-synchronising pulses). The performance of a colour-television channel is usually defined in terms of the ratio of luminance signal power to noise power (this also defines the signal-to-noise power ratio in the chrominance channel because the peak amplitude of the chrominance signal is fixed relative to that of the luminance signal).

The three systems used for terrestrial television broadcasting are the NTSC system (developed in the USA), and the PAL and SECAM systems (developed in Europe). The NTSC system is a 525/60 system (i.e. 60 frames are transmitted each second and each picture is made up of 525 lines); the PAL and SECAM systems are 625/50 systems. In all three of these systems the chrominance channel is modulated onto a sub-carrier (at a frequency of about 3·5 MHz or 4·5 MHz depending on the system). The spectra of the luminance and chrominance channels overlap (see Fig. 4.13), the side frequencies being interleaved in such a way as to minimise interference between the two channels. This type of signal is known as a composite television signal.

The NTSC, PAL and SECAM systems are used for the relay of television pictures by satellite to terrestrial stations (for re-broadcasting) or to the head ends of cable-television networks. On the other hand the multiplexed-analogue-components (MAC) system will be used for broadcasting television

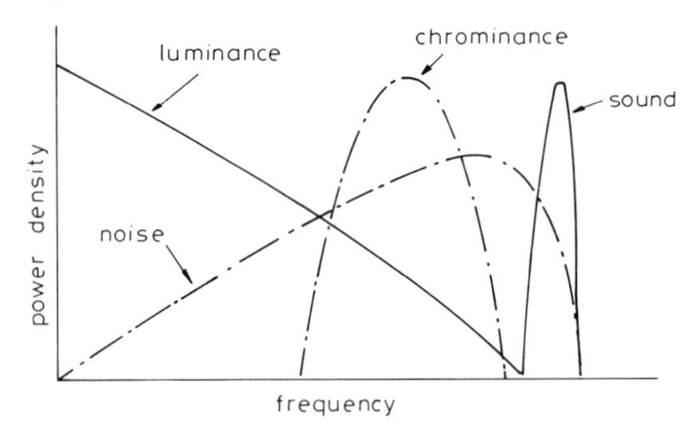

Fig. 4.13 *Spectrum of composite television signal (not to scale)*

programmes via satellites direct to homes (DBS); in the MAC system the luminance and chrominance signals are transmitted in time-division multiplex (see Section 4.7.5.1). Frequency modulation is used with all analogue television transmissions by satellite.

4.7.2 Relay of composite signals

Fig. 4.14 shows the waveform of one line of the luminance channel of a composite signal. The luminance component of the signal is combined with the sync pulses. The total (peak-to-peak) amplitude of the luminance-plus-sync-pulse signal at a point of zero reference level is 1 V and the peak-to-peak voltage of the luminance component (i.e. black level to peak-white level) is 0·7 V.

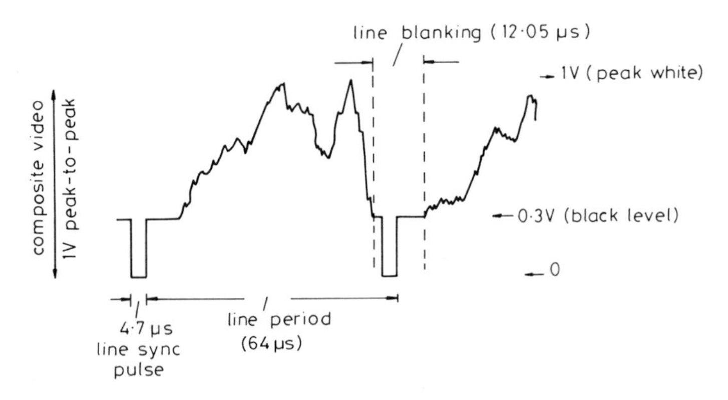

Fig. 4.14 *625/50 television waveform: Luminance component plus sync pulses*

4.7.2.1 Signal-to-noise power ratio

There are several different ways of deriving the signal-to-noise power ratio in a television channel and this can be confusing; the methods most commonly used are therefore summarised in the succeeding paragraphs.

The baseband of a television signal extends from very low frequencies to between 4 and 6 MHz (depending on the system). Thus when the FM equation (expression 4.2*a*) is applied to television transmissions F_l^3 is negligible compared with F_u^3 and the equation becomes:

$$s/n = (C/N) \times 3B \times F_r^2/F_u^3 \qquad (4.9a)$$

where C/N = carrier-to-noise power ratio in the RF bandwidth B
F_r = RMS deviation corresponding to the signal
F_u = top baseband frequency
s/n = ratio of mean signal power to mean noise power, i.e. 20 log[(RMS signal voltage)/(RMS noise voltage)]

However, it is customary to define the signal-to-noise ratio in a video channel as:

$$(s/n) = 20\log[(\text{peak-to-peak luminance voltage})/(\text{RMS noise voltage})]$$

If we use this new definition of signal-to-noise ratio we must replace F_r (RMS deviation) in eqn. 4.9 with the deviation F_{pp} corresponding to the peak-to-peak luminance voltage; thus

$$s/n = (C/N) \times 3B \times F^2_{pp}/F_u^3 \qquad (4.9b)$$

The test tone for a television channel usually has the same magnitude as the luminance-plus-sync-pulse signal (i.e. 1 V peak-to-peak). Thus, if F_t is the peak-to-peak deviation of the test tone then:

$$F^2_{pp} = (0{\cdot}7 \times F_t) = F_t^2/2 \text{ (approximately)}$$
$$\text{and } s/n = (C/N) \times (3/2) \times BF_t^2/F_u^3 \qquad (4.9c)$$

In calculating the s/n of TV–FM transmissions via INTELSAT satellites use is often made of the peak frequency deviation of the test tone F_d (i.e. $F_d = F_t/2$); in this case we must write:

$$s/n = (C/N) \times 6B \times F^2_d/F_u^3 \qquad (4.9d)$$

where F_d is defined by INTELSAT as the peak frequency deviation corresponding to a 1·0 V peak-to-peak test tone at the crossover frequency of the pre-emphasis characteristic. (The deviation may also be quoted as that corresponding to a test tone at a frequency of 15 kHz; the consequences of this are explained in Section 4.7.2.3).

4.7.2.2. *Weighting and pre-emphasis*

In section 4.1.2 the use of weighting to allow for the amplitude/frequency response of the ear was explained. Weighting can also be used to allow for the decrease in sensitivity of the eye to video noise as the frequency of the noise increases. The magnitude of the video-weighting improvement is very dependent on the distribution of the noise power across the video bandwidth. When the noise power density is proportional to the square of baseband frequency (as it is in an FM system working well above threshold) the improvement ranges from about 10 to 16 dB (depending on the television system used) but for white noise (i.e. constant noise power density across the baseband) the improvement is more likely to be in the range 5–9 dB.

Emphasis is used with TV–FM transmissions in order to reduce the noise density at the higher video frequencies. Because emphasis changes the distribution of noise power in the video channel it also affects the weighting improvement; it is therefore best to specify the improvement resulting from weighting and pre-emphasis as a single composite improvement applicable to a particular system working under specific conditions (e.g. with specified pre-emphasis and with C/N well above threshold).

There are many variants of the basic television systems; for example, systems B, C, G and H are variants of PAL, Systems D, K and L are variants of SECAM and system M is a variant of NTSC. A number of different pre-emphasis and weighting improvements have been used with these variants and the s/n ratios calculated using these different improvement factors may vary by as much as 6 dB for transmissions with the same C/N and frequency deviation (although viewers can often see little difference between the quality of the channels).

This can lead to problems, particularly in the context of the international exchange of television programmes, and CCIR has recommended a 'unified' system for the calculation of s/n. In this method the highest baseband frequency F_u is taken as 5 MHz for all systems. Separate weighting networks are specified for the 525/60 system and the 625/50 systems (see Fig. 4.15) and the weighting plus pre-emphasis improvement is specified as 14·8 dB for the 525/60 system and 13·2 dB for the 625/50 systems.

Only the unified system is considered in this book. Let us use it to find the s/n ratio corresponding to a 625/50 transmission (with an allocated satellite bandwidth of 20 MHz) between two INTELSAT Standard-A stations via an INTELSAT V transponder. The characteristics, taken from INTELSAT Earth Stations Standards Module IESS–306 (Table 5*a*), are:

C/N $= 17{\cdot}3\,\text{dB}$
Peak frequency deviation (F_d) $= 10{\cdot}5\,\text{MHz}$
Occupied bandwidth (B) $= 18\,\text{MHz}$

Substituting in eqn. 4.9*d* we get:

$$s/n\ (\text{dB}) = 17{\cdot}3 + 10\log[6 \times 18 \times (10{\cdot}5)^2/5^3]$$
$$= 37{\cdot}1 \text{ (without pre-emphasis and weighting improvement)}$$

With the addition of pre-emphasis and weighting improvement of 13·2 dB the s/n becomes 50·3 dB. For a 525/60 system the s/n for the same C/N and deviation is 51·9 dB.

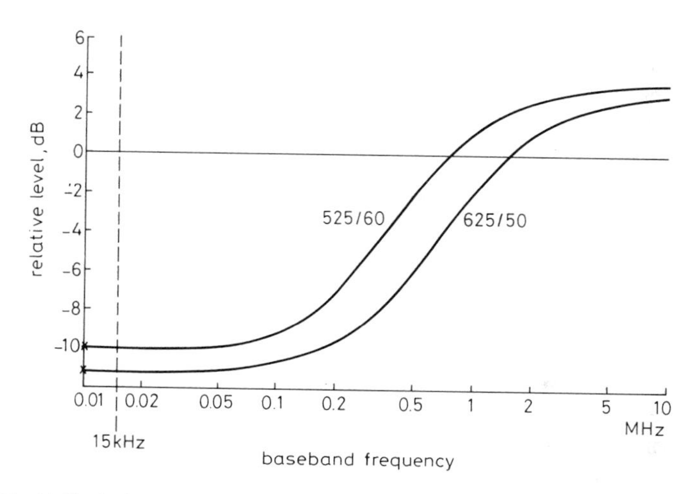

Fig. 4.15 *Unified television weighting networks (CCIR Recommendation 405–1)*

4.7.2.3 *Frequency of test-tone deviation*

There is one more complication which must be mentioned. So far we have been assuming that the test tone is at the crossover frequency of the pre-emphasis network (i.e. the frequency at which the network has no effect on a signal passing through it).

However, the standard frequency for the test-tone deviation for a television channel is 15 kHz. The effect of the pre-emphasis network is to reduce the deviation by the difference in the gain of the network at 15 kHz and at the crossover frequency. As can be seen from Fig. 4.12, this difference is 10 dB for the 525/60 pre-emphasis network and 11 dB for the 625/50 pre-emphasis network. Thus, if the peak deviation at the crossover frequency is 10·5 MHz then the corresponding peak-to-peak deviation at 15 kHz is:

$$2[10{\cdot}5/\text{antilog}(10/20)] = 6{\cdot}64\,\text{MHz}(525/60)$$
$$\text{or } 2[10{\cdot}5/\text{antilog}(11/20)] = 5{\cdot}92\,\text{MHz}(625/50)$$

If the deviation at 15 kHz is substituted in expressions 4.9*b*—4.9*e* the apparent values of the s/n are reduced by 10 or 11 dB relative to those found by using the deviations at the crossover frequencies. This must be corrected by including these factors in the weighting and pre-emphasis improvements. The unified weighting and pre-emphasis improvements therefore become 24·8 dB and 24·2 dB for 525/60 and 625/50 transmissions, respectively. INTELSAT usually quotes both the peak-to-peak deviation at 15 kHz and the peak deviation at the crossover frequency.

4.7.2.4 RF bandwidth and overdeviation

The spectral distribution of RF power in TV–FM transmission is such that it is possible, without introducing unacceptable distortion, to use bandwidths which are significantly narrower than those given by Carson's rule.

It will be recalled that the Carson's-rule bandwidth B is:

$$B = 2(F_p + F_m)$$

where F_p = peak deviation and
F_m = maximum baseband frequency

For the 625/50 TV–FM transmission with an occupied bandwidth of 18 MHz, considered earlier in this Section, INTELSAT specifies the maximum video frequency as 6 MHz (note that the value of 5 MHz used with the unified system of calculating signal-to-noise ratio is not used in calculating RF bandwidth). The peak deviation corresponding to the Carson rule bandwidth is therefore $(18/2 - 6) = 3$ MHz. However, as we saw in Section 4.7.2.2, INTELSAT actually specifies a peak deviation of 10·5 MHz (at crossover frequency). The use of a greater deviation than that given by Carson's rule is described as overdeviation, and the amount of the overdeviation is defined as:

$$20\log\,[(\text{actual deviation})/(\text{Carson-rule deviation})]$$

Thus, the overdeviation for the 625/50 transmission is

$$20\log\,(10{\cdot}5/3) = 10{\cdot}9\,\text{dB}$$

For a 525/60 transmission occupying the same RF bandwidth of 18 MHz INTELSAT again specifies the peak deviation as 10·5 MHz but the maximum video frequency is now 4·2 MHz. The peak deviation is therefore 4·8 MHz and the overdeviation is 6·8 dB.

When overdeviation is used the instantaneous frequency corresponding to the peak deviation is well outside the passband of the band-limiting filters. Thus when the deviation is close to peak the carrier is suppressed and a short burst of noise is generated; this is visible as a spot on the picture. The percentage of time when the carrier is outside the passband is usually very small, but as the amount of overdeviation is increased a point is reached where the quality of the picture deteriorates rapidly.

4.7.2.5 Energy dispersal

The spectrum of a television picture usually has peaks of high power density; the application of energy dispersal is therefore essential for the avoidance of interference by and to television transmissions sharing a frequency band. The usual dispersal signal is a triangular wave form. This is removed as completely as is practicable at the receiving earth station but some trace of the waveform may remain; in order to stop the residual component becoming visible the waveform is synchronised with the field frequency of the video signal and the phase is adjusted so that the points of inflection occur during the field blanking intervals. INTELSAT specifies a symmetrical triangular dispersal waveform producing a deviation of 1 MHz peak-to-peak in the presence of the video signal; when the video signal is absent the energy dispersal is increased to 2 MHz peak-to-peak which ensures that the power in any 4 kHz band of the RF spectrum is at least 27 dB below the carrier power.

4.7.2.6 INTELSAT TV–FM transmissions

The ITU objective for the quality of a long-distance TV circuit is a signal-to-noise power ratio of not less than 53 dB for 99% of the time and not less than 45 dB for 99·9% of the time. Achieving this standard can be expensive and lower standards are often accepted. Many viewers can see nothing wrong with the picture at signal-to-noise ratios down to about 43 dB and notice only slight degradation at ratios of about 40 dB.

With most of the communications satellites now in use the ITU objective either cannot be met or can be met only by allocating a complete transponder to a single TV–FM transmission; INTELSAT calls this 'full-transponder TV'. It may also be necessary to use large (INTELSAT Standard-A) antennas at both the transmitting and receiving stations.

By using overdeviation it is possible to transmit two television programmes through a 36 MHz transponder on an INTELSAT IVA satellite and get acceptable picture quality; alternatively, one half of the transponder can be used for television and the other half can be used for other types of transmission. This is known as 'half-tranponder TV'. Overdeviation also makes it possible to receive pictures of acceptable quality from satellites such as INTELSAT V when using earth-station antennas with a diameter of about 13 m.

All the transponders of INTELSAT IVA satellites have a bandwidth of 36 MHz; these satellites are still being used for half-transponder TV. INTELSAT specifies the allocated bandwidth for this service as 17·50 MHz, the corresponding occupied bandwidth being 15·75 MHz. Overdeviations of 6·2 dB and 12 dB are used with 525/60 and 625/50 transmissions, respectively (an overdeviation of 12 dB pushes the practice to its limits, and some would say beyond acceptable limits).

36 MHz transponders of INTELSAT V and INTELSAT VI satellites may be used for half-transponder TV (with the same bandwidth and deviation as for INTELSAT IVA) but these satellites also have transponders with a bandwidth of 41 MHz. The use of the latter for half-transponder TV enables the allocated and occupied bandwidths to be increased to 20 and 18 MHz, respectively, and the overdeviation associated with 625/50 transmissions to be reduced to 10·9 dB.

Table 4.2 gives some examples of the weighted signal-to-noise ratios available with INTELSAT half-transponder and full-transponder TV transmissions in the 6/4 GHz band. The figures are taken from INTELSAT Earth Station Standards Document IESS–306 (July 1985) which assumed the use of earth stations to the old Standard-A specification, i.e. with $G/T = 40{\cdot}7$ dB(1/K). The new Standard-A specification requires a G/T of only 35 dB(1/K) and few such stations are yet in operation. However, the assumptions made in IESS–306 were pessimistic and, in particular, the EIRPs of INTELSAT satellites have generally been better than specified values; it should therefore be possible to achieve signal-to-noise ratios close to those given in the Table when using the new Standard-A antennas. The G/T of stations using 13 m antennas is generally about 3 dB less than that of the new Standard-A stations.

Table 4.2 *Characteristics of INTELSAT TV transmissions*

Allocated bandwidth (MHz)	30		20		17·5	
Occupied bandwidth (MHz)	30		18		15·75	
Peak deviation (MHz)	9·0		10·5		7·5	
Overdeviation factor (dB)	0		10·9		12·0	
Satellite type	VI	V	VI	V	VI	V
C/N (dB)	19·2	16·5	20·1	17·3	20·7	17·9
Weighted s/n (dB)	53·1	50·4	53·1	50·3	50·2	47·4

(i) Both antennas are assumed to be at an elevation of 10°
(ii) Deviations are at crossover frequency
(iii) Weighted signal-to-noise ratios are those corresponding to the unified system.

4.7.3 TV sound

At least one good-quality audio (programme) channel with a bandwidth of 10–15 kHz is usually provided with every television broadcast. Additional (narrow-band) sound channels are often required, e.g. for commentaries in different languages on sporting events and for management of the broadcasts.

4.7.3.1 FM audio sub-carriers

Probably the most common method of transmitting the programme channel is by frequency modulation onto a sub-carrier which is then added to the video baseband signal. At the receiving earth station the composite video-plus-audio transmission is demodulated, the FM programme sub-carrier is separated from the video carrier by filtering and the sub-carrier itself is then demodulated.

The FM sub-carrier method of programme channel transmission is economical of bandwidth and can give good quality. However, the audio sub-carrier is at the top end of the baseband where it is subject to maximum pre-emphasis on passing through the video network; care must therefore be taken to avoid overdeviation. If a second programme channel is needed (e.g. for a stereo broadcast) a second sub-carrier can be added; it may, however, be difficult to avoid intermodulation between the two sub-carriers.

In order to avoid crosstalk between the audio channels of two television transmissions through the same transponder, INTELSAT specifies two sub-carrier frequencies of 6·60 and 6·65 MHz; only the 6·60 MHz sub-carrier is used in association with full-transponder or single half-transponder transmissions. Other systems may use other audio sub-carrier frequencies such as 5·8, 6·2 and 6·8 MHz.

4.7.3.2 Sound-in-sync

Another method commonly used for the transmission of programme channels is sound-in-sync (SIS). In this method the problems associated with sub-carriers are avoided by multiplexing the audio and video channels in time rather than in frequency. The audio signal is sampled twice during each line and the samples are PCM encoded using 10 bits per sample. The resulting 20 bits of data are stored and inserted in the next sync pulse. The line rate for all three composite television signals is approximately 15 kHz and the corresponding SIS bit rate of approximately 300 kbit/s gives a high-quality sound channel. Sound-in-sync is used in the EUTELSAT system.

4.7.3.3 Supplementary audio channels

Additional audio channels (used, for example, for commentaries and programme management) are usually narrow-band (telephony) channels which may be transmitted as separate SCPC channels or combined with the video channels in similar ways to the programme channel.

4.7.4 TV distribution

Systems for the distribution of television programmes to a large number of earth stations (feeding, for example, cable-television networks, hotels or community antennas) must, if they are to be economic, use receive antennas which are significantly smaller than those used for the international relay of television transmissions. Transponders with high EIRPs are therefore required; for example, a spot-beam transponder of INTELSAT V has an EIRP of about 40–44 dBW at 11 GHz and with this EIRP satisfactory reception of one or two transmissions per transponder can be achieved using antennas with diameters of 3–4 m.

4.7.5 DBS and MAC

The price of the antenna, amplifier and downconverter required to receive television broadcasts from satellites direct to homes (DBS) and make the signals (transmitted at a frequency of 12 GHz) compatible with existing television receivers has to be kept low enough to create a mass market. By using satellite EIRPs of around 65 dBW the diameter of the receiving antennas can be kept down to about 0·5 m.

4.7.5.1 MAC

The use of colour sub-carriers near the top of the video band (as in NTSC, PAL and SECAM) is no disadvantage for terrestrial broadcasting (which uses amplitude modulation). However, with FM transmissions the sub-carriers are situated towards the top of the baseband where the noise density is (despite the use of pre-emphasis) relatively high. This leads to degraded performance in the chrominance channel. In addition it is difficult, using the existing television-broadcast systems, to provide a number of features for which it is expected that there will be considerable demand in the future; these include two or more high-quality sound channels for some video channels, various data services, and methods of ensuring that subscription programmes and services can only be received by those who have paid for them. Digital processing and transmission of television signals is being used increasingly in television studios but digital methods are not, at present, suitable for DBS because they require the use of very large RF bandwidths (or the use of complex methods of reducing bit rate which would make home receivers too expensive).

The MAC system avoids the problems associated with sub-carriers and with the digital encoding of video channels by multiplexing the luminance and chrominance channels in time (see Fig. 4.16) but leaving them in analogue form. In order that the chrominance and luminance signals may be multiplexed they are compressed in time (e.g. by factors of 3 : 1 and 3 : 2, respectively); they are then frequency-modulated onto the carrier. So that MAC signals shall be compatible with existing television receivers, the number of lines and the frame rate are made the same as those of the existing composite systems.

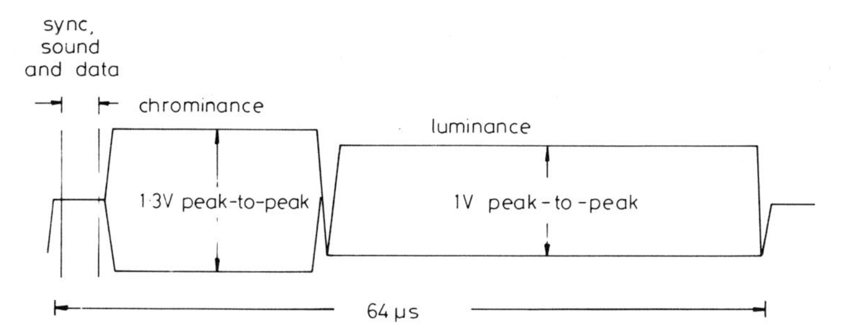

Fig. 4.16 *One line of a MAC signal*

The luminance and chrominance signals do not occupy all of the time allocated for transmission of a line and the remaining period (which, in the composite systems, is used for the blanking waveform and the sync pulses) is used for digital sound and data. The bit rates required for the transmission of sound and simple data services are very modest compared with those required for the transmission of video channels; it is therefore possible to provide a number of high-quality audio programmes as well as data channels and all the synchronising information. Moreover, digital methods allow great flexibility in the way the total bits available are used.

There are a number of different ways of transmitting the digital part of the signal and this gives rise to variants of MAC such as C–MAC, D–MAC and D2–MAC. D2–MAC is being used for DBS transmissions from France and Germany and D–MAC has been recommended for DBS transmissions from the UK. However, one DBS company based in the UK has elected to use PAL for its transmissions and thus reduce the cost of the home receiving equipment by eliminating the need for a MAC receiver.

Chapter 5

Digital satellite communications

5.1 Introduction

The first digital transmission method to be used commercially was pulse code modulation (PCM); this is still the most common method of encoding voice signals and it is the progenitor of most of the more complex coding methods now in use.

PCM was first used for systems carrying time-division-multiplex assemblies of 24 or 30 voice channels over each pair of wires between telephone exchanges. A very large number of these primary-multiplex systems working at approximately 1·5 Mbit/s or 2 Mbit/s have been installed, and these have been joined by many higher-capacity PCM systems working at bit rates of up to 560 Mbit/s.

The rapid reduction in cost and improvement in performance of integrated circuits has led to the widespread adoption of digital switching and the processing of telecommunications signals in ways which would have been completely impractical until recently. The combination of digital transmission, digital switching and digital processing of signals will eventually lead to a world-wide integrated-services digital network (ISDN).

For terrestrial systems one of the major advantages of digital encoding is that the quality of the received signal can be made virtually independent of the length of the transmission path by regenerating the signal at intervals along the path. It is obviously not possible to put regenerators between earth stations and satellites; nor is it practicable, as yet, to include regenerators in commercial communications satellites. Nevertheless, communications-satellite systems are turning to digital methods because:

(i) Digital methods make it possible to use satellites in new and more efficient ways such as time-division multiple access (TDMA) and will, in the future, make it possible to carry out complex processing of signals on board satellites.

(ii) Digital systems can be used for the efficient transmission of both voice and data signals.

(iii) Digital signals can easily be encoded so that they cannot be used by unauthorised persons.
(iv) Digital satellite systems can be designed to be part of national and international integrated service digital networks.

5.2 Pulse code modulation

To convert an analogue signal to PCM form it is necessary:
(i) to sample the signal at regular intervals
(ii) to represent the amplitude of each sample by a binary code.

5.2.1 Sampling

If the highest frequency of the analogue signal is F_m hertz then the waveform can be fully specified by $2F_m$ samples per second and the analogue signal can in theory be recovered without distortion by passing the samples through an ideal low-pass filter with a cut-off frequency of F_m hertz. Telephony signals are usually limited to a maximum frequency of about 3·4 kHz by means of a low-pass filter; the lowest theoretical sampling frequency is therefore 6·8 kHz. In practice a sampling frequency of about 8 kHz is used because the simple low-pass filters used with telephone circuits do not cut off sharply at 3·4 kHz. Fig. 5.1*a* is a simplified representation of the frequency spectrum of a speech signal; Figs. 5*b* and *c* show the process of sampling and the frequency spectrum of the samples. The latter comprises:

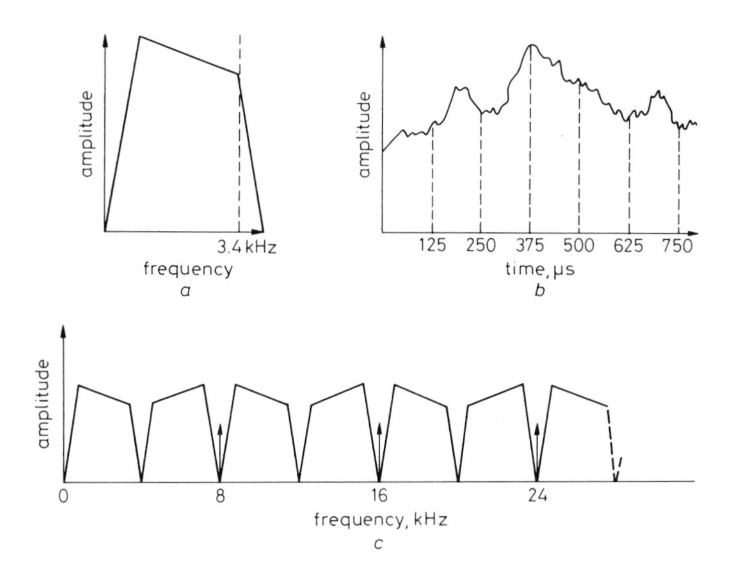

Fig. 5.1 *Sampling*
a Approximate frequency spectrum of speech channel
b Sampling a speech signal
c Spectrum of samples of a speech signal

(i) the original signal
(ii) a component at the sampling frequency with upper and lower sidebands of the same form as the original signal
(iii) harmonics of the sampling frequency, each having sidebands of the same form as the original signal.

It can be seen that if the sampling frequency is not high enough then the lowest sideband will overlap the spectrum of the original signal; in this case it is impossible to recover the original signal from the samples without distortion.

5.2.2 Coding

In order to limit the number of codes required to represent the samples, the amplitude range over which the codec works is divided into a number of sub-ranges or quantisation steps and the amplitude of each sample is represented by the code applicable to the sub-range within which it falls. Fig. 5.2 shows an example of coding using eight quantisation steps, each of which is represented by a 3-digit binary code (this is far fewer quantisation steps than would be used in any practical PCM system). A binary '1' is represented by the presence of a pulse and a binary '0' by the absence of a pulse and, for this example, the codes are represented as a pulse train of the non-return-to-zero (NRZ) type, i.e. there is no gap between successive '1's.

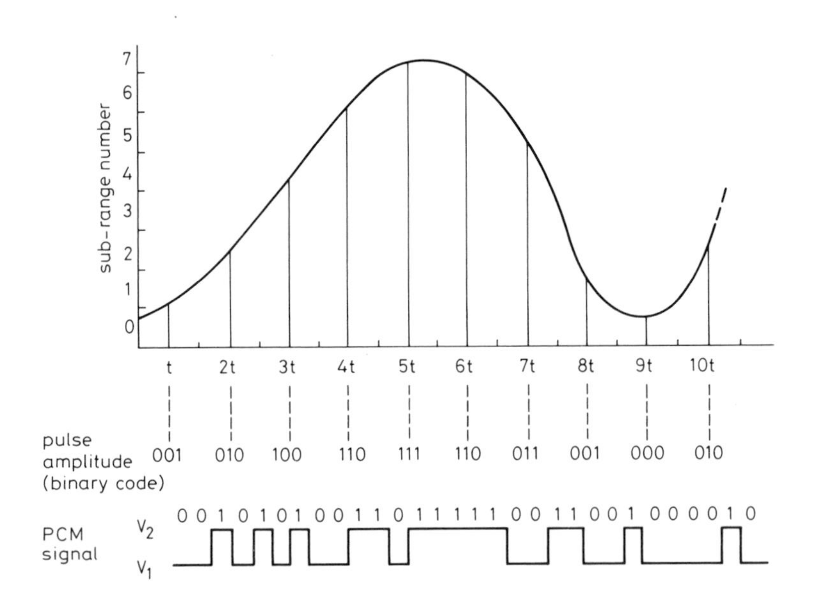

Fig. 5.2 *Pulse-code modulation*

5.2.3 Decoding

When the coded signal arrives at its destination a decoder reconstitutes the samples from the groups of pulses and the reconstituted samples are passed through a low-pass filter to recover the analogue signal. If a code is received correctly then the corresponding quantisation error (i.e. the difference between the amplitude of the original sample and that of the reconstructed sample) will be not more than one half the amplitude of a quantisation step. The effect of quantisation errors is to generate quantisation distortion which appears as noise in the telephone channel. The quantisation noise can be reduced by increasing the number of quantisation steps; however, increasing the number of steps increases the number of bits which have to be transmitted per second and therefore increases the bandwidth required. Digital radio systems are usually designed so that, under normal conditions, only a very small percentage of bits (say 1 in 10^6) are received in error; thus the performance of a digital system is determined by the quantisation noise except for that (usually very small) fraction of time when a relatively large percentage of bits are received in error.

The present national and international networks comprise both analogue and digital transmission links and, particularly on a long-distance call, it may be necessary to make a number of analogue–digital (A/D) and digital–analogue (D/A) conversions; i.e. the channel may include a number of coder–decoder pairs (codecs) in series. CCITT recommends that an 8-bit code (corresponding to $2^8 = 256$ quantisation levels) shall be used for voice signals on international circuits; at a sampling rate of 8 kbit/s this corresponds to a bit rate of 64 kbit/s for a voice channel. When an 8-bit code is used, up to 14 analogue–digital–analogue conversions can be included in a circuit without introducing unacceptable quantisation noise. For most national circuits and for international circuits using communications-satellite links it is unlikely that anything approaching 14 conversions will be necessary; the INTELSAT SCPC system and some national systems therefore use a 7-bit code which reduces the bit rate to 56 kbit/s at the cost of increasing the quantisation noise power per conversion by approximately 6 dB. However, systems using non-standard codes and bit rates are more difficult to fit into international digital networks.

The mean power of speech signals arriving at a PCM encoder from different sources can vary over a very wide range, firstly because some people speak much more loudly than others and secondly because the transmission loss through a switched analogue terrestrial network is dependent on the route; typically 99% of speech signals arriving at an encoder occupy a range of rather more than 30 dB. Thus, if all the quantisation steps are of equal size, the ratio of signal power to quantisation noise power also varies widely from channel to channel. To avoid this effect, the size of the quantisation steps is decreased at the lower signal levels. Two different laws are used to determine the relation between the size of the quantisation steps and the amplitude of the signal; the μ-law, which is used mainly in the USA and Japan, and the A-law. Where necessary, conversion between A-law and μ-law signals (which adds more quantisation noise) is carried out in the μ-law country.

PCM may be used to encode music and video signals as well as speech signals. Typical bit rates are given in Table 5.1.

Table 5.1 *Typical PCM bit rates*

Signal type	Baseband frequency (kHz)	Sampling rate (10^3/s)	Bits per sample	Bit rate (kbit/s)
Speech	0·3 — 3·4	8	8	64
Music	0·05— 15	33	12	396
Video	0 —5500	12 000	10	120 000

5.3 Low-rate encoding (LRE)

5.3.1 Introduction

The RF bandwidth occupied by a digital signal and the power required to transmit the signal are proportional to bit rate. The EIRP available from satellites is limited and frequency spectrum is usually in short supply so it is particularly important that communications-satellite systems should use the lowest practicable bit rates.

Low-rate encoding works by the reduction of the redundancy in signals. When we make notes we often leave out letters or words and yet someone else will usually have no difficulty in understanding our notes; this is because written languages use more symbols than are necessary to convey information and we use the rules and statistical properties of the language both to decide what to leave out and to make an informed guess at what is missing. In the same way, a knowledge of the properties of sound and video signals makes it possible to reduce the redundancy in the signals and thus encode them more efficiently.

Video signals usually have a lot of redundancy because:

(i) In many cases an element of a television picture is very similar to nearby elements in the same line and neighbouring lines
(ii) There is often high correlation between successive pictures, especially when there is little movement in the scene.

Speech and music signals also have regular characteristics and thus contain redundant information but it is not usually possible to reduce the bit rate of data signals as the elements of these signals are generally uncorrelated.

5.3.2 *Differential PCM and delta modulation*

5.3.2.1 *DPCM*

The most widely used method of reducing bit rate is differential PCM (DPCM). A DPCM encoder predicts the amplitude of the next sample of a signal from the preceding sample, or a number of preceding samples. For example, in Fig. 5.3*a* the predictor operates on the three previous (predicted) samples which are fed back through networks giving delays of T, $2T$ and $3T$, where T is the interval between samples. The predicted value is compared with the actual amplitude and the difference between the two (i.e. the error) is encoded and transmitted to the receiver.

If the correlation between the predicted and actual values is good then

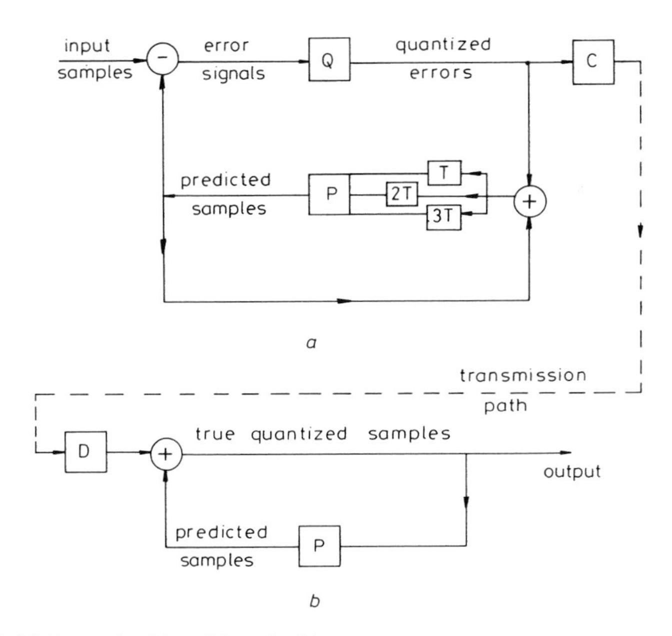

Fig. 5.3 *DPCM encoder (a) and decoder (b)*
Q = quantiser
P = predictor
C = error encoder
D = error decoder
T, 2T, 3T = delay networks giving delays of T, $2T$ and $3T$, respectively

the errors are small and only a few bits are needed to transmit them. Powerful but cheap digital signal processors are now available and complex algorithms can therefore be used to make accurate predictions. (An algorithm is simply a series of precise instructions for making a calculation or solving a problem.)

At the receiver, a decoder (Fig. 5.3*b*) uses the same algorithm as the encoder to predict the values of the samples and then combines the predicted values with the error signals to produce 'true' quantised values. The error signal is also used by the encoder to correct the values used to make the next prediction.

5.3.2.2 *Delta modulation*

The simplest form of differential encoding is delta modulation (DM). In a delta modulator (see Fig. 5.4*a*) the quantiser has only two output states, positive ($+v$) or negative ($-v$). The output of the quantiser is sampled and the predictor integrates these samples. The output of the integrator is compared with the analogue input; if the input is greater than the output of the integrator then the quantiser produces a positive output; if not then the quantiser produces a negative output. Each quantised sample is coded as one bit ('1' for a positive sample and '0' for a negative sample) and transmitted to

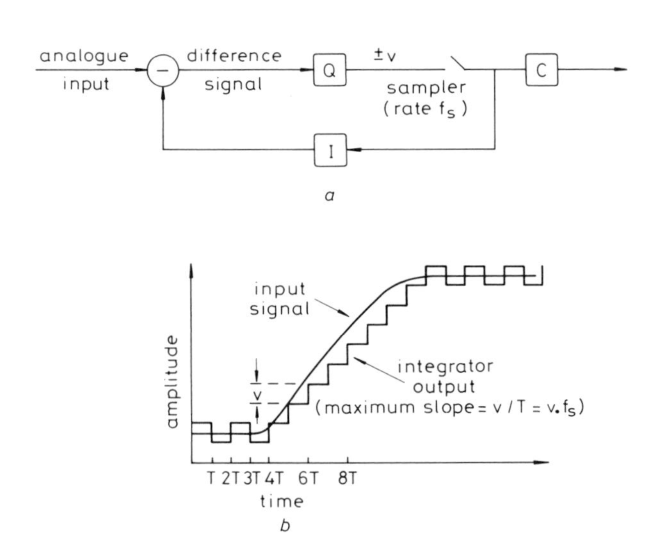

Fig. 5.4 *Delta modulation*
a Modulator
Q = 2-state quantiser
I = integrator
C = 1-bit encoder
b Input waveform and integrator output

the receiver where the codes are converted back to positive and negative pulses which are integrated in an identical circuit to that in the encoder. The output of the decoder is, as usual, passed through a low-pass filter to remove as much as possible of the sampling frequency and its harmonics.

A constant input signal produces an alternating series of positive and negative outputs from the quantiser; this causes the output of the integrator to fluctuate around a constant value (see Fig. 5.4*b*) and produces what is known as 'granular' noise at the receiver. v must be kept small in order to limit the granular and quantisation noise to acceptable levels. On the other hand, the maximum rate at which the output of the integrator can change is v/T where T is the interval between samples; if the input signal has a slope greater than v/T then the integrator output cannot keep up with the signal and this results in overload distortion. To avoid unacceptable overload distortion the delta modulator must use a high sampling rate, say 32 kHz for a voice channel. The bit rate of the transmission is equal to the sampling rate because the error code only requires one bit per sample.

The main advantages of DM are that the modulator and demodulator are simple and cheap to implement, and that transmission errors do not cause as much distortion as they do in PCM systems.

5.3.2.3 Adaptive coding

It is possible to improve the performance of DM by making the size of the quantisation step a function of the characteristics of the signal. For example it can be arranged that the size of the step increases with the number of consecutive positive (or consecutive negative) outputs from the quantiser; this reduces the effects of overload distortion (but may introduce a different type of distortion as the result of large overshoots).

The principle of continually adapting the characteristics of the encoder to the changing characteristics of the signal can be applied to coding methods other than DM. In adaptive DPCM (ADPCM) systems the size of the quantisation step and the parameters of the prediction algorithm are a function of the characteristics of a segment of the signal which is stored in the predictor.

It is not difficult to develop codecs which give acceptable voice quality at 32 kbit/s but nearly all those so far produced have at least one characteristic which makes them unsuitable for integration into the international public switched telephone network (PSTN). For example, some cannot handle in-band signalling tones or voice-band data, some are very susceptible to transmission errors, some introduce unacceptable delays, some are too expensive for commercial use and some do not give good results when used in tandem (i.e. if the signal has to be encoded and decoded several times in succession).

CCITT has, for some time, been close to agreeing on a world standard for a 32 kbit/s ADPCM system which accepts, as its input, 64 kbit/s PCM or voice-band data at speeds of up to 4·8 kbit/s and in which the difference signal is normally encoded as 5 bits for voice-band data and 4 bits for speech. The system continues to work well under poor transmission conditions (i.e. when there are high error rates) but, at the time of writing, there is still some argument as to whether the performance is satisfactory when several codecs are connected in tandem.

5.3.3 Very-low-rate encoding

Very-low-rate encoding is defined here as encoding at bit rates of 16 kbit/s or less. At present these rates are used only for systems (such as mobile or military systems) which would be impracticable or uneconomic without them and users have to accept one or more disadvantages such as poorer speech quality or higher cost.

5.3.3.1 Sub-band coding

ADPCM systems can be designed to operate at bit rates of 16 kbit/s or lower, but the quality of the decoded speech is noticeably degraded. However, good quality speech can be achieved at 16 kbit/s by the use of sub-band coding (SBC). In this method the speech is split into a number of frequency bands (typically eight) by using a bank of filters. Each of these bands is encoded separately (using, for example, ADPCM coders) and the sub-band signals are multiplexed and transmitted. At the receiver the signals are demultiplexed and decoded, and a bank of filters is used to synthesise the signal. Sub-band encoding works by masking the quantisation noise, by allocating more of it to the sub-bands with high speech energy than to the sub-bands with low speech energy.

5.3.3.2 Vocoders

All the coders considered so far are waveform encoders. Bit rates of 4·8 kbit/s or less can be achieved by the use of vocoders; these analyse speech in terms of a simplified model of speech production and continually send the results of the analyses to the receiver where the speech is synthesised using the same model.

Speech sounds may be voiced (e.g. a vowel sound) or unvoiced (fricatives such as 's' and 'f', and plosives such as the sound that begins and ends the word 'pop'); voiced sounds have a fundamental pitch and a number of resonant frequencies, unvoiced sounds may be represented as filtered noise. The parameters transmitted include the type of sound (voiced or unvoiced), the amplitude, the pitch (if the sound is voiced) and certain characteristics of the frequency spectrum. The parameters are updated about 50 to 200 times a second, depending on the speech quality required.

The ear is very sensitive to some of the characteristics of speech and quite insensitive to others; thus it is possible for the output of a vocoder to be perfectly intelligible even though the waveform of the vocoded speech is very different to that of the original. Vocoders can be implemented in a variety of ways; one of the most common is linear predictive coding (LPC) in which the vocal tract is represented as a time-varying linear filter. Vocoders, however they are realised, are relatively complex and expensive devices. Their performance may vary with different speakers and different languages and, up to now, many of them have tended to make speakers sound like robots; their use has therefore been confined mainly to military systems where quality of speech is not usually important (provided that it is intelligible) and high cost is less of a disadvantage than it is in commercial systems.

5.3.3.3 Hybrid speech coders

There are now speech coders which are hybrids of waveform coders and vocoders; these can be used to give bit rates in the range from 16 kbit/s down to 4·8 kbit/s. Much work is going on to establish the advantages and disadvantages of these methods, particularly for use with mobile systems. Acronyms which may be met include RELP (residual-excited linear-predictive encoding), ME–LPC (multipulse-excited LPC) and RE–LPC (regular-pulse-excitation LPC).

5.3.4 Low-rate encoding of television

Full-motion television pictures encoded as PCM require very high bit rates (of the order of 100 Mbit/s or more.

With DPCM coding the bit rate of a full-motion picture can be reduced by a factor of about 2. A typical DPCM coder for television makes predictions using the weighted sum of eight samples. These samples represent neighbouring picture elements (pixels) from the same line and the previous line, and corresponding pixels from the previous picture; it is therefore necessary to provide a picture store at the transmitter.

A further reduction in bit rate can be obtained by conditional element replenishment (CER). In a CER coder the picture is held in a store. When a new picture is generated, each pixel is compared with the corresponding pixel in the stored picture and data is transmitted only where there is a significant difference between the two. The differences are encoded, together with the positions of the pixels to which they refer, and used to update both the picture at the encoder and the picture at the receiver (which is held in an identical store to that at the encoder). The flow of data from the encoder is irregular, so it must be fed into a buffer store from which it is read out to the transmitter at a constant rate. If the picture is changing so much that the buffer store cannot accept all the information that would be generated then the coder has to take action to reduce the bit rate; for example, it may compare only half the pixels at each renewal of the picture.

When conditional encoding is used, the transmission rate required depends very much on the type of pictures to be transmitted. For example, the cameras for video conferencing are usually fixed and the participants in the conference do not move very much; under these conditions it is possible to make very large savings in the data transmitted. Modern codecs designed especially for video conferencing, and using conditional replenishment together with DPCM encoding, can give good picture quality at the standard primary multiplex rates of 2·048 Mbit/s (for Europe) and 1·544 Mbit/s (for North America and Japan). Other systems use even lower rates but give relatively poor picture quality.

5.4 Time-Division Multiplexing (TDM)

Trains of pulses from a number of channels may be interleaved to form a single high-rate bit stream or time-division-multiplex (TDM) assembly and the assembly may be directly modulated onto an RF carrier. This has the advantage that:

(i) Less equipment is required than is needed to modulate each channel onto a separate carrier
(ii) The transmission efficiency of a satellite is usually better when it is carrying a few transmissions (each comprising many channels) than when it is carrying many transmissions (each comprising relatively few channels).

Large multiplex assemblies are usually built up by interleaving a number of channels into a primary multiplex and then combining a number of primary multiplexes to form a secondary multiplex and so on. The series of primary, secondary and higher-order multiplexes is called a digital hierarchy. Unfortunately there are several hierarchies in use (see Table 5.1). All the standard hierarchies are built up from 64 kbit/s channels but those used in North America and Japan start with primary multiplexes of 24 μ-law channels whereas the hierarchy used in Europe starts with a primary multiplex of 30 A-law channels.

Accurate timing is essential to the correct operation of digital systems. The clocks used for encoding and decoding are either:

(i) synchronous, i.e. there is one master clock which controls the timing of all the other (slave) clocks

(ii) plesiochronous, i.e. the clocks run independently but their accuracy is so good that the drift between them can be accommodated by the provision of buffer stores (see Section 5.9).

Table 5.2 *TDM hierarchies*

Number of channels	Bit rate (Mbit/s)	Level
(a) North America		
24	1·544	Primary (DS1, T1)
96	6·312	Second order (DS2, T2)
672	44·736	Third order (DS3, T3)
4032	274·176	Fourth order (DS4, T4)
(b) Japan		
24	1·544	Primary
96	6·312	Second order
480	32·064	Third order
1440	97·728	Fourth order
(c) Europe		
30	2·048	Primary
120	8·446	Second order
480	34·368	Third order
1920	139·264	Fourth order

(i) The level 4 assemblies used in North America and Japan are not international (CCITT) standards (and other higher-order non-standard multiplexes are used in some countries).
(ii) Bit rates of 2·048, 6·312, 44·736 and 139·264 Mbit/s have been agreed for connections between countries using different hierarchies.
(iii) In North America the abbreviation DS1 (digital signal level 1) is used for the primary level and T1 refers to a system carrying a DS1 assembly.

If there were no errors or breaks in transmission it would only be necessary, in a synchronous system, to provide a single marker at the beginning of transmission; the decoder could then identify all the bits which followed from the specified pattern of the transmission. However, in any practical system regular markers must be provided in the bit stream so that the decoder can extract groups of digits representing samples, identify the channels of a TDM assembly and resynchronise the system after errors, breaks in transmission or drift of the clocks. One set of samples from each of the channels in a multiplex is known as a frame and Fig. 5.5 shows one frame of a primary multiplex transmission in the European hierarchy. It will be seen that the frame period is 125 μs (corresponding to a sampling rate of 8 kHz). The frame period is divided into 32 time slots of which 30 slots carry the speech channels, one slot

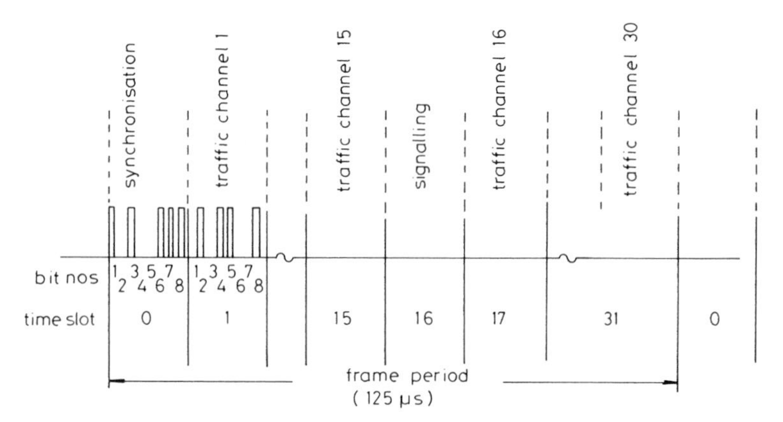

Fig. 5.5 *TDM frame*

(time slot 0) carries the synchonising information and one slot (time slot 16) carries signalling for all 30 channels. As each time slot comprises 8 bits the bit rate is $(8 \times 32 \times 8000) = 2{\cdot}048$ Mbit/s.

When plesiochronous multiplexes are combined to form higher-order multiplexes it is necessary, in order to preserve synchronism over the long term, to allow for the occasional addition of dummy bits (or stuffing bits). This means that the bit rate of the higher-order multiplex will be greater than the sum of the bit rates of the input multiplexes; for example, as will be seen from Table 5.2, the bit rate of the secondary-order multiplex in the European system is 8·446 Mbit/s which is greater than $4 \times 2{\cdot}048$ Mbit/s. The position of stuffing bits must be noted so that they can be removed at the demultiplexer.

5.5 Digital speech interpolation (DSI)

During a telephone conversation it is usual for only one person to speak at a time. Moreover, in any speech there are gaps between sentences, words and syllables. This means that a channel of a four-wire circuit is occupied on average for less than half the time. Speech-interpolation systems exploit the normally unused channel time by allocating a channel to someone only when they start to speak and taking it back immediately there is a pause in the speech (provided that someone else needs a channel and all other channels are occupied). Systems which interpolate analogue speech signals, e.g. TASI (time-assigned speech interpolation), have been in use for many years on transoceanic cable systems. DSI systems operate on voice signals which have been encoded in digital form. Digital-interpolation systems give better performance than analogue systems and the equipment tends to be cheaper.

5.5.1 *DSI gain*

By using DSI with a communications-satellite system, t terrestrial channels incoming to an earth station can be accommodated on s satellite channels where s is usually less than $t/2$. The ratio t/s is known as the DSI gain or DSI advantage. The DSI gain which can be achieved (without unacceptable degradation of the signal) depends on:

(i) The activity factor (i.e. the percentage of time for which the channels are occupied, on average, in the absence of DSI)
(ii) The number of channels available.

It is not possible to interpolate two incoming channels on one outgoing channel because, although each incoming channel is occupied in the long term for only (say) 35% of the time, there is a high probability that both speakers will talk simultaneously and one or other would therefore frequently be shut out of the channel. On the other hand, as the number of incoming channels becomes larger the probability that more than 35% of the channels are occupied at any one time decreases and with a very large number of channels the DSI gain approaches $(1/a)$ where a is the activity factor.

5.5.2 *Freeze-out*

There will be times, in any practical DSI system, when there are more active talkers than there are satellite channels. Thus occasionally when someone starts talking there will not be a channel immediately available and the first part of a 'speech spurt' will be lost. It is possible to calculate the average percentage of time for which speech will be lost ('freeze-out') for any combination of incoming (terrestrial) channels, outgoing (satellite) channels and activity factor. If the percentage freeze-out is sufficiently low (say less than 0·5%) then the number of speech spurts affected and the duration of the clipping will be so small that there will be no perceptible effect on the speech. Fig. 5.6 shows DSI gain as a function of the number of terrestrial channels for an activity factor of 35% and freeze-out of not greater than 0·5%.

In some DSI systems (e.g. the INTELSAT and EUTELSAT systems) the onset of freeze-out is delayed by arranging for the number of bits in each PCM sample to be reduced from eight to seven when a large number of incoming (terrestrial) channels are active; the eighth (least significant) bit from each group of seven channels is used to establish an additional channel (the overload channel). The DSI gain with such a system may appear to be greater than $1/a$.

5.5.3 *DNI*

Some non-speech services (such as data, telex and facsimile) require continuous use of a channel. If channels carrying these services are routed via

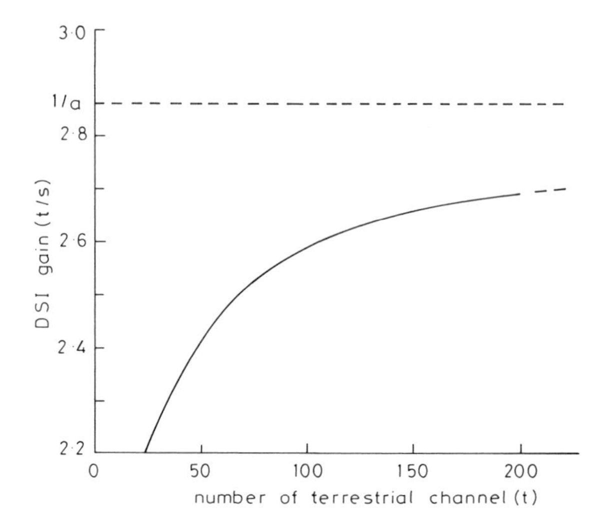

Fig. 5.6 *DSI gain*
$a = 0{\cdot}35$
Freeze-out = 0·5%

a DSI unit they increase the amount of freeze-out. It may pay to arrange for channels which are permanently assigned to non-speech services to bypass DSI equipment, in which case they are referred to as digitally non-interpolated (DNI) channels. Channels which are sometimes used for speech and sometimes for non-speech signals do not usually result in significant degradation of performance provided they do not constitute more than about 10% of the total; where there is a higher proportion of non-speech channels it may be necessary to accept a lower DSI gain in order to keep freeze-out to acceptable levels.

5.5.4 DSI operation

A voice-activity detector (VAD) in the DSI module at the transmitting station looks at each of the incoming channels in turn. If the VAD detects that the power level in a channel has risen significantly above the level of the background noise then the channel is assumed to be active and a processor in the transmit module assigns an outgoing (satellite) channel to the terrestrial channel and sends an assignment message to the distant terminal to tell it which terrestrial channel is connected to the satellite channel. The processor must of course keep continual note of which satellite channels become free and are available for reallocation.

The DSI receive modules at the distant stations use the assignment messages to switch the incoming satellite channels to the correct terrestrial channels. Forward error correction (see Section 5.7) is usually used to ensure that the assignment messages are received virtually error-free.

The various operations, such as detecting the presence of voice signals and making the connections between terrestrial and satellite channels, cannot be carried out instantaneously, and thus there is a danger that the beginning of each speech spurt could be lost (this is called 'front-end clipping'); buffer stores are provided in both the transmit and receive DSI modules to avoid this type of distortion.

5.5.5 SPEC

An alternative form of digital speech interpolation is speech predictive encoded communication (SPEC). In a SPEC system the last encoded sample from each incoming channel is stored at the transmitter and receiver. The current encoded sample for each channel is compared, at the transmitter, with the last sample and if the difference is less than some predetermined amount (called the 'aperture') then the corresponding sample at the receiver is left unchanged; if the difference is greater than the aperture then a new (8 bit) sample is transmitted. A sample assignment word (SAW) is sent in addition to the samples. The SAW comprises one bit for each incoming channel; this bit is a '1' if there is a new sample for the corresponding channel and a '0' if there is not. The SPEC equipment monitors the differences for all the channels (in each sampling interval) and continually adjusts the magnitude of the aperture so that the transmission path is never overloaded.

It is claimed that SPEC is more simple to implement than conventional DSI and that its subjective performance is better than that of DSI at times of high speech activity.

5.5.6 DCME

By the use of 32 kbit/s ADPCM and DSI, the number of channels which can be fitted into a given bandwidth can be increased by a factor of at least five compared with the number possible when using standard (64 kbit/s) PCM equipment. ADPCM and DSI equipment is therefore sometimes referred to as digital-channel multiplication equipment (DCME); this term is sometimes used to refer to the two equipments when they are used in tandem and sometimes simply as a generic term for DSI or ADPCM equipment.

5.6 Modulation

5.6.1 Introduction

5.6.1.1 Modulation methods

A radio-frequency carrier may be written as:

$$A\cos(\omega t + \phi)$$

where A = amplitude
ω = angular frequency
t = time and
ϕ = phase

The carrier may be modulated by causing the amplitude, the frequency or the phase to vary as a function of the amplitude of a baseband signal. When the modulating signal is digital, the three basic modulation methods are known as amplitude-shift keying (ASK), frequency-shift keying (FSK) and phase-shift keying (PSK).

In choosing the most suitable method for any particular system one has to consider what power and bandwidth are available, the performance required, the complexity and cost of the equipment needed, and the effects on performance of distortion, fading and interference. PSK is often used for radio systems because it is relatively simple to implement and, under many practical conditions, it gives a lower BER than ASK and FSK. Communications-satellite systems frequently use binary phase-shift keying (BPSK or 2–PSK), quadriphase phase-shift keying (QPSK or 4–PSK) or variants of these methods.

5.6.1.2 BER

For analogue systems the basic measure of system performance is signal-to-noise ratio in a baseband channel. For digital systems the basic measure is the bit error ratio (BER), i.e. the ratio of the number of bits received in error to the total number of bits transmitted. For example, a performance objective for some communications-satellite systems carrying data is that the error ratio shall be not greater than 1 in 10^6 for at least 99% of any month. The BER is often referred to as the bit error rate although, strictly speaking, this is a misnomer because error rate should have the unit 'errors per second' (whereas a ratio is dimensionless).

5.6.1.3 Eye diagrams

The nearly rectangular pulses which comprise a newly-generated digital signal usually become severely distorted during transmission. Fig. 5.7*a* shows how pulses are distorted when they are passed through a network with restricted bandwidth. Fig. 5.7*b* is an 'eye' diagram; that is it shows the pattern formed by superimposing the waveforms, after transmission, of many random sequences of pulses (all with the same pulse repetition rate). The basic shape of the eyes is the result of filtering whereas the 'fuzziness' (shown by shading) is the result of the noise added during transmission; this varies from instant to instant and therefore from pulse to pulse.

To demodulate a digital transmission it is necessary to sample it at the pulse repetition rate and compare the amplitude of the samples with one or

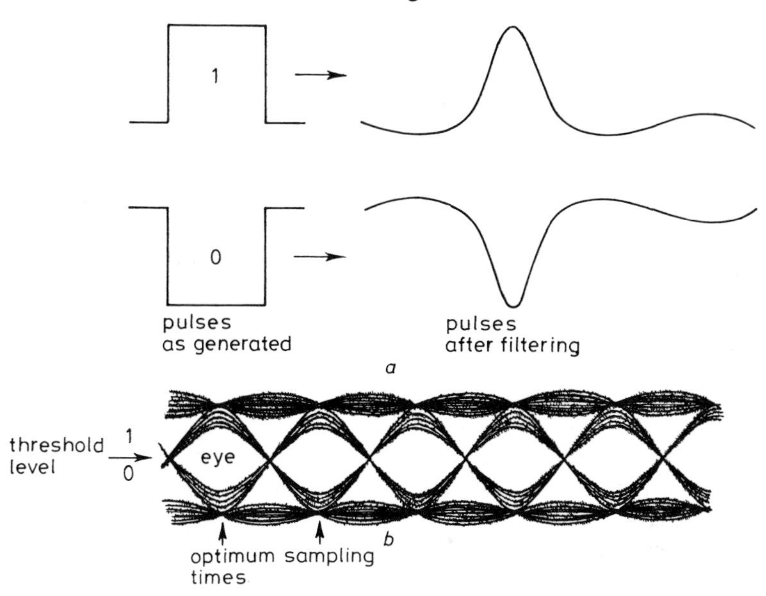

Fig. 5.7 *Filtering and eye diagrams*
a Effect of filtering
b Eye diagram

more reference (or threshold) levels. In Fig. 5.7*b*, for example, there is a single threshold level and a sample is read as '1' if its amplitude is above the threshold level and '0' if it is below the threshold level. The noise varies randomly; thus, even when the signal-to-noise ratio is high, there is a small probability that a '1' will be detected as a '0' or vice versa. When the signal-to-noise power ratio deteriorates the eye becomes smaller and the BER gets larger. The eye diagram shows the importance of sampling the pulses at the right instant, i.e. when the eye is most widely open; for optimum detection it is therefore essential that the clock at the receiver shall be locked to the frequency of the incoming pulse train (so that it follows any relatively slow 'wander' of the frequency) and that it (and the clock at the transmitter) shall be as free as is practicable from 'jitter' (i.e. random variations in phase).

5.6.2 *PSK*

5.6.2.1 *BPSK*

The simplest form of PSK is binary PSK (BPSK or 2–PSK). A BPSK signal can be written as:

$$A\cos(\omega t + \phi)$$

where $\phi = 0°$ when the baseband signal is '0'
and $\phi = 180°$ when the baseband signal is '1'

or as:

$$A\cos\omega t$$

where $A = +V$ when the baseband signal is '0'
and $A = -V$ when the baseband signal is '1'

Fig. 5.8*a* shows how, in theory, the phase of the carrier changes instantaneously by 180° when the baseband signal switches from '0' to '1'; Fig. 5.8*b* represents the two states of the carrier by two vectors with a phase difference of 180°. (This phasor representation may not be of help in understanding BPSK but will be found to be useful when considering more complex modulation methods.)

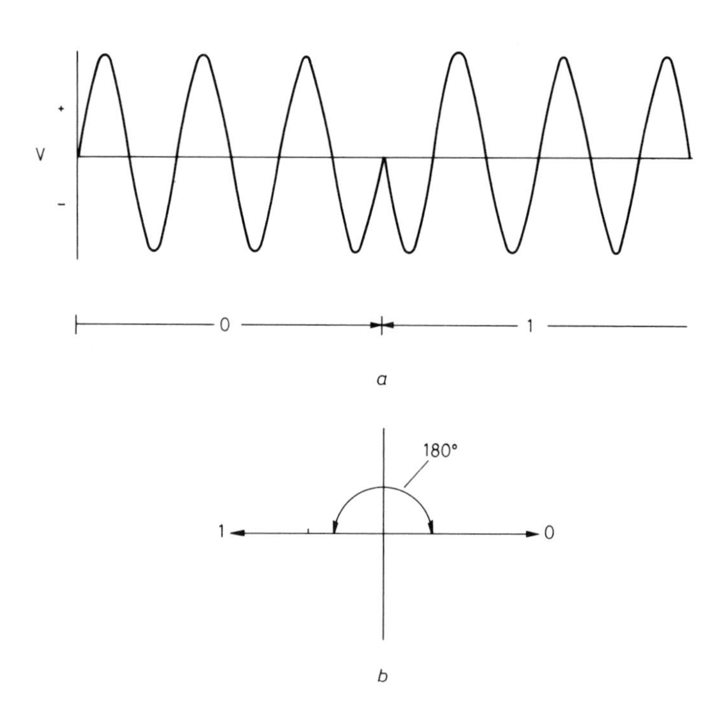

Fig. 5.8 *BPSK*
a Waveform
b Phase states

5.6.2.2 *QPSK*

A slightly more complex form of PSK is quadrature PSK (QPSK or 4–PSK). Two binary digits are needed to describe four states and we can write a QPSK signal as:

$$A\cos(\omega t + \phi)$$

where $\phi = 45°$ corresponds to '00'
$\phi = 135°$ corresponds to '01'
$\phi = 225°$ corresponds to '11'
$\phi = 315°$ corresponds to '10'

Thus each state of the signal carries two bits of information. A combination of two bits (or more) which corresponds to a discrete state of a signal is called a symbol; in the case of QPSK the symbol rate is half the bit rate. Thus, for example, a bit rate of 120 Mbit/s corresponds (with QPSK) to a symbol rate of 60 Msymbols/s or 60 Mbauds (one symbol per second is called a baud after J.M. Baudot, one of the pioneers of telegraphy).

Fig. 5.9 shows the phasor diagram for QPSK. If, at the instant that the signal is sampled, the resultant of the signal voltage (V_s) and the noise voltage (V_n) lies in a quadrant other than that of the signal then an error will occur. The symbols allocated to adjacent quadrants differ by only one bit; thus only peaks of noise high enough to carry the resultant vector into the opposite quadrant cause both of the bits comprising a symbol to be in error.

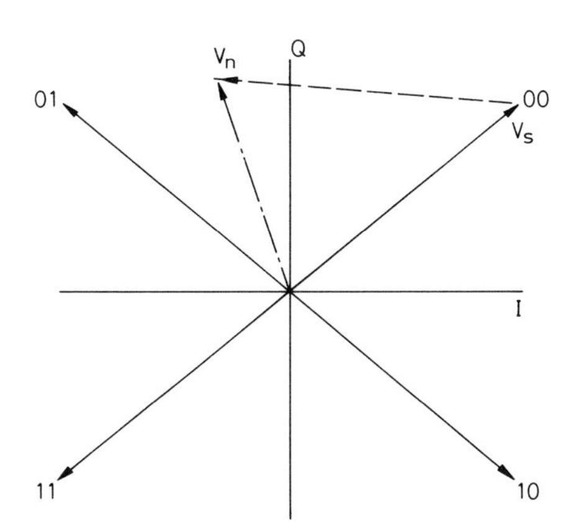

Fig. 5.9 *QPSK phase states*

As can be seen from the diagram, each of the QPSK phase vectors may be generated by adding together two equal orthogonal vectors. An alternative way to represent a QPSK signal is therefore as the sum of two BPSK signals: an 'in-phase' signal $I\cos\omega t$ and a 'quadrature' signal $Q\sin\omega t$, viz:

$$I\cos\omega t + Q\sin\omega t$$

where $I = +V,\ Q = +V$ for '00'
$I = -V,\ Q = +V$ for '01'
$I = -V,\ Q = -V$ for '11'
$I = +V,\ Q = -V$ for '10'

Fig. 5.10*a* is a block diagram of a QPSK modulator. The input data (at rate F_b) is split into two streams (each of rate $F_b/2$) by diverting the first, third and all odd bits into one leg of the modulator and the second, fourth and all even-numbered bits into the other leg of the modulator. The two streams of bits modulate two orthogonal carriers and the two carriers are then combined to form the QPSK signal. Fig. 5.10*b* shows an input bit stream, the two derived bit streams and the phase of the QPSK carrier corresponding to each symbol; in practice, the derived bit streams are delayed relative to the input bit stream but these delays are not shown in the diagram.

The bandwidth occupied by a PSK transmission is proportional to the rate at which the state of the carrier is changed (irrespective of the amount of information carried by each state). Thus the bandwidth required by a QPSK transmission carrying a bit stream at rate F_b (symbol rate $F_b/2$) is half that required by a BPSK transmission carrying the same bit rate.

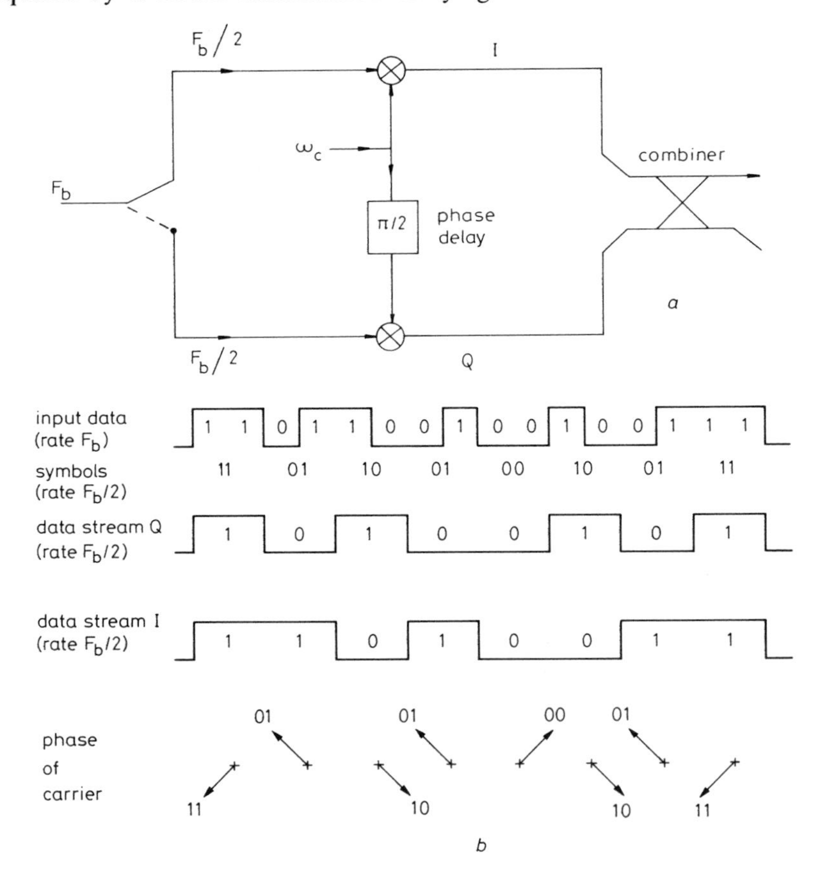

Fig. 5.10 *QPSK modulator*
a Block diagram
b Operation

5.6.2.3 BER, E_b/N_0 and C/N

The theoretical bit error rate p for a BPSK transmission in the presence of Gaussian (i.e. white) noise is:

$$p = (1/2)\text{erfc}(E_b/N_0)$$

where $\text{erfc}(x) = (2/\sqrt{\pi}) \int_x^{\infty} \exp(-t^2)dt$ [erfc(x) is known as the complementary error function]

E_b = energy per bit = (carrier power)/(bit rate)

N_0 = spectral noise density (= kT)

Fig. 5.11 shows p plotted against (E_b/N_0). It will be seen that, in theory, the E_b/N_0 required for BERs of 1 in 10^6 and 1 in 10^3 are 10·5 and 6·8 dB, respectively.

In any practical system the (E_b/N_0) required to achieve a specified BER will be significantly higher than the theoretical value because of the practical limitations of the equipment.

The BER of a QPSK transmission as a function of E_b/N_0 is the same as that of a BPSK transmission. Consider a QPSK transmission with carrier power C and bit rate R (i.e. with $E_b = C/R$). The QPSK transmission may be thought of as two BPSK transmissions in quadrature, each with a carrier power of $C/2$ and a bit rate of $R/2$ (i.e. each BPSK transmission also has $E_b = C/R$). The noise power density (N_0) remains constant and the ratio E_b/N_0 is therefore the same for each of the two BPSK carriers and the QPSK carrier. As the QPSK carrier is formed by adding the two BPSK carriers it has twice the digit rate and twice the error rate of the BPSK carriers; i.e. the error ratio for the QPSK and BPSK carriers is the same. Thus BER and E_b/N_0 are the same for the QPSK transmission as for its component BPSK transmissions.

The carrier-to-noise ratio C/N is related to E_b/N_0 by the relation:

$$C/N = (E_b/N_0)\ (R/B)$$

where B = bandwidth, Hz

The carrier-to-noise ratio of a QPSK transmission is therefore twice that of a BPSK transmission with the same bit rate and E_b/N_0.

5.6.2.4 QAM

As we have seen, combining two bits into a symbol halves the RF bandwidth required for transmission at a given bit rate. Further reductions in bandwidth can be achieved by combining more than two bits into each symbol.

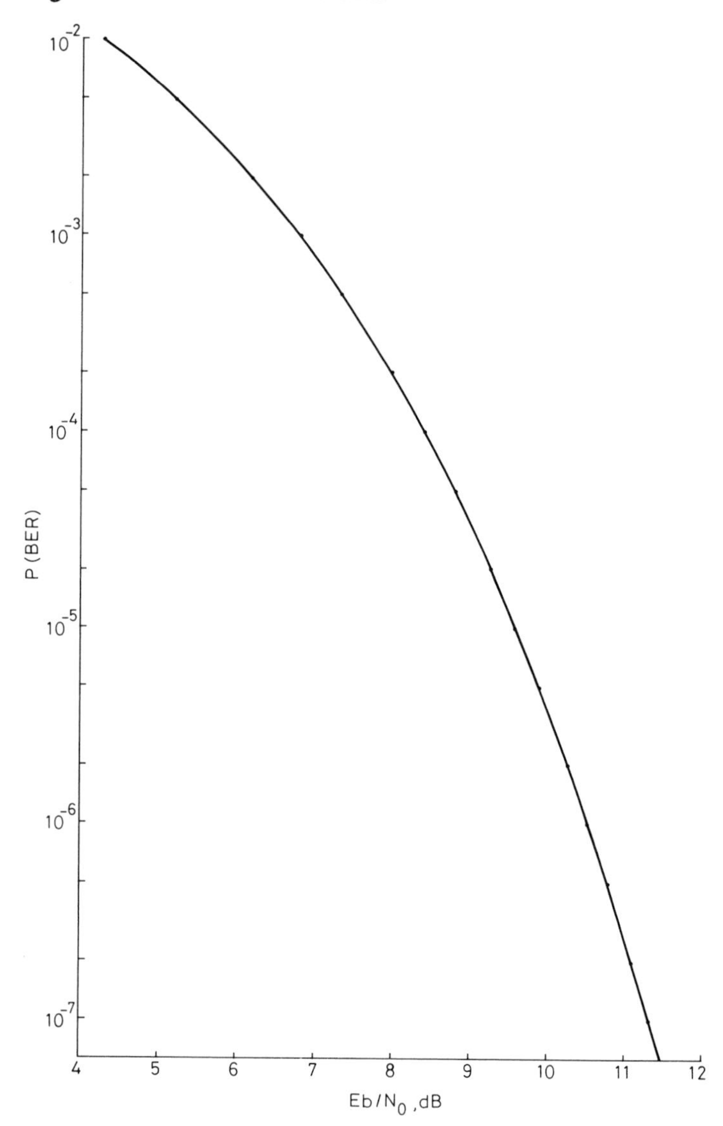

Fig. 5.11 *BER versus E_b/N_0 for BPSK or QPSK*

For example, if the phase of the carrier is allowed to asume any of eight states (each separated from the adjacent states by 45°) then each symbol may comprise three bits (because $2^3 = 8$), and so on. Alternatively, the concept of QPSK may be extended by assigning several possible values of amplitude to each of the orthogonal carriers of a QPSK signal; this type of modulation is known as quadrature amplitude modulation (QAM). If each carrier is assigned four possible values of amplitude then the number of states in each quadrant is 16 (see Fig. 5.12) and the total number of states is 64.

No matter whether the modulation is PSK or QAM, it becomes more difficult to distinguish between adjacent states, in the presence of noise, as the number of possible states increases. Thus the more complex the modulation system the greater is the value of E_b/N_0 required to achieve a given error ratio; that is, a reduction in bandwidth can only be achieved in exchange for an increase in power. (QPSK, which requires half the bandwidth of BPSK but the same E_b/N_0, is an exception to this rule because it is possible to use two, and no more than two, orthogonal carriers in the same bandwidth without interference between them.)

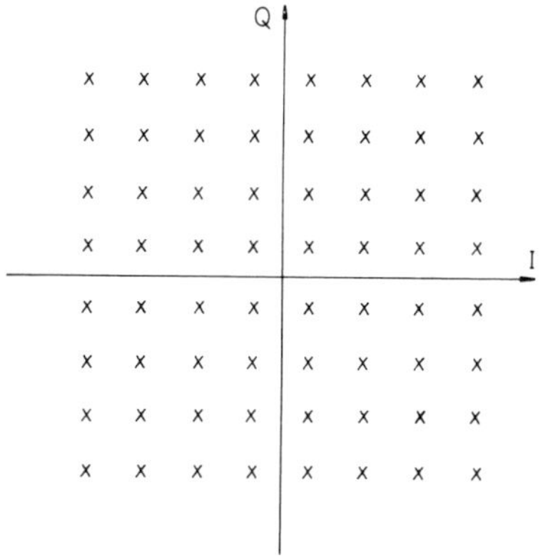

Fig. 5.12 *64-QAM state diagram*

Although 64–QAM is widely used for terrestrial systems (particularly in the USA) it requires amplifiers with better linearity and higher EIRPs than those available with the present generation of communications satellites; BPSK, QPSK and some variants of these modulation methods are therefore normally used with communications-satellite systems.

5.6.2.5 Demodulation

The simplest way, in theory, of demodulating a PSK transmission is to compare it with the unmodulated carrier; this is known as coherent demodulation. For example, Fig. 5.13 shows a BPSK transmission and the original (unmodulated) carrier; in this example state '0' is defined as that where the modulated and unmodulated carriers have the same phase and it follows that a difference in phase of 180° corresponds to state '1'.

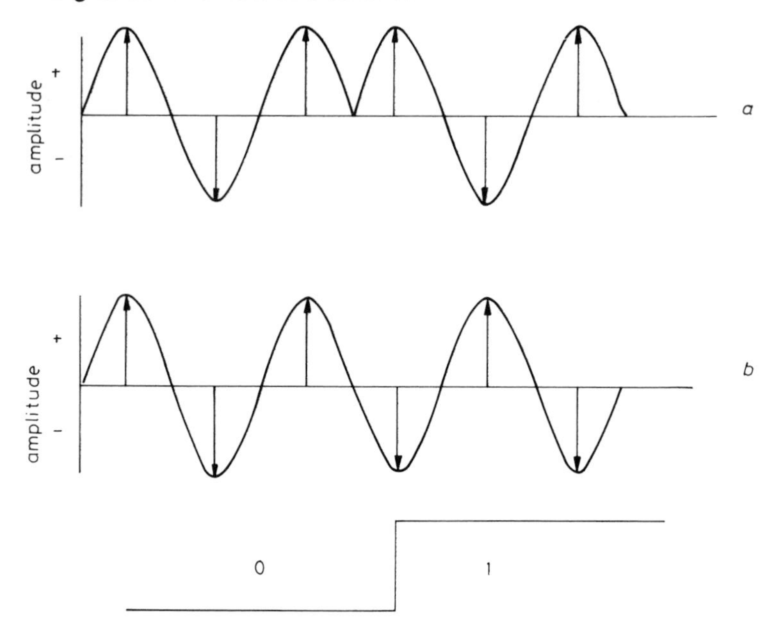

Fig. 5.13 *Coherent demodulation*
a Modulated carrier
b Reference carrier

In practice it is usually more practicable and efficient to recover the carrier from the signal than to transmit a separate unmodulated reference carrier. One way of recovering the carrier from a BPSK signal is to square (or rectify) the modulated signal, extract the second harmonic and divide the frequency by two (see Fig. 5.14). This gives rise to a problem of phase ambiguity; the recovered carrier may have the same phase as the original carrier or it may be 180° out of phase in which case the received bit stream will be inverted (i.e. '1's will be detected as '0's and vice versa). The difficulty may be overcome by starting every message with a specified code; if this is demodulated correctly then the reference carrier is in the right phase; if not then the receiver inverts the bit stream. An alternative solution to the problem is 'differential' encoding (also called 'transition' encoding) which represents a '1' as a change of phase and a '0' as an absence of change (see Fig. 5.15) or vice versa.

A differentially-encoded signal may be recovered by coherent or differential demodulation. A differential demodulator uses a network giving a delay of one bit and compares the phase of the transmission during each bit with the phase during the preceding bit. A differential demodulator is comparatively simple because it does away with the need to recover the carrier but the performance of a differential demodulator is about 1 dB worse than that of a coherent detector for BPSK modulation and 2·5 dB worse for QPSK; this is because both the signals compared in the differential demodulator are noisy whereas it is usually possible to recover a comparatively 'clean' reference carrier for use in a coherent demodulator.

We have only discussed the demodulation of BPSK transmissions but the same principles may be applied to the demodulation of a QPSK transmission which may be treated as two BPSK transmissions.

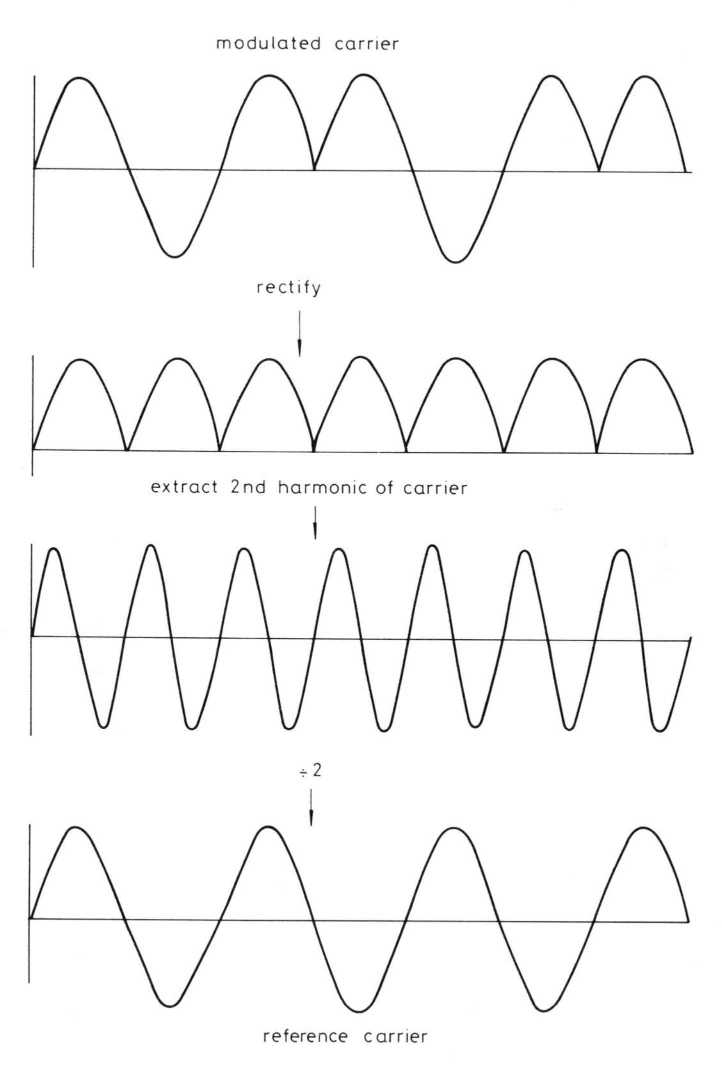

Fig. 5.14 *Carrier recovery*

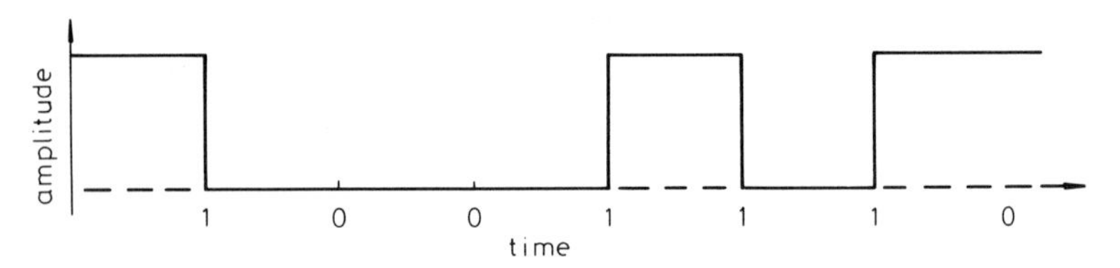

Fig. 5.15 *Differential encoding*

5.6.2.6 *Frequency spectrum*

Fig. 5.16*a* shows a train of rectangular pulses of duration T (= $1/F$) occurring regularly at intervals T_r. Fig. 5.16*b* shows the corresponding frequency spectrum. It will be seen that the spectrum has components spaced at frequency intervals F_r. The envelope of the relative power of these

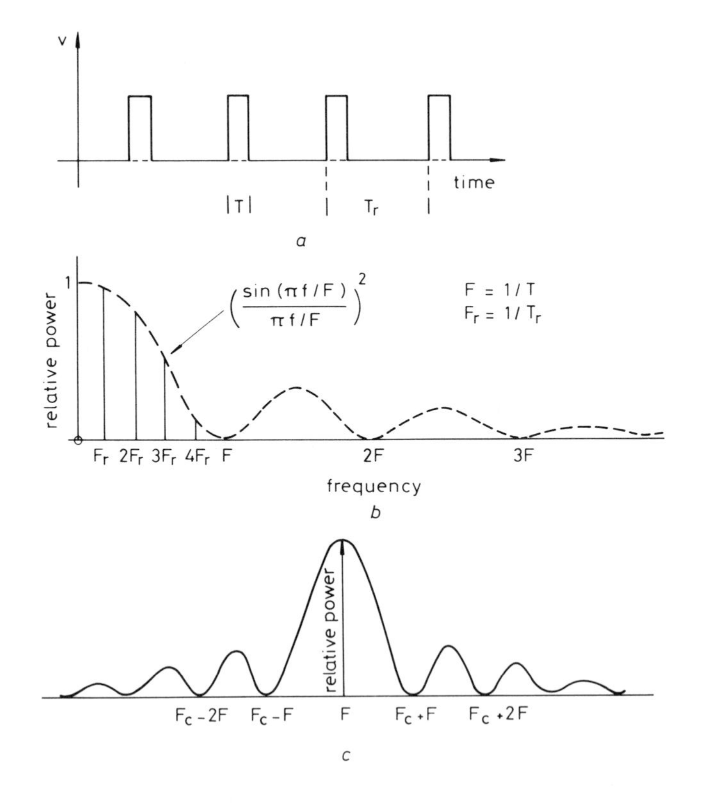

Fig. 5.16 *Power spectrum of a pulse train*
a Regular pulse train
b Spectrum of regular pulse train
c Corresponding RF spectrum after modulation

components has the form $(\sin x/x)^2$, where $x = \pi f/F$, and this envelope is zero at frequencies F, $2F$, $3F$ etc. (It should be noted that Fig. 5.16*b* is not to scale as the successive maxima of the function $(\sin x/x)^2$ decrease rather rapidly; for example the maximum of the first sidelobe, which occurs at a value of x just less than $3\pi/2$, is approximately 0·05.)

Now consider a random train of NRZ pulses with duration T. The envelope of the relative power spectrum of this random train will have the same form as that of Fig. 5.16*b*. However, the spectrum is now continuous (because the interval T_r is random).

So far we have been talking of the frequency spectrum of a baseband signal. The RF spectrum of a BPSK signal is symmetrical about the carrier frequency F_c and the relative distribution of the power in each half of the spectrum is the same as in the baseband. Thus Fig. 5.16*c* shows the RF spectrum corresponding to the baseband spectrum of Fig. 5.16*b*. The power spectrum is dependent on symbol rate; thus the spectrum of a QPSK signal is compressed by a factor of two relative to that of a BPSK signal, i.e. the zeros and maxima occur at half the frequency intervals.

5.6.2.7 *Filtering*

The transmitted RF bandwidth must be limited by filtering in order to avoid adjacent-channel interference (ACI). Filtering is also necessary at the receiver in order to limit the noise power. However, filtering causes distortion of the pulses. In choosing the characteristics of band-limiting filters it is therefore necessary to compromise between reducing ACI and limiting distortion to a level where it does not result in an unacceptable increase in BER.

One of the major sources of distortion which can be caused by filtering is intersymbol interference (ISI). Some understanding of ISI can be gained by considering what happens, in theory, if an impulse (i.e. an extremely narrow pulse) is passed through an ideal filter with the characteristics shown in Fig. 5.17*a* and *b*. The shape of the impulse after it has passed through the filter is shown by the solid line in Fig. 5.17*c*; it has been spread in time and there are oscillating tails both before and after the main lobe. These tails have zero amplitude at times T, $2T$, $3T$ etc. (where $T = 1/F$) before and after the main pulse reaches its maximum amplitude. A second identical pulse (occurring at time T after the first) is shown by the broken line in Fig. 5.17*c*. It will be seen that, provided the pulses are sampled at the right instants (i.e. at maximum amplitude), there is no interference between pulses separated by an interval T (or some integral multiple of T). On the other hand, if the time interval between the pulses is not a multiple of T or if the pulses are not sampled at the correct instant then ISI will occur.

No practical filter has a rectangular amplitude/frequency characteristic

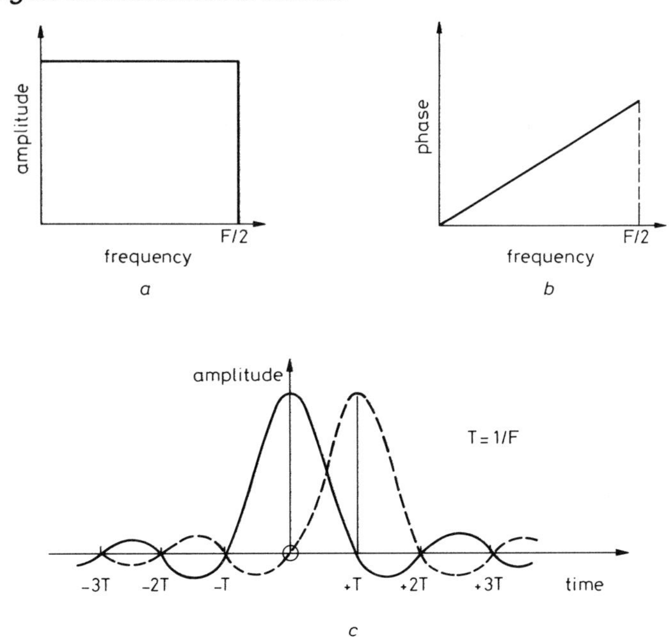

Fig. 5.17 *Filtering an impulse*

such as that shown in Fig. 5.17*a*. However, it is possible to design practical filters which keep ISI to very low values.

Fig. 5.18 shows the normalised amplitude/frequency response of the 'raised-cosine' type of filter which is often used in digital systems.

The 'roll-off factor' α of a raised-cosine filter determines the bandwidth of the filter relative to the Nyquist bandwidth $F/2$, where F is the symbol rate. Thus, for example, roll-off factors of 0·5 and 1 correspond to filters with bandwidths 50% and 100% greater than the Nyquist bandwidth. Increasing the roll-off factor reduces the rate of change of the amplitude of the output pulse at the zero crossings and this reduces the ISI caused by jitter. The optimum roll-off is dependent on a number of factors including the modulation method, the linearity of the transmitters and whether or not the signal is subject to multipath fading.

At this stage it may be useful to recapitulate by considering the bandwidth occupied by a typical PSK signal. Consider a QPSK signal with a bit rate of 120 Mbit/s; the symbol rate is 120/2 = 60 Mbit/s and the Nyquist bandwidth is 60/2 = 30 MHz. If the signal is band-limited by means of a raised-cosine filter with a roll-off factor of 40% then the corresponding baseband width is 1·40(30) = 42 MHz and the (double-sided) RF bandwidth occupied by the signal is 84 MHz. The efficiency of use of bandwidth by a digital radio transmission is sometimes expressed in terms of bits/s per Hz; for the QPSK transmission just considered the efficiency is 120/84 = 1·4 bits/s per Hz.

In discussing the effects of filters on pulse shape we have been talking in terms of the response to a very narrow pulse (having a virtually flat frequency spectrum). In practice most systems use NRZ pulses with a $(\sin x/x)^2$ power spectrum. To compensate for this the amplitude/frequency characteristic of the chosen filter (e.g. a raised-cosine filter) is multiplied by $(x/\sin x)$.

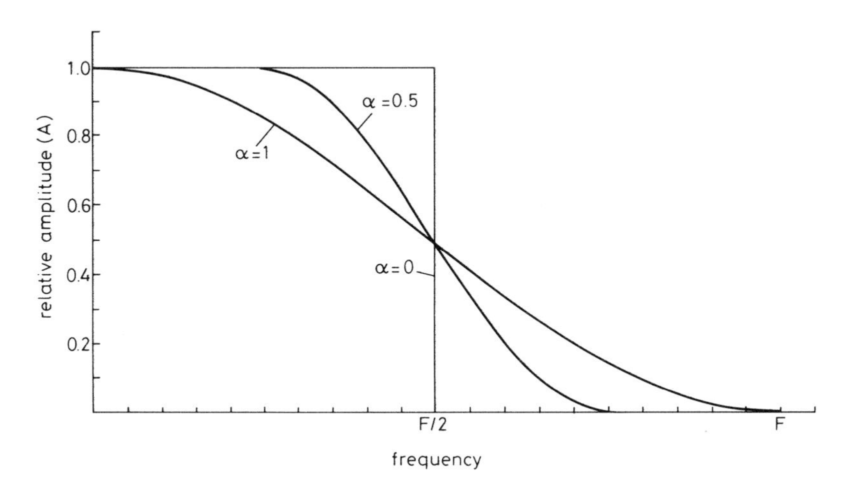

Fig. 5.18 *Raised-cosine filter characteristics*

$A = 1 \quad \text{if } 0 < f < (1-\alpha)\,F/2$

$$A = \tfrac{1}{2}\left[1 + \cos\left\{\frac{\pi}{2}\left(\frac{f-(1-\alpha)\,F/2}{\alpha\,F/2}\right)\right\}\right]$$

$\text{if } (1-\alpha)\,F/2 < f < (1+\alpha)\,F/2$

α = roll-off factor

$A = 0 \quad \text{if } f > (1+\alpha)F/2$

If the transmission channel is linear then the ratio of signal power to thermal-noise power is greatest when the transmit filter and receive filter have the same amplitude characteristic; thus, if the overall characteristic is to be raised cosine then the transmit and receive filters must both have a characteristic which is the square root of the raised-cosine function. Practical systems usually include non-linear amplifiers (e.g. TWT or klystron output amplifiers at the earth station or in the satellite) and in this case some other distribution of the filtering may be optimum. Filtering may be done at baseband, IF or RF (or it may be distributed between these frequencies).

In theory the changes of phase of a PSK transmission take place instantaneously and every cycle of the carrier has the same amplitude. In practice, however, the phase transitions of band-limited PSK signals are relatively slow, a large number of carrier cycles are affected and significant amplitude modulation of the carrier can occur. Fig. 5.19 shows, for example, how the envelope of a QPSK transmission may vary with phase transitions; it will be seen that a transition between adjacent sectors of the phasor diagram (e.g. from symbol '00' to symbol '10') causes only a relatively minor variation in amplitude whereas a transition between diametrically opposite sectors of the phasor diagram (e.g. from '00' to '11') causes the amplitude of the carrier to pass through zero.

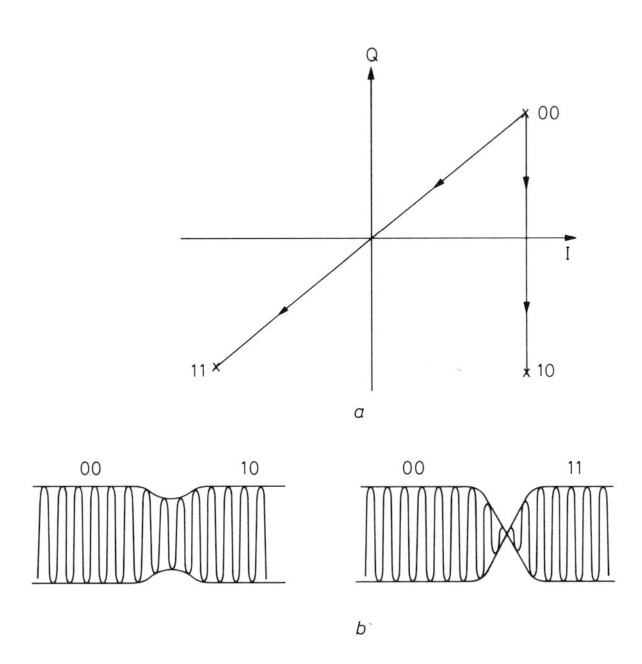

Fig. 5.19 *Envelope of QPSK transmission*
a QPSK phase transitions
b Envelopes

If a PSK signal with significant amplitude modulation is passed through a non-linear amplifier, the AM–AM and AM–PM characteristics of the amplifier interact with the amplitude variations to produce spreading of the spectrum beyond the filtered bandwidth and ISI. Klystron and TWT amplifiers in the earth station or satellite are therefore usually backed off to limit these effects; an input back-off of 2 dB (corresponding to an output back-off of 0·2 dB) is typical.

There are a number of modulation methods which are especially designed to produce RF transmissions with little or no amplitude variation. A few of the more common methods are described briefly in Section 5.6.3.

5.6.2.8 *Scrambling*

Many digital transmissions (such as speech or video signals) are far from random and their RF spectra may therefore have peaks at certain frequencies. These peaks of power density increase the probability of interference between transmissions. It is therefore often necessary to 'scramble' the signal (i.e. make it appear approximately random).

The usual method of scrambling a digital signal is to combine it in some way (e.g. by modulo-2 addition at baseband) with a pseudo-random pulse train. (A pseudo-random sequence is a sequence that can be generated by means of a set of instructions but which has so few regular features that it appears random for the purposes for which it is used.) The receiver is provided with a generator which produces a pseudo-random pulse train identical to that used by the scrambler and is thus able to restore the signal to its original form.

5.6.2.9 *Optimising performance*

In practical digital systems there may be a large number of sources of degradation, some of which may interact. Choosing the optimum system for a particular set of constraints may therefore be difficult and it is often necessary to use a combination of theory, computer simulation and measurement in designing practical systems.

5.6.3 *O-QPSK, MSK and 2,4-PSK*

It was noted in Section 5.6.2.7 that spreading of the spectrum of band-limited BPSK and QPSK signals occurs when they are passed through a non-linear amplifier. This effect is associated with sudden changes in phase of the carrier (and particularly with phase changes of 180°). There are a number of modulation methods which are designed to lessen these sudden phase changes and thus reduce spectrum spreading; three of these methods are offset-QPSK (O-QPSK), minimum shift keying (MSK) and 2,4-QPSK.

5.6.3.1 *O-QPSK*

O-QPSK is also known as offset-keyed QPSK (OK-QPSK). Aviation QPSK (A-QPSK), which is used with the INMARSAT aeronautical mobile system, is also a form of O-QPSK.

Fig. 5.20*a* is a block diagram of an O-QPSK modulator; the only difference between this and the QPSK modulator of Fig. 5.10*a* is that the bit stream in the Q arm of the modulator is delayed by half a symbol (one bit) relative to that in the I arm. Fig. 5.20*b* shows that this delay results in intermediate carrier phase states which eliminate all 180° transitions. It is instructive to compare Fig. 5.20*b* with the corresponding Figure (Fig. 5.10*b*) for the QPSK modulator (the Figures are drawn for identical input data streams).

The BER of an O-QPSK transmission is the same as that of the corresponding QPSK transmission; the frequency spectrum of an unfiltered O-QPSK transmission is also the same as that of a QPSK transmission.

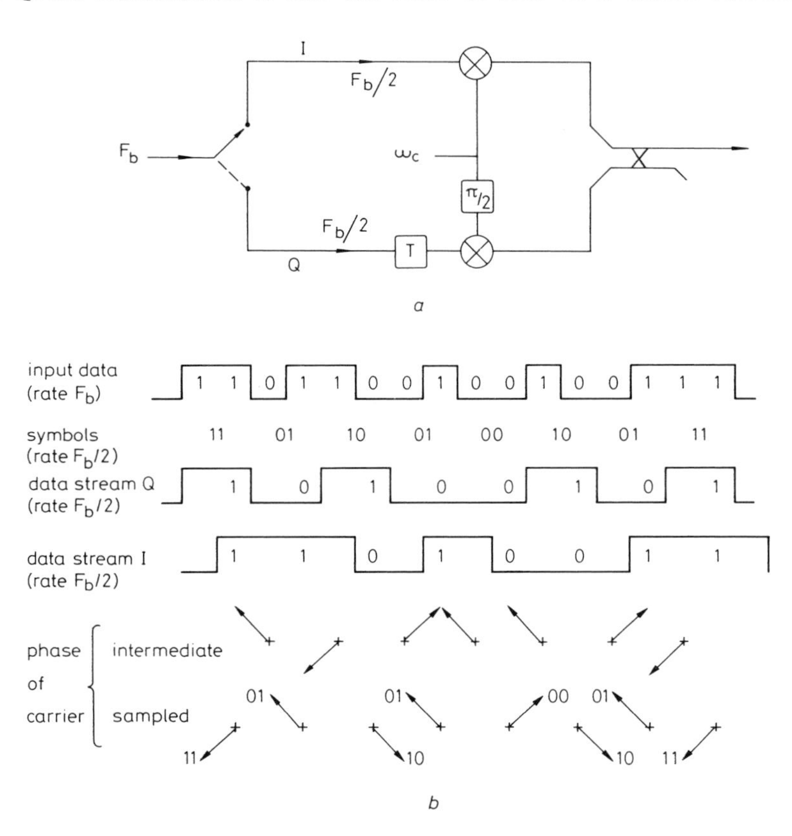

Fig. 5.20 *O–QPSK modulator*
a Block diagram
b Operation

5.6.3.2 2,4-QPSK

Fig. 5.21*a* is a block diagram of a 2,4-PSK modulator. The 2,4-PSK modulator is the same as the O-QPSK modulator except that the output of the modulator is taken alternately from the two arms (i.e. it is not the sum of the outputs of the I and Q arms). Fig. 5.21*b* is drawn for the same data input as Fig. 5.10*b* (QPSK) and Fig. 5.20*b* (O-QPSK) and should be compared with them. It will be seen that the output of the modulator now switches between the two quadrature BPSK carriers at bit rate and there are two states (one for each carrier) corresponding to '1' (and, of course, two states corresponding to '0'). Once again there are no 180° transitions of the carrier. Aviation-BPSK) which is used in the INMARSAT aeronautical mobile system is the same as 2,4-PSK. The BER and unfiltered spectrum of a 2,4-PSK transmission are the same as those of a BPSK transmission.

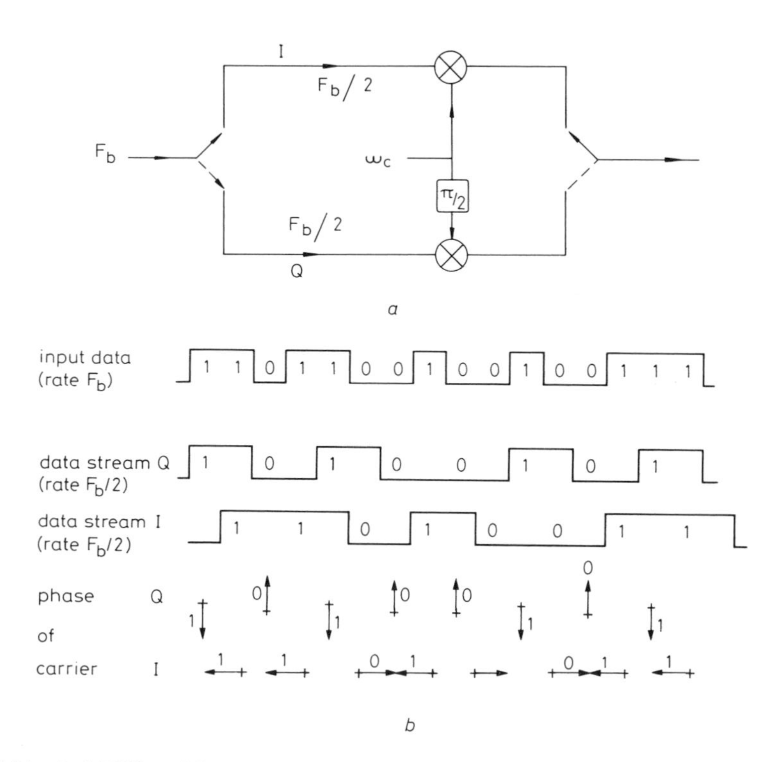

Fig. 5.21 *2, 4-PSK modulator*
a Block diagram
b Operation

5.6.3.3 *MSK*

Steps in phase are avoided completely when minimum shift keying (MSK) is used. With MSK the phase of the carrier is increased or decreased steadily according to whether the signal is '1' or '0' (see Fig. 5.22), the rate of change being such that the carrier phase changes by 90° for each bit. A steady increase in relative phase of $\pi/2$ radians in a bit period of T seconds is equivalent to an increase in frequency of $1/(4T)$ hertz; thus MSK is equivalent to frequency shift keying (FSK) with the two frequencies corresponding to '1' and '0' separated by $1/(2T)$ hertz. The BER of a MSK transmission is the same as that of a QPSK transmission but the envelope of the spectral power density of an unfiltered MSK transmission falls off much more rapidly with increasing frequency offset from the carrier than the corresponding envelope of a QPSK transmission.

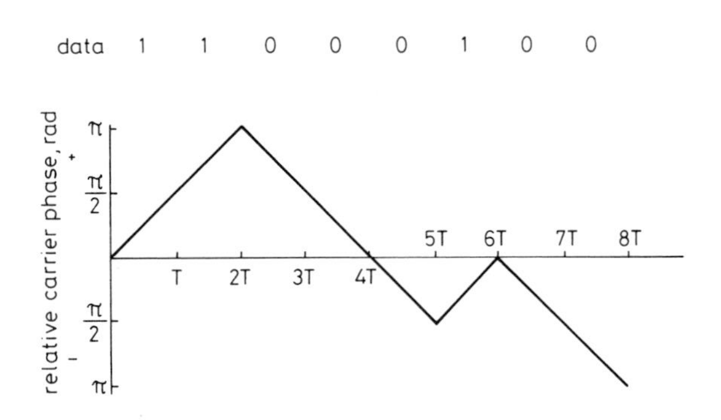

Fig. 5.22 *MSK phase versus time*

5.7 Error correction

The BER of a digital system may be improved either by increasing E_b/N_0 or by detecting and correcting some of the errors in the received data.

Earlier in this Chapter we discussed methods (such as ADPCM) which reduce the number of redundant bits in speech, music and video signals in order to make more economic use of bandwidth. It may seem strange that we are now about to consider methods which require the deliberate addition of redundant bits to messages. However, the bits which are added are very carefully chosen and error-correction systems make it possible to achieve large savings in the power required to achieve low BERs.

There are two ways of correcting errors: automatic repeat request (ARQ) and forward error correction (FEC).

5.7.1 ARQ

ARQ has been used with terrestrial systems for many years. It requires the transmitted data to be divided into blocks which are encoded in some way which makes it practicable for the receiver to detect if there is an error in a block. It also requires a return channel which the receiver can use to send messages to the transmitter.

With 'stop-and-wait' ARQ the transmitter waits at the end of each block for a message from the receiver; if the block has been received correctly the transmitter sends the next block, if the block was in error then the transmitter sends it again.

With continuous ARQ the transmitter keeps on sending until it receives an error message from the receiver. The blocks have to be numbered so that the receiver can tell the transmitter which block was in error. The transmitter may repeat just the block in error or the block in error plus all the succeeding blocks.

ARQ is suitable for improving the BER of transmissions which have a low inherent error ratio (especially if the errors tend to come in bursts) but it quickly becomes inefficient as the error ratio increases and many blocks have to be retransmitted. Long transmission delays such as are associated with systems using geostationary satellites:

(i) reduce the efficiency of ARQ systems
(ii) require the provision of large data stores at terminals working at high data rates and using continuous ARQ.

Continuous ARQ is, however, used in some satellite systems: in the INMARSAT Standard-C system for example, where it is used in conjunction with FEC which ensures that relatively few blocks have to be repeated.

5.7.2 FEC

Forward error correction depends on the use of codes which enable the receiver not only to detect errors but also to correct them. The two basic types of FEC code are block codes and convolutional codes. The theory of FEC is complex and a large number of coding and decoding methods have been proposed; in this book it is practicable only to glance at the subject very briefly.

5.7.2.1 Block codes

For block coding the input data is divided into blocks of k bits and each block is independently encoded as n bits, i.e. there are $(n-k)$ 'parity' bits in each encoded block. This is described as an (n,k), or rate k/n, block code. If the message (input) bits appear as part of the code (output) bits then the code is called a systematic code.

As an elementary example consider a (5,2), or rate 2/5, systematic code in which two bits of information appear as the first two bits of a 5 bit code and bits three, four and five (the parity bits) are derived as follows:

(i) Bit 3 is '0' only for information symbols '11' and '00'
(ii) Bit 4 is '1' if the first information bit is '1' and bit 5 is '1' if the second information bit is '1'.

The code words are:

00000, 01101, 10110, 11011

and they and the code rules can be represented diagrammatically as follows:

00	01	10	11
0 | 00	1 | 01	1 | 10	0 | 11

The bits within the boxes are the information bits, the bit to the left of the box is (parity) bit 3, and the bits above the boxes are (parity) bits 4 and 5. It will be seen that, before transmission, the number of '1's in the row containing the information bits and each of the columns is always even. If, after transmission, one of the information bits is in error then there will be an odd number of '1's in both the row and the column containing the bit in error; if one of the parity bits is in error then there will be an odd number of '1's in either the row or the column containing the bit in error (but not in both). Thus a single error can be corrected.

The code we have just considered was chosen solely to illustrate the basic principle of block codes and without regard to its efficiency. The code with the best performance (for given values of k and n) is that in which the allowable code words (corresponding to messages) are chosen from the possible code words so as to give the highest probability that when an incorrect word is received it will be possible to decide which of the allowable code words was sent.

Practical codes are constructed by using algebraic methods such as the manipulation of matrices. Some of the more frequently used forms of code are:

(i) Hamming, which can correct one error per block
(ii) BCH (Bose–Chauduri–Hocquenghem), which can correct several errors per block)
(iii) Reed–Solomon, which can correct several symbols per block.

Block coding is well suited to systems, such as TDMA systems, where the data is necessarily transmitted in discrete packets. The INTELSAT 120 Mbit/s TDMA system uses a (128, 112) BCH code; this code can correct two errors per block.

There are several ways of decoding block codes. The simplest method to understand is decoding by correlation; in this method the decoder compares the received code word with all the permissible code words and selects the permissible word which is the closest match; this method is impracticable for large values of k and n. Some other methods of decoding require the use of complex processors.

5.7.2.2 *Coding gain*

The performance of an error-correction system is usually stated as the increase in E_b/N_0 required to give a specified BER without the error correction relative to that required with the error correction. Note that we are interested in the energy required to pass a given amount of information; E_b is therefore always taken as the energy per information bit (i.e. C/R, where C is the carrier power and R is the bit rate at the input to the encoder) never as the energy per coded bit, i.e. $[(C/R)(k/n)]$, which is less than C/R.

Coding gain usually varies with error rate, being high at low BER's and decreasing with increase in error rate until, at some 'cut-off' point the coded transmission has a higher BER than the uncoded transmission. The theoretical coding gain for the (128, 112) BCH code used in the INTELSAT TDMA system is about 2·5 dB at a BER of 1 in 10^5.

5.7.2.3 *Convolutional codes*

The essential elements of a convolutional encoder are a shift register and one or more modulo-2 adders (a modulo-2 adder gives an output of '1' when an odd number of the inputs to it are '1's). The inputs to the modulo-2 adders are taken from selected points in the register and the code is formed from the outputs of the adders. The coding algorithm depends on the length of the register, the number of adders and the way in which the inputs to the adders are selected. The INTELSAT IDR and IBS systems use (1/2)-rate and (3/4)-rate convolutional encoding respectively.

Fig. 5.23 shows a simple rate-(1/2) convolutional encoder. The input data

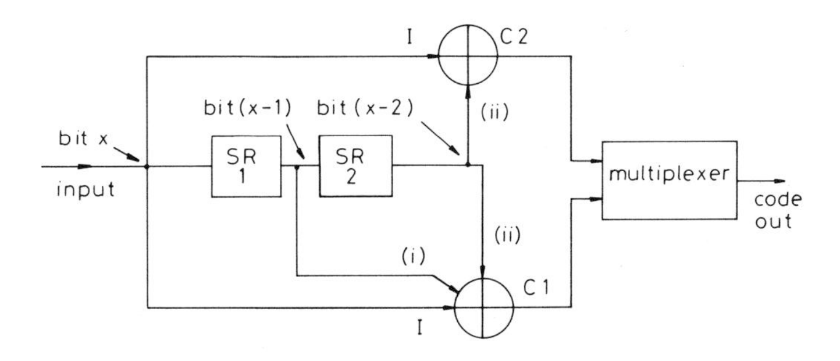

Fig. 5.23 *Rate-1/2 convolutional encoder*

is fed serially to the two-stage shift register; thus, when bit x arrives at the input, bit x-1 appears at the output of stage 1 of the register (SR1) and bit x-2 appears at the output of stage 2 (SR2). Bits x, x-1 and x-2 are fed to a modulo-2 adder, and bits x and x-2 are fed to a second modulo-2 adder. The outputs from the two modulo-2 adders are interleaved in a multiplexer to give an encoded bit stream at twice the bit rate of the input.

Table 5.3 shows the outputs, *C*1 and *C*2, from the two modulo-2 adders for all combinations of x, x-1 and x-2.

Table 5.3 *Outputs of encoder*

x	x-1	x-2	C1	C2
0	0	0	0	0
0	0	1	1	1
0	1	0	1	0
0	1	1	0	1
1	0	0	1	1
1	0	1	0	0
1	1	0	0	1
1	1	1	1	0

The encoder is said to have a constraint length of 3 because each pair of bits (C1, C2) emerging from the encoder is a function of three consecutive input bits.

Another way of representing the action of the encoder is by means of a tree diagram (see Fig. 5.24). The upper or lower branch of the tree is followed according to whether the input is '0' or '1', respectively, and the corresponding output symbol (pair of digits) is shown on the branch. From Table 5.3 we see that the encoder output for a '1' preceded by two '0's is '11'; the same information is represented in Fig. 5.24 by the fact that any downward branch preceded by two upward branches leads to the symbol '11'. Because the constraint length of the code is 3, the pattern of the code tree is repetitive after the first three sets of branches.

As an example of the coding process, suppose that the input data is the sequence:

101101 . . .

By tracing this path through the tree it will be seen that the output from the coder is:

11 10 00 01 01 00 . . . . .

Now suppose that when this sequence arrives at the receiver the fourth symbol is in error and the encoded message reads:

11 10 00 11 01 00 . . . . . .

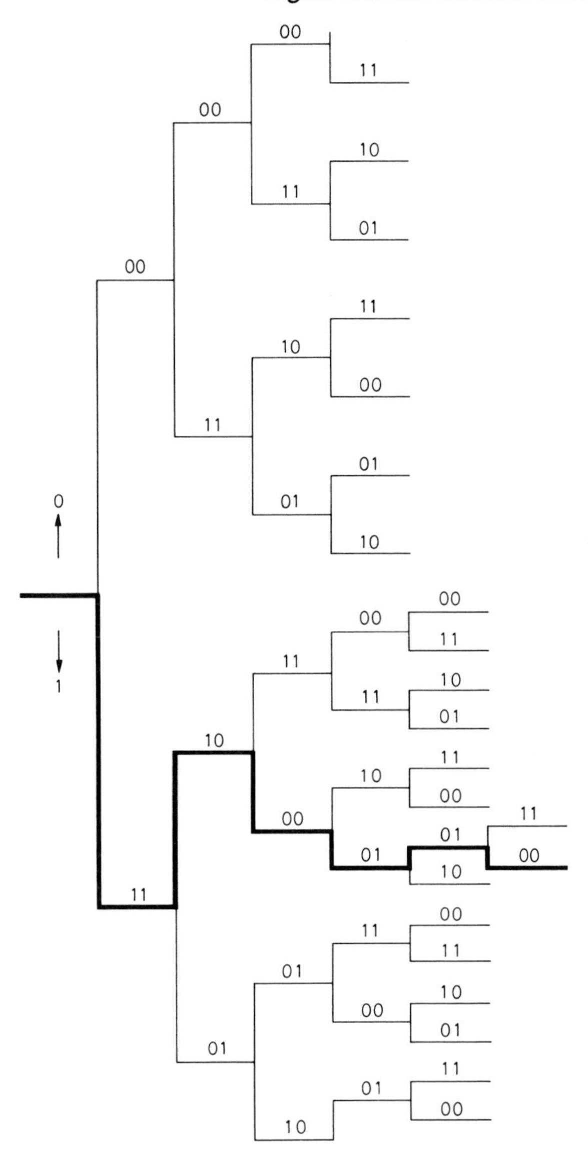

Fig. 5.24 *Code tree*
The heavy line shows the route for the input sequence 1 0 1 1 0 1

This path cannot be traced through the tree and the decoder will therefore know that an error has occurred. One way of proceeding is to guess what correction is needed and then try once again to follow the path, using the new sequence. However, for a long sequence of symbols containing several errors it may be necessary to go a long way back in the sequence and to make a very large number of tries before the most likely sequence is discovered. It is therefore necessary to find a decoding method which does not require the provision of an unacceptable amount of data storage and processing. An efficient and frequently used decoding algorithm is the Viterbi algorithm.

5.7.2.4 *Soft decisions*

In demodulating an uncoded digital signal it is necessary to make a 'hard' decision, e.g. in the case of a QPSK signal we have to decide in which quadrant of the phasor diagram the transmission falls at the time at which it is sampled. However, because of the random nature of noise (and variations in the power of the received signal) the vector representing the signal plus noise will sometimes fall well away from the boundaries between sectors of the phasor diagram (which means that the decision of the demodulator is very probably correct) and will sometimes be close to a boundary (in which case the decision may well be in error). If the signal is coded it is possible to make use of any information we have about the probability that a particular bit of the code is correct. Demodulators for FEC data may therefore be designed to make 'soft' decisions which include information as to the probability that the decision is correct. Coding gain using soft decisions can be up to about 2 dB greater than that achievable with hard decisions. Rate-1/2 convolutional encoding in association with soft decisions and the Viterbi algorithm can give coding gains of about 6 dB at a BER of 1 in 10^6.

Where very low error rates are required it is often more efficient to achieve the required correction in two stages, using two relatively simple processes rather than by using a single very complex decoder.

5.7.2.5 *Interleaving*

FEC methods work well when the BER is low and the errors are randomly distributed. However, under some circumstances bursts of consecutive errors may occur and this can render the encoding useless. To overcome this problem the symbols must be rearranged in such a way that the periods between transmission of successive related symbols are greater than the duration of the great majority of the bursts of errors; this process is called 'interleaving' and it should ensure that, in most cases, enough related symbols are left uncorrupted to enable the original message to be reconstructed. The data must, of course, be interleaved in accordance with a prescribed set of rules so that it can be reassembled in the original order after reception.

Consider, for example, the INMARSAT Standard-C system which will be introduced in 1989. This system uses 1/2-rate convolutional encoding with a constraint length of 7, i.e. each bit entering the encoder influences fourteen consecutive output bits. The transmissions between ships and satellites will be subject to slow fading as a result of multipath transmission (see Section 6.6.1) and these fades can introduce long bursts of errors. Blocks of data (comprising between 2176 and 10 368 bits) are therefore interleaved before transmission.

5.8 Multiple access

There are three ways in which transmissions may share a satellite transponder:

(i) Each transmission may be allocated a different carrier frequency; this is frequency-division multiple access or FDMA.
(ii) Transmissions may occupy separate (non-overlapping) time intervals; this is time-division multiple access or TDMA.
(iii) All transmissions are simultaneous and use the same bandwidth; this is code-division multiple access (CDMA) or spread-spectrum multiple access (SSMA).

5.8.1 CDMA

In the case of CDMA or SSMA, interference between transmissions cannot be avoided because they use the same frequency band at the same time. The interference is kept to an acceptable level by spreading the power from each transmission as evenly as possible over a wide bandwidth.

If the RF power is to be spread evenly over the spectrum then the modulating signal must be virtually random; on the other hand the transmissions must carry a message (which may have regular features) and each transmission must have a unique signature by which it can be recognised and retrieved. In order to reconcile these requirements it is necessary to scramble each transmission by means of a pseudo-random sequence. Only receiving stations which can generate the pseudo-random sequence in synchronism with the transmission can recognise and recover the message.

One way of generating a CDMA transmission (known as 'direct sequence') is to combine a train of pseudo-random pulses with the message signal. The pulse repetition rate of the pseudo-random signal must be high enough to spread the signal over the whole of the available bandwidth and one message bit is therefore combined with a train of many pseudo-random bits (the latter are usually called 'chips' to distinguish them from the message bits); for example, the message might have an information rate of 20 kbits/s and the pseudo-random sequence might have a chip rate of 2 Mbit/s so that every message bit is combined with 100 chips. The receiving station knows the pseudo-random sequence and can generate a chip train which is coherent with that of the wanted transmission; it can therefore unscramble the message by a process analogous to coherent detection. On the other hand all the other transmissions, which are combined with different (uncorrelated) pseudo-random sequences, look like noise. The signal-to-noise power ratio is proportional to the number of chips per bit and the quantity [10log(chips per bit)] is known as the processing gain.

Another method of CDMA (frequency-hopping CDMA) requires the frequency of the transmission to be changed as a function of a pseudo-random sequence.

In general, CDMA systems have the following characteristics:

(i) They are resistant to interference (whether deliberate or accidental) and new users can join the system without danger that they will cause serious interference to existing users.
(ii) They are secure (provided that unauthorised users do not know what pseudo-random code is in use).
(iii) They do not make efficient use of the frequency spectrum.

CDMA systems are widely used for military communications where the first two characteristics are very advantageous and efficient use of the frequency spectrum is not of prime importance. Until recently little consideration has been given to CDMA for civil communications-satellite systems; it is, however, a serious contender for use with systems using VSATs and other systems where interference is a serious problem.

5.8.2 FDMA

Two types of digital transmission are used in association with FDMA: single-channel-per-carrier (SCPC) transmissions and carriers modulated with time-division-multiplex (TDM) asssemblies of channels. TDM assemblies are usually the most efficient means of transmission for routes carrying heavy traffic.

Any FDMA system has the disadvantage that intermodulation will occur between transmissions sharing the same satellite transponder; the most common method of achieving linear amplification is to back off the transponder output amplifier (i.e. to work it at a reduced output power) and this reduces the traffic which the satellite can carry.

5.8.2.1 SCPC

An SCPC transmission carrying a 64 kbit/s PCM voice channel and using QPSK modulation requires a bandwidth of about 38 kHz. Carrier frequencies in such a system are spaced at intervals of about 45 kHz and a satellite transponder with a usable bandwidth of 36 MHz can therefore support up to 800 channels. If a more efficient encoding method (e.g. delta modulation or ADPCM) is used it is practicable to reduce the bit rate of a voice channel to

32 kbit/s and the carrier spacing to 22·5 kHz. Even lower bit rates and spacings can be used in special circumstances where lower voice quality is acceptable (e.g. the INMARSAT Initial Aeronautical system will use a bit rate of 21 kbit/s). SCPC systems are also used for the transmission of a variety of data channels. Channels may be either demand-assigned or pre-assigned.

Instability of oscillators associated with frequency changers in earth stations and satellites can result in errors in the frequency of received transmissions of as much as ± 50 kHz; frequency errors of this order are unimportant in the case of wideband transmissions but are unacceptable with transmissions having RF bandwidths of tens of kilohertz. Automatic frequency control (AFC) is therefore required in SCPC systems. A typical AFC system is described in Section 6.4.5.

Carriers in SCPC systems are usually voice activated, i.e. transmission of the carrier ceases during pauses in speech; this results in a reduction in the required transponder power of more than 50%. The automatic gain control at the receiving station cannot work when there is no incoming carrier and a 'squelch' circuit must therefore be provided to suppress the background noise when the carrier is cut off. The output power of TWT or klystron transmitters in the transponder must be backed off (usually by about 4 — 6 dB relative to single-carrier saturation power) in order to reduce intermodulation between SCPC transmissions to acceptable levels. However, a significant reduction in intermodulation is becoming practicable with the development of solid-state power amplifiers suitable for use with communications satellites.

An advantage of SCPC operation is that the power of each transmitted carrier can be adjusted individually to take account of factors such as the size of the receiving antenna, the geographic location and position in the satellite beam of the receiving earth station, rain attenuation and the required BER.

5.8.2.2 TDM–PCM–PSK

Carriers modulated with TDM assemblies of PCM channels are particularly useful on routes carrying heavy traffic. The modulation method is usually PCM and the transmission method is then known as TDM–PCM–PSK. TDMA is usually more efficient than TDM–PCM–PSK for serving a number of routes carrying relatively light streams of traffic but TDM equipment has the advantage of being relatively cheap and simple.

The INTELSAT IDR (intermediate date rate) system is a TDM–PCM–PSK system which uses QPSK and a 3/4-rate FEC code with soft decision decoding. The recommended information rates for transmissions using the system include 64 kbit/s and all the first-order, second-order and third-order multiplex rates (see Table 3.2), but any information rate between 64 kbit/s and 44·736 Mbit/s may be used if required.

TDM carriers with a large capacity are often called 'digital pipes' and data and video services may be multiplexed in with voice channels. Digital pipes used in conjunction with ADPCM and DSI offer very large traffic capacities.

5.8.3 *Time-division multiple access (TDMA)*

All the stations in a TDMA system transmit on the same carrier frequency. In some relatively lightly-loaded systems the earth stations transmit at random (see Section 5.8.4) but TDMA systems serving fixed stations carrying relatively heavy traffic are mainly ordered systems in which each station waits for its turn to transmit a burst of data in accordance with a predetermined time plan. In a practical TDMA system there are usually many earth stations but, for simplicity, Fig. 5.25*a* shows only two earth stations transmitting in TDMA to a satellite transponder. The bursts of transmission from each earth station are timed so that they are interleaved at the transponder and thus the transmission from the satellite is the ensemble of the bursts arriving at the transponder. Short guard intervals (of the order of 1 μs) are left between bursts to allow for small timing errors. Since the transponder never has to amplify more than one carrier at a time there is no intermodulation and no significant back-off is necessary; it is therefore possible to use virtually all the power which the output amplifier can deliver. (A small amount of back-off may be necessary for reasons which are explained in Section 5.6.2.)

Each earth station must:

(i) Store the innformation arriving from the terrestrial network which will comprise its next burst of transmission and transmit the burst at the precise time necessary to ensure that, when it arrives at the satellite, it does not overlap the preceding or subsequent bursts.

(ii) Store the information arriving from the satellite and addressed to the station and read this information out to the terrestrial network at the appropriate rates and to the correct circuits.

5.8.3.1 *TDMA frame*

A complete set of transmissions from all the stations using a transponder is known as a TDMA frame. A new frame is commenced as soon as the last one is finished. The start of a frame is determined relative to a reference burst which is transmitted by a station nominated as a reference station; the reference burst occurs at the beginning of each frame in the INTELSAT system but, in general, it may be located anywhere in the frame. Fig. 5.25*b* shows the structure of a typical frame.

Part of each burst is occupied by a preamble of fixed length which does not carry traffic; the percentage of the frame occupied by the preambles and the guard intervals (for a fixed number of bursts) becomes smaller as the frame length is increased, i.e. the efficiency of the system is improved.

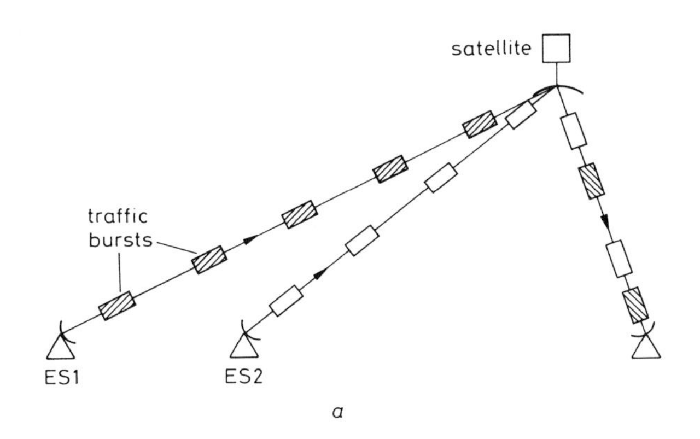

a

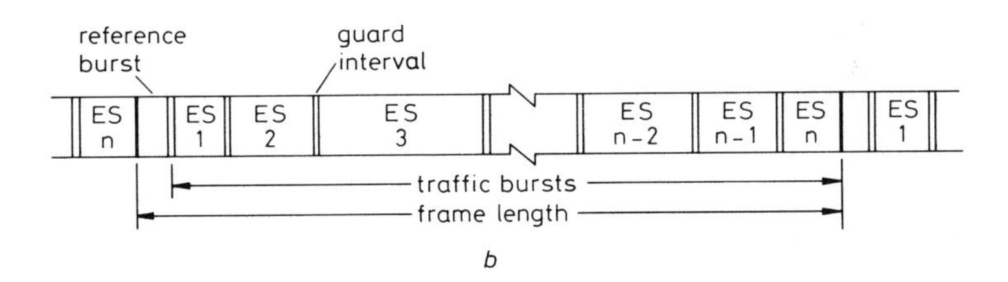

b

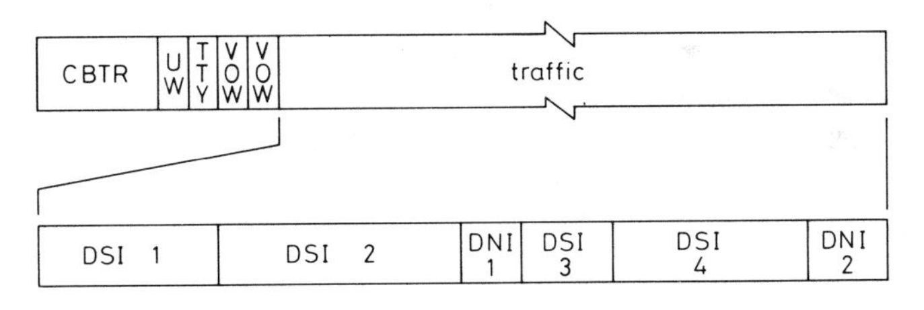

c

Fig. 5.25 *TDMA*
a System
b Format of frame
c Format of traffic burst
ES = earth station
CBTR = carrier and bit timing recovery sequence
UW = unique word
TTY & VOW = teletype & voice order wires
DSI & DNI = interpolated and non-interpolated bursts

However, if the length of the frame is doubled the amount of data which each earth station must store during the frame is also doubled. The choice of frame length is therefore a compromise between efficiency and the size of the data stores required at the earth stations.

In the INTELSAT TDMA system, for example, the frame length is 2 ms and thus a station using this system must be able to store 128 bits for each 64 kbit/s channel which it is transmitting or receiving (in the case of a voice channel these 128 bits comprise sixteen 8 bit codes representing sixteen samples taken at a sampling rate of 8000 per second).

The bit rate of the INTELSAT system is 120 Mbit/s so that the 128 bits corresponding to a 64 kbit/s channel are transmitted in just over 1 μs. The number of 64 kbit/s satellite channels carried by the INTELSAT 120 Mbit/s system would be 1875 (= $(120 \times 10^6)/(64 \times 10^3)$) if the frame efficiency were 100%. In practice the capacity is about 1700 satellite channels, i.e. frame efficiency is about 90%.

The system uses DSI so that 1700 satellite channels can carry well over twice that number of terrestrial channels (the precise capacity in terms of terrestrial channels depends on the proportion of data channels which the system is carrying, since these cannot be interpolated).

5.8.3.2 CBTR

There are slight differences in frequency and bit rate between successive bursts of transmission, because they emanate from different earth stations. In order to demodulate the transmissions both the frequency and the bit rate of each new burst must be accurately established by the receiving station. This cannot be done instantaneously and each burst must therefore start with a carrier and bit timing recovery (CBTR) sequence (see Fig. 5.25*c*); the exact form of this sequence depends on the modulation method used. In the INTELSAT system the modulation is QPSK and the CBTR sequence comprises unmodulated carrier for a period of 48 (2 bit) symbols followed by a period of 128 symbols in which the phase of the carrier is reversed by 180° every symbol (in the latter period the symbols are alternately '00' and '11'); thus, in this system, both the carrier frequency and the clock rate must be established in less than 3 μs.

5.8.3.3 UW

The CBTR sequence is immediately followed by a sequence of digits known as the unique word (UW). Recognition of the UW serves to confirm that a burst is present and gives an accurate timing marker by which the position of each bit in the rest of the burst can be determined; such a marker is essential if the start and finish of individual messages in the burst are to be identified and the messages correctly decoded. The UW must be chosen to be easily recognisable and must comprise a long train of bits in order to reduce the frequency with which it occurs by chance. In the INTELSAT system the UW comprises two consecutive groups of 12 symbols, viz:

P bit stream: 011110001001 011110001001
Q bit stream: 011110001001 011110001001

(Note: The system occasionally makes use of three other UWs; these are derived from the basic UW by inverting either or both of the bit streams comprising the second group of symbols.)

Because the INTELSAT system uses QPSK with absolute (not differential) encoding it is necessary to resolve the phase ambiguity of the carrier. One of the groups of twelve symbols is used for this purpose, and one or both of the bit streams are inverted if necessary. Until the UW has been received the earth-station receiver has no means of synchronising its operations with the received transmission; scrambling must not therefore be applied to any part of the transmission up to the end of the unique word.

5.8.3.4 Preamble

The CBTR sequence and UW form part of the 'preamble' of a burst; additional items which may be included in the preamble are:

(i) Voice and teletype order wires (engineering service circuits) which are used for the exchange of messages concerning the operation and maintenance of the system.
(ii) A station identification code.
(iii) Control channels which may be used for messages which require immediate attention (e.g. alarm messages indicating loss of synchronism or messages used to aid acquisition).

5.8.3.5 Traffic burst

The traffic burst follows immediately on the preamble. In the INTELSAT system, the traffic bursts are divided into sub-bursts, each of which is intended for a particular station (or stations). There are two kinds of sub-bursts: DSI sub-bursts comprising channels all or most of which have been subject to digital speech interpolation and DNI sub-bursts comprising only non-interpolated channels. (There are usually only a few DNI sub-bursts.)

Traffic channels at multiples of the basic channel bit rate can usually be provided on a TDMA system. The INTELSAT system specification allows for channels at bit rates up to 8·192 Mbit/s (unfortunately this is just below the European second-order multiplex rate of 8·446 Mbit/s).

7/8-rate BCH coding is, at present, applied to all traffic bursts in the INTELSAT system but the design allows the coding to be applied to designated bursts only. TDMA systems are well suited to the selective use of FEC: with other systems different modems are required for coded and uncoded channels (because the coded channels have a higher bit rate); with TDMA the coded channels are transmitted at the same rate as the uncoded channels and simply occupy more frame time.

5.8.3.6 Acquisition

Each burst is allocated a time slot in which it must fall on arrival at the satellite. The start and finish of the slot is usually defined in terms of numbers of symbols from the start of the frame. The earth stations are, in general, all at different distances from the satellite and the corresponding differences in transmission time must be taken into account in determining when each station shall transmit its burst. These differences in transmission time can be quite large. For example, an earth station which sees the satellite at 5° elevation is more than 5000 km further from the satellite than a station which sees the satellite at an elevation of 80°. A transmission in free space takes 16·7 ms to travel 5000 km and, at a bit rate of 120 Mbit/s, this corresponds to 2 Mbits (i.e. more than eight frames). The earth station has thus to ensure that its burst arrives in the correct frame as well as in the correct position in the frame. Since the guard time allowed between bursts is only of the order of 1 μs it is clear that transmitting a burst at precisely the right time is not easy.

The method used is typically as follows:

(i) The earth station transmits a short burst comprising just the CBTR sequence and the UW. The timing accuracy necessary to place this short burst in the relatively long time slot is much less than that required in normal operation; it is therefore practicable to calculate the time at which the burst should be transmitted from the range of the satellite (which can be derived from the position of the earth station and a knowledge of the satellite orbit).

(ii) If the earth station can receive its own burst then its TDMA terminal notes the position of the short burst in the time slot and makes a correction to the burst transmission time. After each correction the terminal must wait for a period equal to the return transmission time (i.e. approximately 0.25 s) to see the effect of its action before making a further correction. When the short burst is in the correct position (i.e. at the start of the time slot) the rest of the burst is added and the station can start transmitting traffic.

(iii) If the earth station cannot receive its own burst (which is often the case when spot beams are in use) it is necessary for a second earth station to determine the error and advise the transmitting station of the necessary correction. (This method is used by all stations in the INTELSAT TDMA system, whether or not they can see their own bursts.)

(iv) Because of the cyclic movement of the satellite (which can be as much as two or three hundred kilometres a day) the burst transmission time must be continually updated; the more often this is done, the smaller is the required guard time between bursts. In the INTELSAT system the minimum guard time beween bursts is 64 symbols.

The process of placing a burst in the right position in the frame in order to start transmission of traffic is called 'acquisition'. Maintaining the burst in the right position after acquisition is 'synchronisation'.

5.8.3.7 *Transponder hopping*

A TDMA system may operate to more than one transponder. For example, stations in a west spot beam might be transmitting to stations in an east spot beam via one transponder while channels in the opposite direction are carried by another transponder. In the INTELSAT system reference bursts are provided in every transponder carrying TDMA transmissions (this simplifies operation but is not absolutely necessary).

It is also possible for an earth station to transmit a burst to one set of stations via one transponder and then, in the same frame, to transmit a burst to a second set of stations via another transponder. This is known as transponder hopping.

5.8.3.8 *Burst time plan*

For any TDMA system a burst time plan (BTP) is necessary. The BTP must specify, for each transponder, the frequency and polarisation of the transmissions as well as the position and duration of each burst in the frame, and the name of the originating station. The INTELSAT BTP also records details of the channels (e.g. destination and whether the channel is part of a DSI group), allocations of order wires etc. Each terminal in the system has to store the relevant part of the time plan in its processor. As traffic grows the BTP has to be updated. Terminals are therefore provided with a second store in which the revised BTP can be loaded; all stations switch to the revised BTP simultaneously (under the control of the reference station) and thus the new plan is brought into use without any loss of traffic. Errors in the time plan can lead to serious degradation of system operation; great care must therefore be taken to ensure that the plan loaded into the terminal is absolutely correct. In the INTELSAT system, an earth station which has received data for a new BTP retransmits it for check by the originating station before it is used.

5.8.3.9 *TDMA–DA and TDMA–FDMA*

The INTELSAT and EUTELSAT TDMA systems are virtually identical. Both these systems are intended to provide medium-capacity international trunk routes and both use the total bandwidth of a transponder. Different types of TDMA system may be advantageous for the provision of business and specialised services. For example:

(i) Capacity may be demand assigned (DA); this necessitates continual revision of the BTP and adds considerably to the complexity of the system.
(ii) Several TDMA carriers with relatively narrow bandwidths may be accommodated in a single transponder (TDMA–FDMA); this provides additional flexibility for tailoring a system to any special requirements but intermodulation between carriers results in a large reduction in the capacity of the transponder.

5.8.3.10 SS–TDMA

Each succeeding generation of communications satellites tends to make more use of multiple beams with relatively small coverage areas (zone or spot beams). Zone or spot beams make it possible to reuse frequency spectrum; moreover, the relatively high gains of the satellite antennas needed to form zone or spot beams facilitate the use of smaller antennas and/or lower transmitter powers at earth stations. A reduction in the required earth-station EIRP is particularly useful in TDMA systems as the need to transmit short bursts (at high bit rate) leads to a requirement for high peak power.

However, as the number of satellite antenna beams is increased it becomes more difficult to provide connections between all the earth stations. In INTELSAT V a complex array of channel-switching matrices allows a range of interconnections to be made between the receive and transmit beams of the satellite. The switches in the matrices can be operated by command from the ground but changes are made only infrequently (e.g. if there is a major change in the pattern of traffic carried by the satellite). INTELSAT VI carries additional matrices which enable TDMA traffic from any of the (hemi or zone) uplink beams to be switched rapidly between downlink beams during the course of a TDMA frame (according to a predetermined sequence). Thus, for example, Zone 1 might be connected to Zone 2 for the first 80 μs of the 2 ms frame, then Zone 3 connected to Zone 4 for the next 60 μs and so on until every interconnection required by the pattern of traffic has been made. This is known as satellite-switched TDMA (SS–TDMA).

In a SS–TDMA system, the arrival time at the satellite of the bursts must be synchronised with the operation of the on-board switching matrix. Extra guard time must be provided at each rearrangement of the matrix firstly because operation of the switches is not instantaneous and secondly to allow for small differences in timing between the satellite and earth-station clocks.

When on-board processing becomes practicable for commercial systems it will be possible to store data arriving at the satellite and switch it to a new time slot in the outgoing TDMA transmission. This will simplify the equipment at the earth stations; it will also result in a slight improvement in the efficiency of TDMA systems since it will no longer be necessary for an earth station to send more than one burst per frame.

5.8.3.11 Terminals

TDMA terminals usually comprise a common TDMA terminal equipment (CTTE) and terrestrial interface modules (TIMs). The TIMs are responsible for multiplexing and demultiplexing channels, for the generation of signalling and supervisory information, and for digital speech interpolation. The CTTE contains a processor which executes the system protocols and is responsible for acquisition and synchronisation.

5.8.4 Random-access TDMA

Random-access TDMA is a simple multiple-access method which may be used for the transmission of packets of data. Random-access TDMA is sometimes called time-random multiple access (TRMA).

The simplest type of random-access TDMA system uses a protocol which was first used with the ALOHA radio network at the University of Hawaii. When using the ALOHA protocol an earth station transmits a short burst of data whenever it wishes to use the system without worrying about what the other stations may be doing. If the burst collides with a burst from another station then this is detected and both bursts are retransmitted; random delays are introduced before the retransmissions to minimise the risk of another collision. As the traffic generated by the stations increases so does the probability of collisions and eventually the system may get to a point where there are so many collisions that little or no useful data is transmitted.

Consider a channel with a capacity of R symbols per second (if fully occupied by a continuous transmission). If the total (random) traffic originating from the stations averages NS where N is the number of packets per second and S is the number of symbols per packet, then the traffic passing through the channel will increase with NS up to the point where $NS = R/2$. At this point the probability that a packet does not collide with any other packet is $1/e$ ($= 0{\cdot}368$) so that the total traffic getting through the system will be $0{\cdot}368R/2$; i.e. the maximum throughput of the system is, on average, about 18% of the channel capacity. If the traffic originated by the stations becomes greater than $R/2$ then the probability of collisions increases faster than the traffic, and throughput decreases.

In the INMARSAT Standard-A system ships send requests for calls by means of random-access TDMA messages. Each message occupies 35 ms and the capacity of a channel on a continuous basis would therefore be 28·6 messages per second. The maximum number of randomly occurring requests which can be handled by a single channel is therefore, on average (0·18 × 28·6) or just over five requests per second.

There are various methods of improving the efficiency of the ALOHA protocol at the cost of some added complication; for example:

(i) *Slotted ALOHA (S–ALOHA)*: Stations must place their bursts in any of a series of fixed time slots; in any collision the bursts will now overlap completely and the elimination of partial overlaps reduces the total number of collisions by 50%. An S–ALOHA channel has a maximum throughput of 0·36R bits/s (i.e. twice that of the corresponding basic ALOHA channel) and this throughput occurs when the total traffic originated by the earth stations is R bits/s.

(ii) *Implicit slot reservation*: Whenever a station is successful in transmitting a packet (in one of a number of fixed slots) this slot is reserved for that station for as long as it continues to use it. The trouble with this system is that there is no way of preventing a station from capturing most or all of the slots.

(iii) *Explicit slot reservation*: Part of the capacity of the channel is used to allow stations to request slots for future transmissions; a record is kept of the slots in use and of requests for reservations, and as data slots become free they are allocated to waiting packets on a 'first-come, first-served' basis; the control of the system may be vested in a single station or it may be distributed (i.e. all stations keep a record of both free slots and requests for reservations). The INMARSAT Standard-C system uses an S–ALOHA reservation system.

5.9 Interfaces with terrestrial networks

A number of problems may arise in interfacing digital satellite systems with terrestrial networks.

If the terrestrial network uses analogue methods then it is necessary (in the case of satellite systems using TDM or TDMA transmissions) to convert the multiplexes from frequency division to time division and vice versa. It would be possible to use standard FDM and TDM equipment to demultiplex to channel level and remultiplex but the most convenient and economical way is to use a digital transmultiplexer. A typical transmultiplexer converts a FDM supergroup directly to two (30-channel) primary time-division multiplexes by the use of a digital processor and digital filters.

For any international digital channel there is a problem of synchronisation, whether or not the channel includes a satellite link. Each national digital system works to its own master clock. In an international circuit there must therefore be some interface where there is a difference between the rates at which information is arriving and leaving and a buffer store must be provided at this point. If the store becomes completely full or completely empty it is necessary to accept a discontinuity in the flow of information across the interface. By providing clocks with high stability and sufficiently large stores it is, however, possible to ensure that discontinuities are very infrequent. CCITT recommends that digital networks which work to an international interface should have a clock stability of not worse than 1 in 10^{10} (this is plesiochronous, i.e. nearly-synchronous, working).

A TDMA system has its own clock; thus, in the case of international transmissions, buffer stores must be provided at both earth stations to allow for slip between the TDMA clock and the clocks of the two national systems. Where the satellite system does not have its own clock (e.g. when the transmission is TDM–PCM–PSK) only one buffer store is required and this may be situated at the earth station or in the terrestrial network of the destination country.

The clock rate of a digital transmission by satellite is affected by Doppler shift caused by the movement of the satellite relative to the earth stations. Sufficient storage must be provided at earth stations to allow for this but, for a geostationary satellite, this relative movement is mainly cyclic (over 24 hours) and slip should never occur.

5.10 Error performance objectives

What matters to the user of a communications service is the overall performance of the channel or circuit. Thus it is necessary to set error performance objectives for terminal-to-terminal connections. On the other hand the designer of a satellite-communications system needs to know the performance objectives for that part of the channel which he is providing (i.e. a link between two earth stations via a satellite). In the case of a public service there are usually long terrestrial connections between the customers and the earth stations, whereas in the case of specialised services provided by satellite communications the earth stations are likely to be at or near to the users' premises. In the former case the terrestrial connections may generate a large proportion of the allowable errors whereas in the latter case it may be possible to allow the satellite system to generate virtually all of the errors.

Different causes may give rise to different distributions of errors in time. Thus thermal noise gives rise to errors which are randomly distributed in time but there are many sources of degradation (such as bad connections, fading and impulsive noise arising from electrical equipment) which may give rise to bunches of errors.

The effect of a particular distribution of errors frequently depends both on the design of the system and the type of signal. For example, in a data service using ARQ errors which occur in clusters usually have less effect than randomly distributed errors; this is because a block of data must be retransmitted whether it contains one or many errors but fewer blocks will need to be retransmitted if the errors are bunched. The performance objectives for the ISDN therefore include a recommendation that at least 92% of seconds shall be free of errors.

The maximum acceptable BER is determined by the type of signal (speech or data). For speech carried as a 64 kbit/s PCM transmission the effect of error ratios less than 1 in 10^5 is not usually noticeable, but BERs of 1 in 10^8 or less are required for some data services; these very low BERs are usually obtained with the aid of error-correction systems.

In most systems it is necessary to accept that there will be a small percentage of the time when degraded performance must be accepted and an even smaller percentage of the time when the error count will be so high (as a result of fades for example) that channels are not usable. Except in some mobile and military systems, channels are usually considered to be unavailable from the moment when error rate rises above 1 in 10^3 for 10 consecutive seconds until such time as error rate has fallen below 1 in 10^3 for 10 consecutive seconds.

In practice, error-performance objectives are usually a compromise between the performance that the user wants and what he is prepared to pay.

Early error-performance objectives set by CCIR for an international telephony channel between two earth stations were that BER should be better than:

1 in 10^6 (10 min mean value) for at least 80% of any month
1 in 10^4 (1 min mean value) for at least 99·7% of any month
1 in 10^3 (1 s mean value) for at least 99.99% of any year

A typical objective for business systems (e.g. the INTELSAT IBS and the EUTELSAT SMS systems) is that the BER shall be better than 1 in 10^{10} for at least 99% of the year; it is usually necessary to provide error-correction in order to achieve this objective.

Objectives set by CCITT for the end-to-end performance of a hypothetical long-distance 64 kbit/s circuit-switched connection in the ISDN (integrated-services digital network) are:

(a) *Degraded minutes*: Fewer than 10% of 1 min intervals shall have a mean BER worse than 1 in 10^6
(a) *Severely-errored seconds*: Fewer than 0·2% of 1 s intervals shall have a BER worse than 1 in 10^3
(c) *Errored seconds*: Fewer than 8% of 1 s intervals shall have any errors

The complete hypothetical ISDN connection is considered to comprise:

(i) Two 'local-grade' links between users premises and the local exchanges
(ii) High-grade links which are long-haul national and international connections
(iii) Two 'medium-grade' national links from the local exchanges into the national system, i.e. those parts of the link not included in (i) and (ii).

15% of the degraded minutes and errored seconds are allocated to each of the two local-grade links and each of the two medium-grade links (making 60% in all). This leaves 40% for the long-haul national and international links; half of this (i.e. 20%) is usually allocated to a link via a satellite system where this forms the international connection.

Thus two of the objectives for a 64 kbit/s satellite channel forming part of an ISDN connection are:

> That BER should be not worse than 1 in 10^6 (1 min mean) for at least 98% of the time
>
> That 98·4% of seconds should be free of errors.

A third objective for a satellite channel is derived in the following way. The 0·2% of severely-errored seconds are divided into two lots of 0·1%. One of these lots is divided up between the various links of the overall connection in the same ratio as for the first two objectives, i.e. a total of 0·06% to the local-grade and medium-grade links and the remaining 0·04% to the high-grade connection (of which 0·02% is usually given to the satellite link). The other 0·1% is given entirely to the long-haul and international link in recognition that this may include a number of terrestial radio links and a satellite link, each of which may suffer severe degradation for a small percentage of the time when propagation conditions deteriorate. After consideration of the fading statistics of terrestrial radio systems and satellite systems CCITT has given one-tenth of this allowance (i.e. 0·01% of the time) to the satellite system. Thus the error ratio for the connection via the satellite system should be worse than 1 in 10^3 for fewer than (0·02 + 0·01)% of 1 s intervals and the third objective may be restated as:

> The BER should be better than 1 in 10^3 (1 s mean) for more than 99·97% of the time.

The satellite system considered above may be assumed to be between two International Switching Centres (ISCs) and the objectives may well be modified. Objectives for satellite links between other points in the ISDN network are still to be determined.

Chapter 6

Maritime and other mobile services

6.1 Growth of maritime radio communications

Radio was first used for commercial maritime communications in 1899 when a message (in Morse code) was transmitted by a Marconi station on the Isle of Wight (UK) to the US liner *St Paul.* This was only two years after Marconi first demonstrated radio communication to the British General Post Office.

During the first decade of the twentieth century the use of radiotelegraphy to communicate with ships became well established. The value of radio in emergencies at sea was quickly recognised and, in 1906, the signal SOS was adopted as the international distress signal.

Two-way voice communication between a ship and a shore station was first established in 1922 and 'Mayday' (from the French 'm'aidez') was adopted as the distress call for use with radiotelephony in 1927. However, radiotelephony using the high-frequency (HF) band has serious disadvantages; the vagaries of propagation via the ionosphere and the limited bandwidth available can result in serious congestion, and delays of hours or even days in establishing voice communication with a ship are by no means uncommon when using the HF service. HF (Morse) telegraphy transmissions require only very narrow bandwidth and are much less liable to meet with congestion, but telegraphy has the disadvantage of slow speed of information transfer. For ships within 30 to 40 miles of the coast a reliable service of good quality can be provided by the use of VHF radio but there was no really satisfactory means of communicating with ships over long distances until the advent of the first maritime communication-satellite system (MARISAT) in 1976.

The number of ships using satellite communications is now (1988) about 5000. However, a ship earth station (SES) for the original (Standard-A) system costs about $30 000; because of this relatively high price about 90% of deep-sea ships still rely on HF radio for long-distance communications and most of these ships use morse telegraphy. INMARSAT (the International Maritime Satellite Organisation) is therefore introducing a new system, the Standard-C system (see Section 6.6), which will provide a store-and-forward message service. The new system should become fully operational in 1989 and it is expected that a Standard-C SES will cost $5000 or less and that several tens of thousands of ships will be fitted with these SESs in the early 1990s.

INMARSAT also proposes to introduce a Standard-B system which will use all-digital communication methods and will be better matched to modern digital-communication networks than the Standard-A system. When the Standard-B system becomes operational (probably in the early 1990s) no more Standard-A SESs will be installed, although Standard-A service will continue to be available for a number of years.

6.2 First steps towards an international maritime satellite service

It is not surprising, considering the limitations of HF radio, that the possibility of using satellites to communicate with ships received early study. In 1966, the year after INTELSAT I began service, IMCO (the International Maritime Consultative Organisation which later dropped the word 'Consultative' from its title and became IMO) appointed a panel of experts to study the technical, economic and organisational problems associated with establishing a maritime satellite system. It did not take long for the panel to decide that maritime satellite communications:

(i) Was feasible with existing technology
(ii) Could provide reliable circuits of good quality
(iii) Would do away with the delays associated with the existing long-distance HF telephony service
(iv) Would make it possible to provide new services to ships at sea.

It was not so easy to decide whether a maritime satellite system could pay its way at that stage of development of communications satellites. The main problem was, as is usual in trying to assess the demand for a new service, that the potential customers wanted to know what the costs would be before they would say if they were likely to use the service and conversely the potential suppliers wanted to know the size of the market before they were prepared to make a guess at what they might charge.

Most people were convinced that a maritime communications-satellite system ought to be run by an international body along the lines of INTELSAT. However, it usually takes a long time to launch a new international organisation and INMARSAT was no exception. It was 1973 (seven years after the panel of experts was formed) before the IMCO Assembly convened the first of a series of international conferences to consider the matter and 1976 before the last of these conferences adopted a convention and operating agreement for INMARSAT. The convention said, amongst other things, that:

(i) 'The purpose of the Organisation is to make provision for the space segment necessary for improving maritime communications, thereby assisting in improving distress and safety of life at sea communications, efficiency and management of ships, maritime public correspondence services and radio-determination capabilities.'

(ii) 'The Organisation shall seek to serve all areas where there is a need for maritime communications.'

(iii) 'The Organisation shall act exclusively for peaceful purposes.'

(iv) 'The INMARSAT space segment shall be open to use by ships of all nations on conditions to be determined by the Council. In determining such conditions, the Council shall not discriminate among ships on the basis of nationality.'

(v) 'The Council may, on a case-by-case basis, permit access to the INMARSAT space segment by earth stations located on structures operating in the marine environment other than ships, if and as long as the operation of such earth stations will not significantly affect the provision of services to ships.'

(*Note:* INMARSAT has gone further than was envisaged at the time when the Convention was written. It has approved the use of SESs on offshore drilling rigs and oil production platforms, and by disaster relief agencies; it is carrying out trials of an aeronautical communications-satellite service and it is considering the possibility of providing land-mobile communications.)

The adoption of the convention did not mean that INMARSAT was now a reality because, as is customary in international matters of this sort, a lengthy period (in this case up to three years) was allowed for potential participants to make up their minds to commit themselves and it was agreed that at least 95% of the investment shares must be taken up before INMARSAT could start work.

6.3 MARISAT, the first maritime satellite-communications system

While IMCO was studying the question of an international maritime system a number of other developments were taking place. Firstly, several countries were experimenting with maritime satellite communications, secondly the World Administrative Radio Conference for Space Telecommunications (WARC–ST) 1971 allocated frequency bands for the use of maritime communications-satellite services and lastly COMSAT developed the first operational maritime communications-satellite system (MARISAT).

MARISAT came into existence when it did because the US Navy decided to use satellites for communications with its ships and then ran into serious delays in its programme. The Navy asked COMSAT to provide temporary facilities and COMSAT's studies showed that the Navy's requirements would not make full use of the satellite capacity available. It was therefore decided to develop a system offering commercial as well as military communications. This was, for COMSAT, a very satisfactory way of side-stepping the question of whether a purely civil system could pay its way but it was not a solution that was likely to commend itself to those who wanted an international system in the tradition of INTELSAT.

The MARISAT system became operational in 1976 when three satellites were placed in position over the Atlantic, Pacific and Indian Oceans. Three shore stations were also established, at Southbury and Santa Paula (USA) to communicate with ships via the Atlantic and Pacific satellites, respectively, and at Yamaguchi (Japan) to communicate with ships via the Indian Ocean satellite. The system gave excellent service from the start, the number of civil ships equipped to use the system grew rapidly and the success of MARISAT encouraged many hitherto undecided nations to sign the INMARSAT convention.

6.4 INMARSAT Standard-A system

6.4.1 INMARSAT becomes operational

The INMARSAT Convention finally entered into force at midnight 15/16 July 1979.

Although it took such a long time and such a lot of effort to establish INMARSAT there is no doubt that it was worthwhile and that those who overcame the many problems of founding and developing the organisation deserve our gratitude. INMARSAT has been very successful and is an excellent example of how nations with diverse systems of government and beliefs can co-operate. There were twenty-eight founder members including some from every continent; those members with an investment share of over 5% being USA, USSR, UK, Norway and Japan. Over 50 countries are now members.

The first Director General of INMARSAT was appointed in January 1980. The most pressing task facing the organisation was the procurement of the space segment. A request for proposals was issued with commendable speed and in May 1980 INMARSAT received offers of the lease of satellites from MARISAT and the European Space Agency (ESA), and an offer from INTELSAT of the lease of maritime communications subsystems (MCSs) which would be added to INTELSAT V satellites.

It did not make sense to have two competing maritime communications-satellite systems so there was little doubt that, if suitable terms could be agreed, the INMARSAT system would start by taking over the MARISAT satellites (which by now were serving only civilian ships as the US Navy had overcome the difficulties of establishing its own system). However, additional satellites were required as spares in orbit and also to replace the MARISAT satellites, because the latter had limited capacity and would be approaching the end of their design life by the time INMARSAT commenced operations in early 1982. The outcome was that INMARSAT placed contracts for the immediate lease of the three MARISAT satellites and provided for the future by ordering two MARECS satellites (from ESA) and taking an option on up to four MCSs.

The earth segment of the INMARSAT Standard-A system comprises the ship earth stations (SESs), the coast earth stations (CESs) which provide the interfaces to the public telephone and telex networks, and the network co-ordinating stations (NCSs) which are responsible for network management and act as intermediaries in the allocation of communication channels. All these stations must comply with INMARSAT technical specifications and approval procedures.

About 30% of the (approximately) 5000 Standard-A SESs in service are on oil tankers and about 70% are on ships over 10 000 tons (this includes virtually all the tankers); at the other end of the scale there are several hundred Standard-A SESs on ships of under 1600 tons (many of these are on specialised ships such as seismic and research vessels) and a few hundred terminals on platforms and vessels engaged in off-shore oil production. There are, however, many tens of thousands of ocean-going ships (nearly all under 10 000 tons) which are not yet equipped for satellite communications; most of these ships should be able to afford Standard-C SESs and it is expected that many large ships which already have Standard-A terminals will add a Standard-C SES.

In the MARISAT system there were only three CESs, one for each ocean area; this meant, for example, that all calls between ships in the Atlantic and customers in Europe had to be routed via the CES at Andover (USA) thus adding the cost of a transatlantic call to the cost of the call between the CES and the ship. In the INMARSAT system there are (1988) over 20 CESs and more are planned; the terrestrial part of the circuit on most calls via this system is therefore very much shorter than was the case with the MARISAT system.

Maritime satellite communications has been allocated exclusive use of frequency bands at 1·5 and 1·6 GHz (L-band) for transmissions from satellites to SESs and SESs to satellites, respectively. Frequency bands at 4 GHz and 6 GHz (C-band) are used for transmissions from satellites to CESs and from CESs to satellites; these bands have to be shared with the fixed satellite service. Each INMARSAT satellite contains two transponders: one for receiving transmissions from the CESs in the 6 GHz frequency band and down-converting them to 1·5 GHz for transmission to the ships (the C-to-L transponder) and the other for receiving transmissions from the ships at 1·6 GHz and up-converting them to 4 GHz for transmission to the CESs (the L-to-C transponder).

L-band is better than C-band for transmission between satellites and ships for two reasons, viz:

(i) Free-space path loss and rain attenuation are lower at L-band than at C-band and this partly offsets the low G/T of SESs (which have to use small antennas)

(ii) An L-band antenna has a wider beamwidth than a C-band antenna of the same collecting area; L-band antennas do not therefore have to be steered as accurately as the equivalent C-band antennas.

SESs work only to CESs and are therefore equipped only for operation at L-band (ship-to-ship calls can be made but they have to be relayed by a CES). CESs have to be able to work at both C-band and L-band; they use C-band frequencies for communication with ships, L-band frequencies for communication with NCSs and other CESs, and they transmit and receive pilot signals in both bands for purposes of automatic frequency control (AFC).

6.4.2 Services available

The types of call available via the INMARSAT Standard-A system include:

(i) Telephony
(ii) 50 baud telex
(iii) Data at up to 9600 bit/s, facsimile and slow-scan television (all via telephony channels)
(iv) High-speed (56 kbit/s) data (HSD) in the ship-to-shore direction only, and very high-speed data (VHSD) using rates up to 1 Mbit/s; these services require greater EIRP than is needed for transmission of a voice channel
(v) 'Broadcast' telex or voice messages to groups of ships (e.g. to all ships in a particular ocean area, all ships of a specific nationality or all ships in a specific fleet).

Most of the services available via the network of a country owning a CES are available to ships through that CES (provided that the INMARSAT channels will support them) and some additional services are provided specifically for ships. Examples of services available through most CESs are:

(*a*) Automatic calls using the international number
(*b*) International operator assistance
(*c*) Collect calls
(*d*) Person-to-person calls
(*e*) Technical assistance
(*f*) SES commissioning tests
(*g*) Advice on navigational hazards and
(*h*) Medical advice or assistance

Services are available by dialling a two-digit code which is the same in every ocean area.

A SES can make a very important contribution to crew welfare and to safety at sea. Every SES is fitted with a distress facility which gives it immediate access to a CES and, through the CES, to a Rescue Co-ordination Centre (RCC). In most maritime communications systems messages may be classed as distress, urgent or safety and all these categories take precedence over ordinary traffic. In the Standard-A service, delays are not expected to occur under normal circumstances and traffic is therefore divided into only two categories: priority and ordinary. Priority traffic pre-empts ordinary traffic and will seize a busy circuit if necessary.

As well as providing run-of-the-mill communications, the INMARSAT system is being used in ways that are impractical with other maritime communications systems. For example:

(i) Ships can be diverted, at short notice, to pick up cargoes from ports near their route; delay in getting in touch with a vessel without a SES can lead to loss of potential spot contracts.
(ii) Instant communication with ships carrying out operations such as seismic surveys allows data to be sent back to headquarters for analysis as soon as it is collected; this means, for example, that an unprofitable line of investigation can be dropped much more quickly than is practicable with other means of communication.
(iii) Data on the performance of a ship can be gathered by an automatic monitoring system and relayed back to shore, where it can be used to improve the scheduling of maintenance and the efficiency of fleet management.

6.4.3 Setting up a call

6.4.3.1 Ship-to-shore calls

Most ships have only occasional need for telecommunications services; circuits via the INMARSAT system are therefore usually assigned only on demand. When a ship wants to make a call it sends a message on the request channel. The message is in the form of a short burst of data (see Fig. 6.1) which includes the identity of the ship, the type of channel required and the identity of the CES through which the ship wants to make its call. All the CESs in the ocean area monitor the request channel but only the station addressed responds to a specific request.

The request channel is a random-access time-division multiple-access (TDMA) channel (see Section 5.8.4). If request messages from two ships overlap in time they will interfere with one another. An error-detection code is therefore added to the messages so that CESs are able to detect these overlaps. When CESs detect an overlap they do not reply to the requests and each ship repeats its message (after waiting for an interval, which it selects at random in order to avoid a further collision). Two request channels are provided and ships use them alternately; this ensures that failure of a request channel does not put the system out of action. The random-access request channels work well provided that each channel does not have to handle more than about five requests per second; however, channels which are overloaded

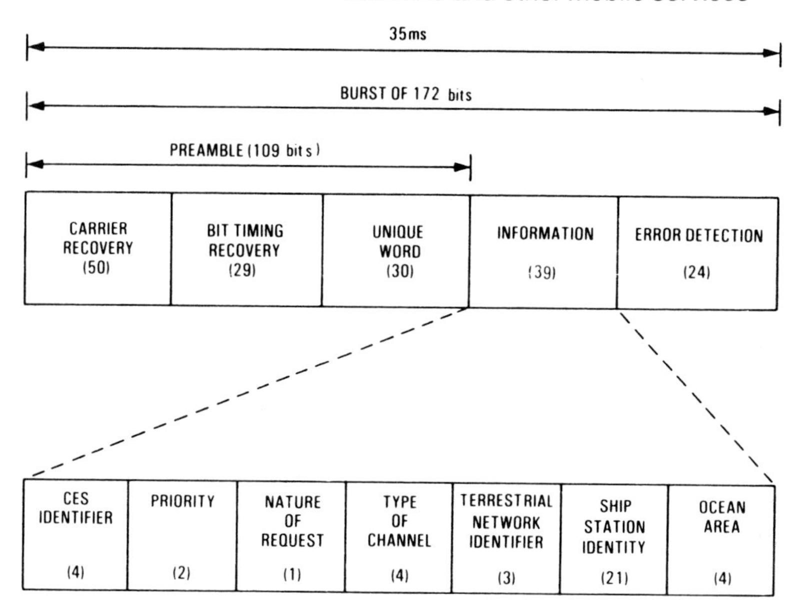

Fig. 6.1 *Format of a request message (Reproduced from British Telecom Engng, 1982,* **1***)*

1 The data rate is 4800 bit/s
2 The error detection coding is BCH 63,39
3 The figures in brackets indicate the number of bits allocated
4 The terrestrial network identifier is necessary only where there is a choice of common carrier

become inefficient because of the large number of collisions, and long delays may occur.

6.4.3.2 Assignment of channels

A request message correctly received by a CES is passed to the area network co-ordinating station (NCS) and the NCS sends a reply to the ship. This reply, called an assignment message, gives details of the channels to be used for the call and is transmitted on a common TDM channel. This common TDM channel is monitored by all ships and all CESs in the ocean area.

In the case of telex calls the CES determines what message channels are to be used but passes the information to the ship via the NCS; in the case of telephone calls the NCS allocates the message channels as well as sending the information to the ship. The reason for the different procedure in the two cases is that different methods of transmission are used for the two types of call.

6.4.3.3 *Telex calls*

Each CES has its own TDM carrier (or carriers) for telex transmissions to ships; each of these TDM carriers provides twenty-two 50 baud telex channels and a signalling channel (see Fig. 6.2).

The telex transmissions in the return direction (i.e. from ships to a CES) form a TDMA assembly at the satellite (see Fig. 6.3). These TDMA transmissions are not made at random; each SES is allocated a time slot to which it has access at regular intervals and synchronism of the transmissions is maintained by means of a timing signal transmitted by the CES in the TDM carrier.

Each frame of a TDMA telex carrier is divided into 22 time slots and each of these slots is paired with one of the 22 time slots of a corresponding TDM carrier; one of these pairs of time slots is allocated to a ship on receipt of a request for a telex call.

6.4.3.4 *Telephone calls*

A telephone call uses two narrow-band frequency-modulated (NBFM) radio-frequency carriers, one for the ship-to-shore transmission and one for the shore-to-ship transmission. If all the bandwidth and power of a MARECS satellite were to be devoted to telephone channels it could carry about 60 channels in the shore-to-ship direction or 80 with partial carrier suppression (see Section 6.4.6), the corresponding figures for a MCS package on an INTELSAT V satellite being 35 and 50. Some of the power and bandwidth must, however, be reserved for TDM and TDMA transmissions, so the actual number of telephone circuits available is significantly reduced.

With so few channels available it is impracticable to give each CES exclusive use of a group of channels. When a CES receives a request from a ship for a telephone call it must therefore ask the NCS to allocate a pair of channels from the pool. It does this by means of a message on a TDM channel and the NCS replies with an assignment message on its common TDM channel. On receipt of this message both ship and CES tune their equipment to the allocated channels and, after an automatic continuity check, the CES sends a 'proceed-to-select tone' which tells the caller that the number may now be dialled. Any further signalling is in-band (i.e. sent in the telephone channel) and the CES translates the signals used on the INMARSAT system to those required by the terrestrial network to which it is connected (and vice versa).

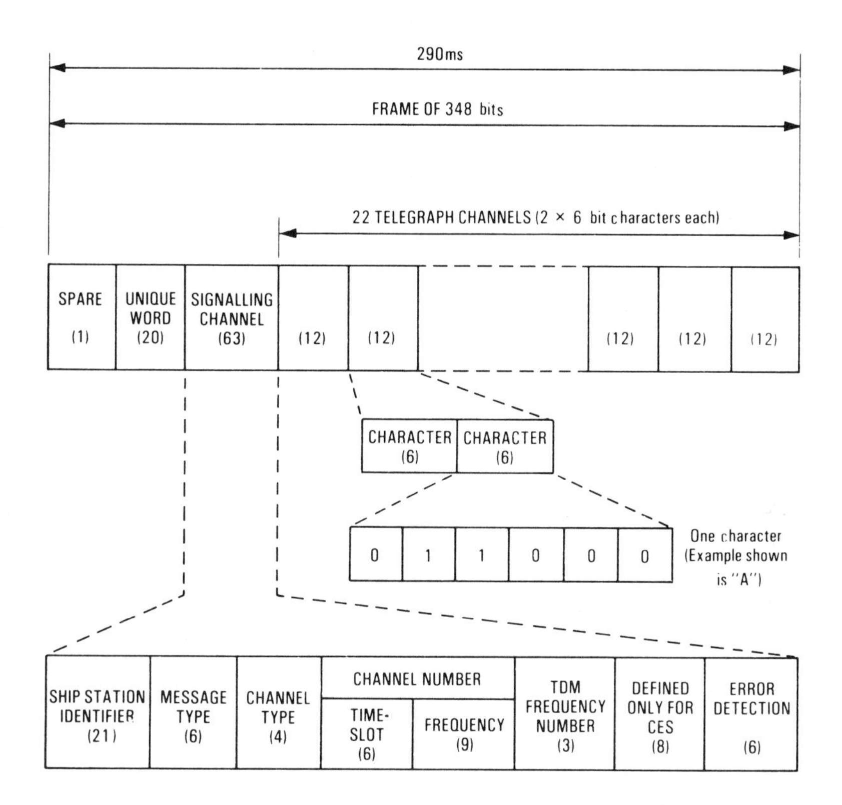

Fig. 6.2 *Format of a TDM carrier (Reproduced from British Telecom Engng., 1982,* **1***)*

1 The data transmission rate is 1200 bit/s; the telex character rate is 151 character/s
2 In the telegraph channel, the first bit transmitted indicates the type of character field. When the first bit is 0, the subsequent 5-bit character field represents an international telegraph alphabet (ITA) no. 2 character; when the first bit is 1, the subsequent 5 bits represent line conditions for signalling
3 The figures in brackets indicate the number of bits allocated
4 The error detection coding is BCH 63, 57
5 The signalling channel is used in the CES's own TDM channel for *request-for-assignment* messages to the NCS. In the common TDM channel transmitted by the NCS it is used for the *channel-assignment* message to the SES and the CES
6 The unique word is inverted every sixth frame

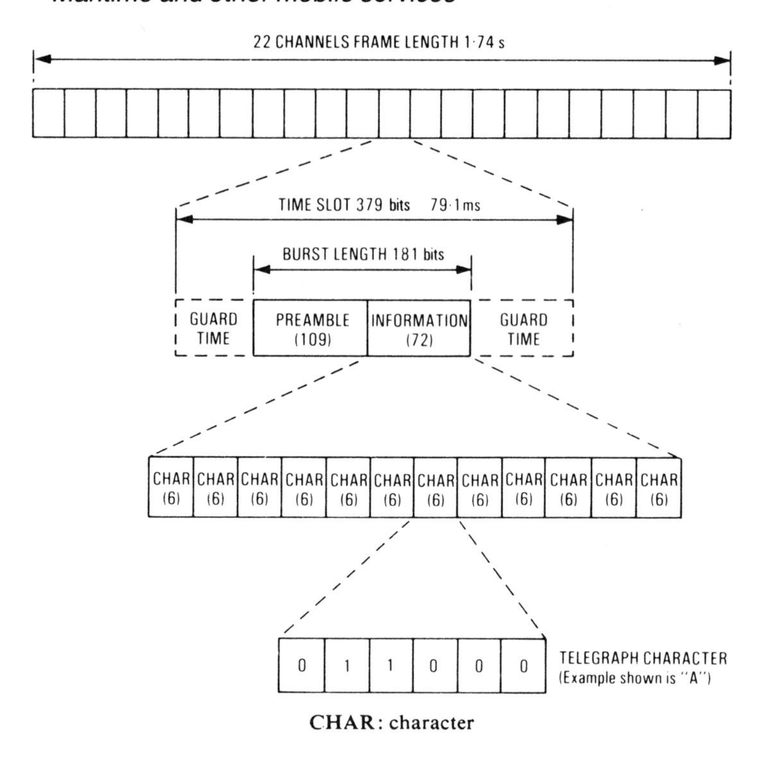

Fig. 6.3 *Format of a TDMA transmission (Reproduced from British Telecom Engng., 1982,* **1***)*

1 The data transmission rate is 4800 bit/s; the telex character rate is 151 character/s
2 In a telegraph channel, the first bit transmitted indicates the type of character field. When the first bit is 0, the subsequent 5-bit character field represents an international telegraph alphabet (ITA) no. 2 character. When the first bit is 1, the subsequent 5 bits represent line conditions for signalling
3 The preamble includes carrier recovery, bit timing recovery and the unique word, in a similar manner to the request message

6.4.3.5 Shore-to-ship calls

For calls from shore to ship the procedures are very similar. For telex calls the CES allocates the time slots to be used in its TDM and TDMA carriers; for telephone calls the request is passed to the NCS which allocates SCPC channels. Assignment messages are transmitted by the NCS on the common TDM channel (in the same way as for calls initiated by ships).

6.4.4 Ship earth stations (SESs)

6.4.4.1 ADE and BDE

The components of a ship's terminal (see Fig. 6.4) are usually categorised as above-deck equipment (ADE) and below-deck equipment (BDE). The ADE comprises the antenna and its mounting, the radome (which protects the equipment from the weather and the corrosive effects of salt spray), the high-power amplifier (HPA) and the low-noise amplifier (LNA). The latter two units have to be mounted on or adjacent to the antenna in order to minimise losses and thus get good performance at an acceptable price. The BDE comprises up-converter and down-converter units, frequency synthesisers, equipment for processing the signals (e.g. IF amplifiers, modems and timing circuits), a display and control unit, and one or more terminal

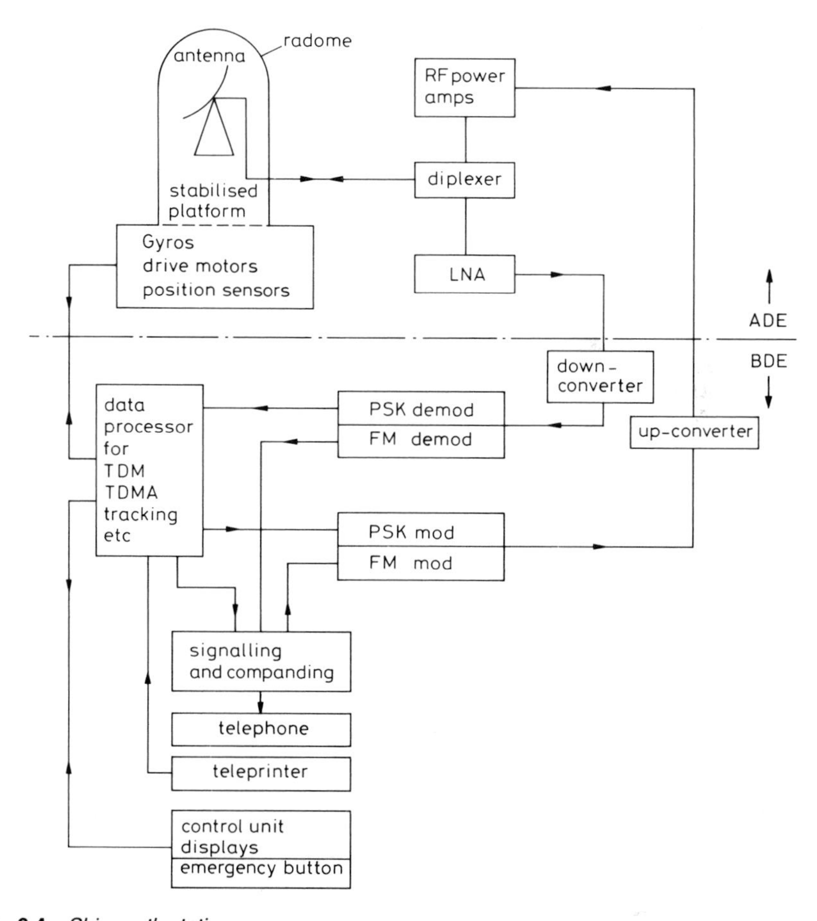

Fig. 6.4 *Ship earth station*

equipments (e.g. a telephone or telex machine, and ancillary units such as facsimile machines). Emergency push buttons for activating the priority distress channel can be mounted wherever they may be needed.

It is an advantage if all the equipment is compact since space is often at a premium even on relatively large ships. The antenna must be mounted in a position where, as far as possible, it has an unobstructed view of the satellite whatever the ship's course and even when the vessel is rolling and pitching in heavy seas. This usually means putting the ADE high up on the superstructure and it is therefore essential that this equipment shall be small and lightweight but robust; if the unit is too bulky or heavy to be mounted on the existing superstructure then installation may well prove expensive. Finding a good position for the antenna can be difficult but the problem may sometimes be simplified by taking account of the routes used by the ship and choosing a position which is satisfactory for present usage but may have to be changed if the ship sails to new destinations.

6.4.4.2 *Tracking and stabilisation*

It is not always easy to keep the antenna of a Standard-A SES pointing at the satellite because of the effects of:

(i) Roll, pitch and yaw which, for small ships in rough seas, may reach as much as $\pm 30°$, $\pm 20°$ and $\pm 5°$, respectively
(ii) Changes in the ship's heading
(iii) Changes in azimuth and elevation of the satellite with movement of the ship.

A common method of counteracting the effects of roll and pitch is to mount the antenna on a platform which is gyroscopically stabilised by two flywheels spinning on orthogonal axes; this can reduce the angular excursions by a factor of 50. The changing heading of the ship may be counteracted by linking the drive of the platform to the ship's compass and the antenna may be made to follow the changing direction of the satellite resulting from the changing latitude and longitude of the ship (and satellite motion) by the provision of a step-track system (see Section 7.3.5).

The INMARSAT system uses circular polarisation so that ships' antennas do not have to track the plane of polarisation.

6.4.4.3 *Characteristics*

The use of antennas with diameters as small as 0·9 m has been made possible by the availability, at reasonable cost, of gallium arsenide LNAs with good performance; this has led to reductions of up to 50% in the size and weight of ADE relative to early generations of equipment.

Table 6.1 gives typical characteristics of a Standard-A SES.

Table 6.1 *Characteristics of a Standard-A SES*

Antenna	
Diameter	0·9–1·2 m
Gain	21–23 dB
Half-power beamwidth (±)	6–7°
EIRP	36 dBW
G/T (at 5° elevation with a clear sky)	− 4 dB/K

6.4.4.4 Types of SES

There are three types of standard-A SES:

(a) Class 1 stations, which carry telephone and telex traffic
(b) Class 2 stations, which carry telephone traffic only
(c) Class 3 stations, which carry telex traffic only.

6.4.4.5 Maintenance and repair

It is essential that terminals shall be as reliable as practicable but it is inevitable that they will need maintenance and repair from time to time. The provision of a maintenance and repair service for SESs is a particularly difficult problem because stations in need of urgent repairs may turn up at harbours anywhere in the world. Most large manufacturers of SESs have arranged with local companies in the larger ports and harbours to provide service using personnel trained in the manufacturer's factory and this service is backed up, where necessary, with a flying squad of the manufacturer's own technicians.

6.4.5 Coast earth stations (CESs)

Fig. 6.5 is a block diagram of a CES.

6.4.5.1 Antenna(s)

The figure of merit *G/T* required at the C-band frequencies is specified as not less than 32 dB/K. This requires an antenna with a diameter of at least 10 m. It may, however, be necessary to use a larger antenna in order to achieve the required EIRP, taking into account the maximum number of carriers to be transmitted, the output power of available transmitters, the required back-off (if a number of transmissions are to be amplified by a single output stage), the

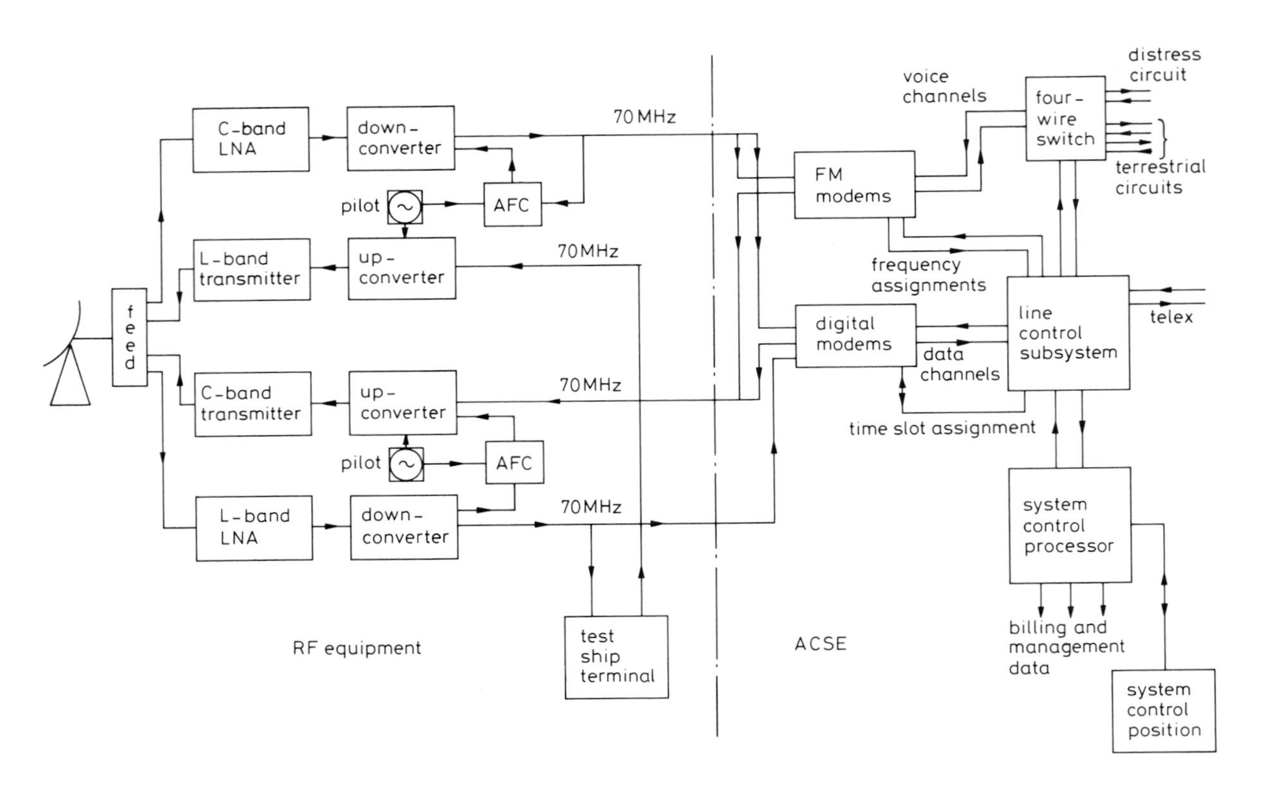

Fig. 6.5 *Coast earth station*

losses introduced by any transmitter combining networks which may be necessary and the diameter of existing antenna designs. For example, the antenna of the British Telecom International CES at Goonhilly Earth Station (which works at both C-band and L-band) has a diameter of 14·2 m; this enables the simultaneous transmission of up to about 40 C-band carriers through a single 3 kW HPA but the precise size of the antenna was determined by the fact that it is an adaptation of an existing design.

A single antenna may be used to transmit and receive the L-band as well as the C-band signals or a separate L-band antenna can be used. The provision of a separate L-band antenna avoids the need for a relatively complex feed system (to combine and separate the outgoing and incoming L-band and C-band signals) but this advantage must be weighed against the cost of procuring and installing a second antenna.

6.4.5.2 L-band functions

The CES needs to be able to receive and transmit L-band signals for three reasons:

(i) For communication with the NCS and other CESs; this requires the ability to transmit and receive L-band signals because the satellite transponders convert C-band signals to L-band (and vice versa) and it is therefore impossible to provide direct connection between two C-band terminals.
(ii) Because the CES provides automatic frequency control (AFC) for both ship-to-shore and shore-to-ship transmissions. Making the CES responsible for AFC in the ship-to-shore direction, as well as the shore-to-ship direction, helps to keep the SESs as simple and as cheap as is practicable.
(iii) So that a complete test of the CES equipment can be carried out without the co-operation of a ship earth station (a test terminal is provided at each CES for this purpose).

6.4.5.3 AFC

The INMARSAT system requires automatic frequency control (AFC) to correct for Dopper shift (the chief cause of which is inclination of the satellite orbit) and errors in frequency translation in the satellite and earth stations. The total frequency shift from these causes without AFC could be more than 50 kHz; this is of the same order as the spacing between the NBFM channels and would be enough to cause failure of the system.

The AFC reduces the frequency shift to a few hundred hertz by comparing pilot carriers transmitted via the satellite with reference oscillators at the CES and using the difference signals to control the frequencies of the local oscillators associated with the up-converter and down-converter (see Fig. 6.5). A pilot transmitted at C-band and received at L-band is used to control the up-converter and thus offset the frequencies of the operational carriers to compensate for Doppler shift and satellite frequency translation errors in the shore-to-ship direction. Similarly, a pilot transmitted at L-band and received at C-band is used to control the downconverter and correct for Doppler shift and errors in translation in the ship-to-shore direction. All frequency errors are corrected except those arising from frequency instability of the SES up-converter and down-converter, and Doppler shift resulting from the relative velocity of satellite and ship.

6.4.5.4 ACSE

A CES must include access, control and signalling equipment (ACSE). The main purposes of the ACSE are:

(a) To recognise requests for calls, and to set up and release circuits; this requires response to and initiation of in-band and out-of-band signalling over satellite and terrestrial paths.
(b) To recognise distress calls and pre-empt channels for them when necessary.
(c) To check that ships are on the list of authorised users and to bar calls (except distress calls) from or to unauthorised ships.
(d) To switch voice circuits between terrestrial circuits and the CES FM channel modems.
(e) To switch telex circuits between terrestrial channels and the TDM and TDMA time slots.
(f) To determine the transmit and receive frequencies used by the FM channel units in accordance with the channel allocations made by the NCS
(g) To allocate TDM and TDMA timeslots
(h) To collect statistics for billing, space-segment accounting, international accounting (for transit calls), traffic analysis, management and maintenance purposes.

The ACSE is sometimes specified to include all the communication equipment other than the RF and IF equipment (e.g. in Fig. 6.4 the ACSE includes the modulators and demodulators).

6.4.6 The space sector

The MARISAT satellites have been retired from service. Two MARECS satellites and four Maritime Communication Subsystems (MCSs) on INTELSAT V satellites now provide operational and spare transponders for the Atlantic, Indian Ocean and Pacific Ocean Regions.

Fig. 6.6 shows the coverage given by the system. Virtually all major shipping routes are served but geostationary satellites do not give coverage of

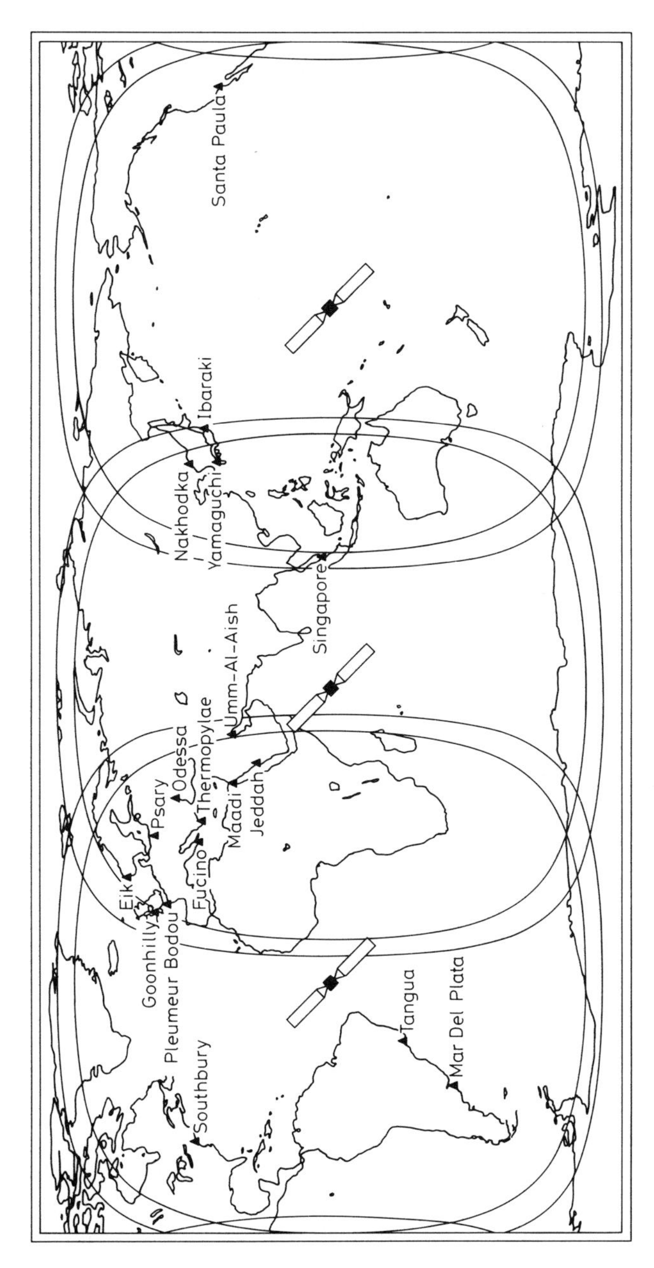

Fig. 6.6 *Areas served by the INMARSAT system showing 0° and 5° elevation contours (Acknowledgment to INMARSAT)*

Arctic and Antarctic waters. Coverage at all latitudes would be possible if satellites in highly-inclined orbits were used. Such satellites are cheaper to launch than geostationary satellites but more of them would be required to give continuous communication and all stations in the system would periodically have to transfer service from one satellite to another.

Both the MARECS satellite and the MCS comprise a single transponder in each direction, i.e. a ship-to-shore (L-band to C-band) transponder and a shore-to-ship (C-band to L-band) transponder.

The transponders are associated with two C-band horn antennas, one for reception of the 6 GHz transmissions from the CESs and one for transmission to the CESs at 4 GHz. A single L-band antenna is used for both reception and transmission; MCS uses a conical helix array and MARECS uses a parabolic antenna with a diameter of 1·8 m. The C-band antennas and the helix have a gain at beam edge of about 16 dB. The MARECS L-band antenna uses beam-shaping techniques to give maximum gain (about 18 dB) at beam edge, i.e. where the path loss is greatest; this reduces the variation in flux density over the coverage area.

Some characteristics of the transponders are given in Table 6.2.

Table 6.2 *Characteristics of MARECS and MCS transponders*

		MARECS	MCS
C-band to L-band repeater			
Receive band	(MHz)	6420·0–6425.0	6418·0–6425·0
Transmit band	(MHz)	1537·5–1542·5	1535·0–1532·5
G/T	(dB(1/K))	−15	−12
EIRP	(dBW)	34·5	33·0
Capacity (voice channels)		60	35
L-band to C-band repeater			
Receive band	(MHz)	1638·5–1644·0	1636·5–1644·0
Transmit band	(MHz)	4194·5–4200·0	4192·5–4200·0
G/T	(dB(1/K))	−11·2	−13·0
EIRP	(dBW)	16·5	20·0
Capacity (voice channels)		90	120

Although both the MARECS satellites and the Maritime Communications Subsystems have been designed so that their EIRP at L-band (i.e. in the direction of the ships) is much greater than their EIRP at C-band (i.e. in the direction of the CESs) the capacity of the system is still limited by the satellite-to-ship link; this is mainly because of the small collecting aperture of the SES antennas.

Better use can be made of satellite power if telephony carriers are suppressed when subscribers are not speaking (i.e. the carriers are voice-activated). However, the automatic gain control of a receiver does not work while the carrier is suppressed and channel noise can therefore rise to an annoying level during pauses in the speech unless the receiver includes a muting circuit. Many existing SESs do not have such a circuit and it is therefore impracticable to use full carrier suppression in the Standard-A system. Nevertheless, it is possible to partially suppress the carriers (i.e. reduce their level when speech is absent) without perceptible degradation of channel quality and this can give a useful gain in traffic capacity. By this means it is practicable to raise the capacity of the MARECS C-band to L-band repeater from 60 to 80 channels and the capacity of the corresponding MCS repeater from 35 to 50 channels. The second-generation satellite has been designed so that its capacity in both directions is approximately equal when carrier suppression is used (see Section 6.5).

6.4.6.1 Quality of the channels

Syllabic companding (see Section 4.5) is used to provide a subjective improvement in the signal-to-noise ratio of the INMARSAT telephone circuits.

CCIR recommends that the noise power in a telephone circuit of a maritime communications-satellite system shall be not greater than the subjective equivalent of 25 000 pW0P when there is no fading on the link and the satellite is at an elevation of 10° relative to the SES and states that this criterion can be achieved with a carrier-to-noise density ratio (C/N_0) of of 52–53 dB(H_3). The corresponding recommendation for telegraphy is that the satellite link shall not contribute more than eight errors in 100 000 characters, with a confidence level of 99%, when the SES is at the edge of the satellite coverage area; this corresponds to a bit error ratio (BER) of 1 in 100 000. The INMARSAT system is designed to meet these CCIR objectives.

6.5 INMARSAT second-generation and third-generation satellites

6.5.1 Second generation

INMARSAT had hardly settled down to using the MARECS satellites and the MCSs before it realised that they were likely to become overloaded in the late 1980s and it was therefore already time to think seriously about a new generation of INMARSAT satellites.

The main requirements for the second-generation satellites were summarised as:

(i) A considerable increase in capacity relative to MARECS and MCS.
(ii) The ability to operate satisfactorily with a range of mobile terminals and

with EPIRBs (emergency position-indicating radio beacons, see Section 6.7)

(iii) Compatibility with at least two launch vehicles.
(iv) Long life and the utmost reliability.
(v) Availability by 1988.

A contract for three new satellites (at a cost of about $150M) with an option for up to six more was placed in early 1985. The first of the new satellites should be delivered in 1988 and come into service in 1989. The requirement for the utmost reliability together with the relatively short period available for development and production dictated the use of technology already proven in the space environment; the second-generation satellite is therefore based on the bus used for MARECS and ECS and carries a conventional communications package with global-coverage antennas and no signal processing (other than amplification and frequency conversion). The satellite is comparable in size and power with INTELSAT V and will have a design life of ten years. It was specified that the satellite should be compatible with at least two of Ariane, Atlas–Centaur, the Shuttle, Proton, Thor–Delta and Titan, and it was designed to be compatible with the first three of these.

In the shore-to-ship direction there is a single transponder, as for MARECS and MCS. In the ship-to-shore direction, however, there are four transponders and the gain of each of these transponders may be adjusted independently, to adapt them for use with different services using different types of mobile terminal, e.g.:

(i) The Standard-A, Standard-B and Standard-C maritime services
(ii) High-speed data transmissions, the number of which is expected to grow rapidly
(iii) Aeronautical mobile service
(iv) EPIRBs.

The total EIRP of the L-band to C-band transponders (at 39 dBW) is about 4·5 dB more than that of MARECS. This gives a capacity equivalent to about 125 Standard-A voice channels (or about 250 channels using voice-activated carriers). For later satellites, ordered under the option, it will be possible to have a communications package with a mass about 50% greater than that on the first three satellites; this can be used to increase the capacity of the satellite or to introduce new features (or both). Ordering only three satellites in the first place has given INTELSAT further time to think about the best way to use this opportunity for further development of the payload.

INMARSAT will be responsible for control and monitoring of the satellites, and for monitoring of transmissions via the satellites. It has procured TTC&M equipment which will be installed at CESs and maintained by the CES owners; under normal conditions, however, the equipment will be operated by remote control from the INMARSAT satellite control centre (SCC) in London.

Table 6.3 summarises the most important characteristics of the satellite.

Table 6.3 *Characteristics of the second-generation INMARSAT satellite*

C-band to L-band repeater		
Receive band	(MHz)	6435·0–6441·0
Transmit band	(MHz)	1530·5–1546·0
G/T	(dB(1/K))	−14
EIRP	(dBW)	39
Capacity (Standard-A voice channels)		125 (250
L-band to C-band repeater		
Receive band	(MHz)	1626·5–1647·5
Transmit band	(MHz)	3600·0–3621·0
G/T	(dB(1/K))	−12·5
EIRP	(dBW)	24
Capacity (Standard-A voice channels)		250

6.5.2 *Third generation*

In its first twenty years of operations, INTELSAT satisfied the growing demand for satellite circuits between fixed stations by:

(i) Introducing successive generations of larger and more powerful satellites.
(ii) Using more than one satellite to serve each ocean area.
(iii) Reusing frequencies by means of orthogonally-polarised transmissions.
(iv) Using higher radio frequencies (where more bandwidth is available).

Unfortunately, INMARSAT cannot rely on the same methods. The reasons for this are:

(*a*) The second-generation INMARSAT satellites are already large and powerful but, because of the very low gain of mobile-station antennas, the traffic capacity of INMARSAT satellites is only a fraction of that of satellites with comparable power and bandwidth working to fixed earth stations.
(*b*) The small antennas of mobile stations have wide beamwidths and cannot therefore discriminate adequately against transmissions from nearby satellites using the same frequency band; thus the capacity of a mobile system cannot be increased by using a number of satellites to serve the same coverage area.
(*c*) Frequency reuse by means of orthogonally-polarised transmissions is

impracticable for maritime satellite systems. This is because signals are usually depolarised on reflection from the sea and the ship-station antennas could receive sufficient power from these reflected signals to cause unacceptable interference.

(*d*) The use of higher frequencies is probably impracticable in the time scale of the INMARSAT third-generation system because of the costs of developing and manufacturing the mobile terminals, and the effects of the higher precipitation losses; the use of higher frequencies would also raise problems of compatibility with existing terminals.

The most practicable method of providing increased capacity for maritime systems is probably the use of satellites carrying phased-array antennas generating multiple spot beams. This would give the advantages of both frequency reuse and much higher satellite EIRP, but the use of multiple spot beams introduces problems in setting up calls (similar problems have, of course, to be solved in terrestrial cellular mobile systems). On-board processing could be of help in routing and switching calls and could also make it practicable to establish direct links between mobile stations.

6.6 INMARSAT Standard-C system

6.6.1 Introduction

6.6.1.1 Objectives and services

The primary objective of the Standard-C system is to reduce the cost, size and weight of the shipborne equipment required for satellite communications and thus to make it practicable for small ships of all types to work to the INMARSAT system. The system should, for example, prove very useful to trawlers and other small fishing vessels.

An experimental Standard-C SES has been built and tested by INMARSAT. The complete station (excluding the antenna) measures $30 \times 22 \times 11$ cm and weighs about 6 kg (excluding the power supplies). INMARSAT has placed contracts for a test CES and SES, and for network co-ordinating stations and it is hoped that the new system will be operational by early 1989.

The gain of the SES antenna is not specified explicitly by INMARSAT but Fig. 6.7 shows how the minimum allowable G/T varies with elevation. It will be seen that the required gain is very low, varies by only a few decibels over the range of elevation angles from $-15°$ to $+90°$ and is independent of azimuth angle. This means that the ship-borne antenna can be non-steerable and non-stabilised and yet the G/T and EIRP are adequate for communication wherever the ship may be in the satellite coverage area and even in rough

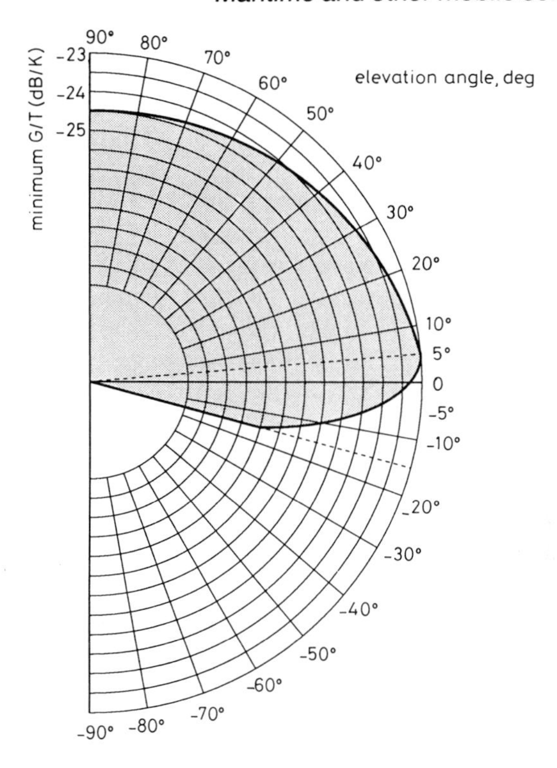

Fig. 6.7 *SES minimum G/T versus elevation (Acknowledgment to INMARSAT)*

1 Gain referred to a right hand circularly polarised isotropic antenna
2 Temperature in dB relative to 1K
3 No azimuth variation
4 G/T not defined for elevation −15 to −90°

seas. Such a low-gain antenna can be very small and light; this, together with the elimination of the requirement for tracking and pointing equipment and a stabilised platform, greatly reduces the cost of provision and installation of the SES, and improves reliability by reducing complexity.

The use of a ship-borne antenna with very low gain makes it impossible to provide voice communication without using uneconomic amounts of bandwidth and satellite power. The Standard-C system is therefore intended only for the transmission of text and data.

The main types of communication via the system will be telex messages to and from individual ships, and group-call messages; the latter may be messages to a selected group of ships (e.g. safety bulletins or weather forecasts directed to ships in a specific geographical area). All CESs must be able to handle telex transmissions and group-call messages, and be able to transmit and receive distress messages. The Standard-C system will also support services such as packet switching, voice-band data and teletext provided that the appropriate equipment and interfaces are available on the ship and at the CES.

All messages, whether received from ships or for transmission to ships, are passed via a store-and-forward message switch at the CES. Distress messages will be allocated the highest priority and the objective is that, under the worst conditions, 99.9% of these messages shall reach a rescue co-ordination centre within 55 s. Other messages are labelled 'immediate delivery', 'delayed delivery' or 'overnight delivery'. Different tariffs will be charged for the different classes of message and INMARSAT aims to give the following standards of service:

(i) *Immediate:* Average delay less than 5 min, 99.9% of messages passed on in less than 10 min
(ii) *Delayed:* Average delay less than 30 min, 99.9% of messages passed on in less than one hour
(iii) *Overnight:* Average delay less than six hours, 99.9% of messages passed on in less than twelve hours.

There are three basic types of Standard-C SES, viz:

(i) *Class 1:* This provides the basic functions of ship-to-shore and shore-to-ship message transfer.
(ii) *Class 2:* This type of SES has two alternative modes of operation which can be selected by the operator. In the first mode it behaves as a Class 1 terminal but can also receive group-call messages when it is not being used for ordinary traffic. In the second mode it can be set to continuous reception of group-call messages.
(iii) *Class 3:* This is Class 1 SES with the addition of a second receiver which allows continuous reception of group-call messages even when the terminal is being used for the transfer of ordinary messages.

All SESs are, of course, able to transmit distress messages.

A Standard-C SES does not need a skilled operator. Terminals can be used for the automatic reception of safety bulletins and commercial information and for the automatic transmission to shore of information such as the position of the vessel, the performance of the ship's engines, the condition of the cargo and meteorological data.

The system will normally provide a simplex connection; i.e. there is no two-way circuit and, once a channel has been set up, a message will be passed from ship to CES or from CES to ship without interaction between the two (apart from requests for retransmission of blocks of data). One advantage of this type of operation is that the SES can use a transmit/receive switch instead of the more expensive diplexer which would be needed for simultaneous transmission and reception. However, it is possible to set up a circuit from a CES to a terrestrial subscriber and keep it open for the duration of a call; information is then passed between the SES and CES as it becomes available (this is known as half-duplex operation). Where there is a special requirement it would be possible, with a suitable SES, to provide both-way circuits (i.e. full-duplex operation).

6.6.1.2 Transmission methods and performance

ARQ and half-rate convolutional encoding (with a constraint length of seven) are used in both directions of transmission. With second-generation satellites the symbol rate will be 1200 per second in each direction but with first-generation satellites the symbol rate in the ship-to-shore direction will be only 600 per second (because of the low EIRP of these satellites at C-band frequencies). The modulation method is binary phase-shift keying.

In addition to the signal arriving at a ship by direct transmission from the satellite, the same signal arrives by other paths after reflection from the sea. The signals arriving by the indirect paths vary in a random manner because the surface of the sea is in motion; the resultant of these signals therefore varies in phase and amplitude and, under some conditions, the variation may be quite slow and the average power of the total indirect signal may be only a few decibels below that of the direct signal. With a narrow-beam antenna the indirect signals would be attenuated to a level where they were negligible but with a Standard-C antenna the combination of the direct and reflected signals may result in severe multipath fading. The same mechanism gives rise to multipath fading on the reciprocal, ship-to-satellite, links.

Slow fades of the type experienced with the Standard-C system can introduce long bursts of errors which disable the convolutional encoder. To overcome this problem the data within blocks is interleaved (see Section 5.7.7). In the shore-to-ship direction the blocks comprise 10 368 symbols and occupy a period of over 8 s; in the ship-to-shore direction the blocks comprise between 2176 and 10 368 symbols (in steps of 2048 symbols).

In systems using ARQ, an increase in channel bit error ratio (BER) does not usually result in a corresponding increase in the number of errors in the received messages; the increase in BER does, however, increase the probability that a package will contain one or more errors and thus increases the number of packages which have to be retransmitted. The performance of the Standard-C system is therefore specified in terms of packet error probability (PEP) rather than BER. For example, INMARSAT states that the PEP for a shore-to-ship channel with a C/N_0 of 35 dB(Hz) is 0.002 for a control packet comprising 10 bits, and 0.020 for a message packet comprising 128 bits.

6.6.2 Operation of the system

The following description is much simplified.

6.6.2.1 Registration of SESs

A network-co-ordination station (NCS) is associated with each active satellite and is responsible for network management in the coverage area of that satellite. When a ship equipped for Standard-C operation arrives in an ocean area (i.e. an INMARSAT satellite-coverage area) it must register with the NCS (which keeps a list of all the SESs which are active in the area). When the ship leaves the area it must register with the NCS for the new ocean area. The registration of the ship with the new NCS is signalled to the other NCSs via a NCS-to-NCS signalling channel and the previous NCS then deletes the ship from its list. Each NCS keeps all its associated CESs informed of arrivals and departures so that the CESs know whether a particlar SES is in their area before they accept a message for transmission to it. (This is essential because subscribers may be billed for message transfer at the time the message is accepted into the CES memory.)

6.6.2.2 NCS common channel

Each NCS transmits a continuous common channel (CC). All SESs registered with a NCS stay tuned to the CC (see Fig. 6.8*a*) except when they are occupied with a message transfer.

Each CES has at least one TDM channel available to it; this channel (and other channels used by the CES) may be permanently assigned or may be shared with other (lightly-loaded) CESs on a demand-assigned basis. In what follows it is assumed that channels are permanently assigned to the CES as this makes the procedure of setting up a call more simple (and thus helps make this explanation more brief). The TDM channels (see Fig. 6.8*b*) carry all messages, and signalling and control information, from the CESs to the SESs.

6.6.2.3 Shore-to-ship messages

When a message for a ship arrives at a CES from the terrestrial network the CES checks both that the ship is authorised to receive calls and that it is in the ocean area. If the ship is not available the CES signals this to the terrestrial source, otherwise it tells the source to go ahead with transmission of its message. When the CES has received and stored the complete message, and is ready to pass it on, it tells the NCS that it has a call for the ship. After some further checks, the NCS initiates a call announcement on the CC; if the SES is busy the announcement is put into a queue, otherwise the NCS makes the

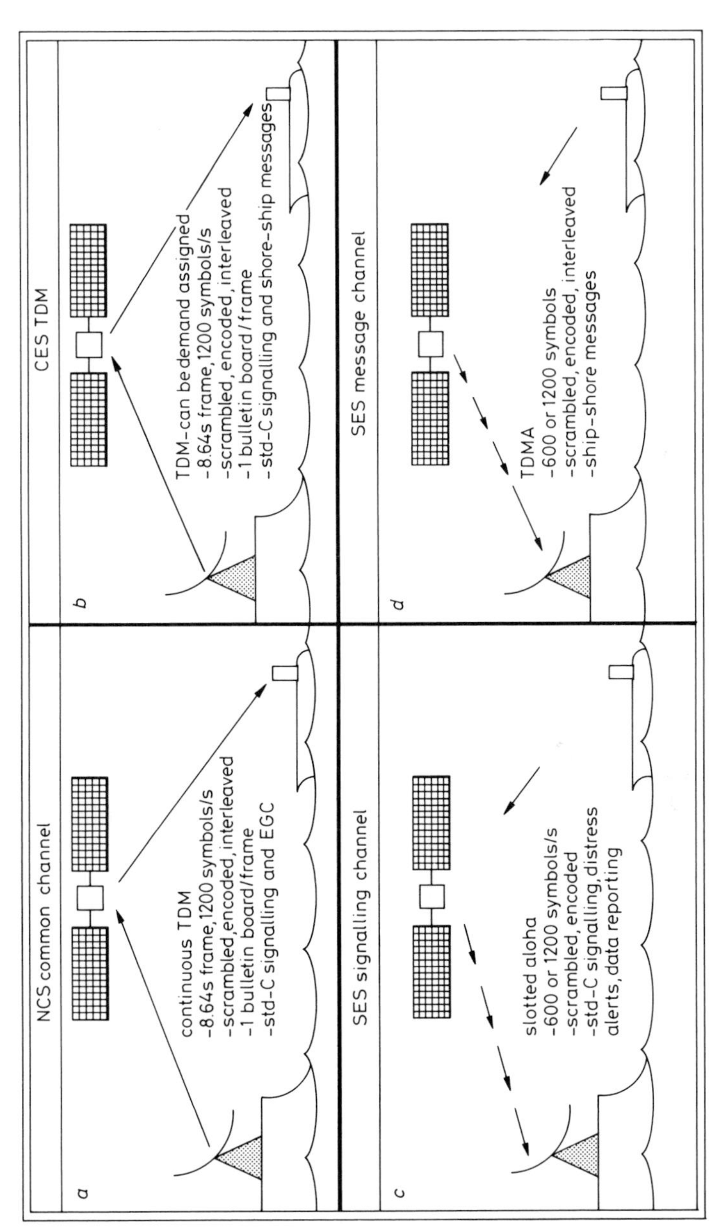

Fig. 6.8 *Standard-C channels (Acknowledgment to INMARSAT)*

announcement as soon as is practicable. The announcement includes the identity of the SES and CES, the priority of the call, the direction of transfer (i.e. shore to ship) and the CES TDM channel which will be used. When it has received the announcement the SES tunes to and synchronises with the CES TDM channel.

Each CES has one or more channels on which it can receive signals from SESs (see Fig. 6.8*c*). These channels are random-access TDMA channels using a form of slotted-ALOHA protocol (see Section 5.8.4) and each frame of the CES TDM transmission includes a bulletin board which tells the SES what time slots in the signalling transmission are free. The SES selects a free slot and transmits an announcement response to the CES (as a short burst of data). It then reverts to listening to the CES TDM channel. It is possible that two ships may use the same slot at the same time; in this case the collision between the announcement responses is detected by the CES, the slot in which the collision occurred is noted on the TDM-channel bulletin board and each SES must retransmit its burst (after a delay which it chooses at random).

When the CES receives a correct announcement response it knows that the SES has heard the call announcement and is ready to proceed; it now sends the SES a channel-assignment packet which includes the length of the message, any routing and information needed to direct the message to a particular terminal on board the ship and a logical channel number (LCN). The message is divided into packets for transmission and each packet is labelled with the LCN; in addition, each packet is given a sequential number (so that the SES can tell if any packets are lost) and each packet includes a check sum (which enables the SES to detect if the packet contains errors). As soon as the SES acknowledges the channel-assignment packet, the CES begins to send the message packets on its TDM channel. The last packet is followed by a request for acknowledgment. The acknowledgment packet from the SES includes a list of packets which are missing or which were received in error. The CES will retransmit these packets and the cycle is repeated until all packets have been received correctly; the CES then disconnects the channel.

6.6.2.4 Ship-to-shore messages

The NCS transmits a bulletin board (in each frame of the CC) which includes the TDM-carrier frequencies for all CESs in the ocean area. When a SES wants to set up a call it looks at the bulletin board and finds the frequency of the TDM carrier being transmitted by the CES it wants; it then tunes to that frequency and synchronises its operation with the CES TDM frames.

The SES now selects a free signalling-channel slot and transmits an assignment request to the CES (using the slotted-ALOHA random-access protocol). The assignment-request packet includes the length of the message, the identity of the SES and the identity of the network to which the message is to be transferred (e.g. the telex network identification code). The SES then reverts to listening to the CES TDM channel.

If the CES can accept the message it tells the NCS, so that the latter can put the SES on the list of busy stations. The CES then sends an assignment packet to the SES; this packet gives the frequency of the channel on which the message is to be passed, a slot number for the start of the message and a LCN; after receipt of this message, the SES starts transmitting in the next frame (at the specified slot number).

The SES message channel is a TDMA channel which may be shared by a number of SESs (see Fig. 6.8*d*). A transmission on this channel must, like all TDMA transmissions, start with a preamble which allows the receiving station to recover the carrier and clock. The preamble is followed by a frame of information and then by further frames of information until the message is complete (see Fig. 6.9). The SES message channel is thus unlike the TDMA channels which we have met previously since it is neither a random-access channel nor a channel of a system such as that described in Section 5.8.3 (where each of a number of earth stations is given a pre-assigned time slot of fixed length in which it transmits a burst of information in every frame).

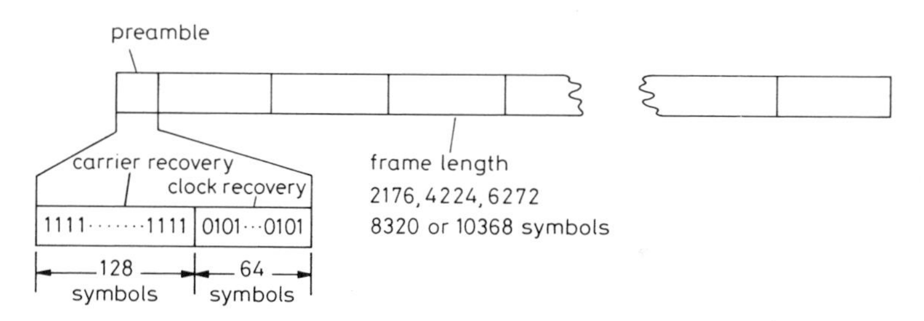

Fig. 6.9 *Format of SES message channel*

6.7 Satellite communications and safety of life at sea (SOLAS)

Reliable and fast methods of maritime communications can do much to improve safety at sea and aid rescue operations, and INMARSAT has always planned and implemented its system with this in mind.

6.7.1 FGMDSS

The International Maritime Organisation (IMO) specifies communications equipment which cargo ships over 300 gross registered tons and all passenger ships must carry in the interests of safety; at present this is VHF and MF/HF equipment. However, the existing maritime distress and safety system has a number of failings and IMO proposes to implement a new system, the FGMDSS (Future Global Maritime, Distress and Safety System) in which

INMARSAT will play a vital part. If the keel of a ship is laid after August 1991 then the ship will have to meet the requirements of FGMDSS from the outset of its operational life; for existing ships IMO has drawn up a schedule for re-equipment which should ensure that all ships meet the new requirements by February 1997.

For the purposes of the FGMDSS the seas are divided into four areas, viz:

(i) A1 — within coverage of VHF coast stations
(ii) A2 — within coverage of MF coast stations (but excluding A1)
(iii) A3 — within coverage of geostationary maritime satellites (but excluding A1 and A2)
(iv) A4 — all areas not included in A1, A2, and A3.

Ships voyaging in area A4 are required to carry equipment operating at all the distress and safety frequencies in the band 1605–27 500 kHz plus general communications equipment for the frequency band 1605–27 500 kHz but IMO has approved the use of an INMARSAT SES on ships in areas A2 and A3 as an alternative to the prescribed MF and HF equipment. The great advantage of the INMARSAT system is that it gives ships reliable and virtually immediate access to rescue co-ordination centres (RCCs) whereas the uncertainties of HF propagation and the limited range of MF radio mean that ships relying on them may be unable to establish direct communication with the shore and may have to rely on other ships to relay their distress messages.

INMARSAT SESs are able to generate distress messages automatically. The distress alert generated by a Standard-C SES comprises 96 bits of information; this is sufficient to specify the identity of the ship, its position (and the time when this was last updated), the course and speed, and a code indicating the nature of the emergency (e.g. fire, flooding, collision, sinking, abandoning ship).

Once the RCC has received a distress message it can use the INMARSAT system to co-ordinate emergency action including search and rescue operations.

The relatively cheap Standard-C SESs will make it possible for many more ships to use the INMARSAT system. The Standard-C system will also make another important contribution to safety at sea because marine safety information (such as navigational and meteorological warnings) will be broadcast to all ships via the NCS CC and can be printed out automatically (this facility can also be added to a Standard-A SES).

6.7.2 EPIRBs

Another way of alerting RCCs in case of emergency, even an emergency which overwhelms a ship without warning, is by means of an emergency position-indicating radio beacon (EPIRB). An EPIRB is a low-power transmitting equipment which can continually send a distress message similar in format to that transmitted by the Standard-C SES. The equipment is usually packaged in a small buoy and an EPIRB can be simple, cheap and reliable. It is activated automatically when it is thrown overboard or when it floats off a sinking ship; it can usually also be activated manually, from the bridge of a ship for example. The distress message is relayed by a satellite and received, decoded and printed out at an RCC.

The COSPAS–SARSAT programme (begun as an experiment in 1979 by Canada, France, the USA and USSR) uses EPIRBs, which can be carried on ships or aircraft, in association with US and Soviet satellites in 12 hour polar orbits. COSPAS–SARSAT has been responsible for the rescue of about 700 people since 1982 and IMO has decided that all ships subject to the FGMDSS regulations must carry a float-off EPIRB working at the COSPAS–SARSAT frequency of 406 MHz (except for ships in area A1 which may carry an EPIRB working at VHF).

6.8 Aeronautical satellite communications

VHF or UHF terrestrial stations can provide good communications with aircraft over line-of-sight paths (e.g. at distances of up to about 100 miles for aircraft flying at 10 000 ft or 200 miles for aircraft flying at 30 000 ft). All countries use VHF or UHF for air traffic control (ATC) where practicable and a public (air-to-ground) telephone service via line-of-sight links has been available to aircraft flying over the USA since 1984.

When aircraft are not within sight of terrestrial stations communication has to be via HF transmissions. However, the unreliability of HF propagation reduces the effectiveness of ATC on long-distance flights and the limited bandwidth available at HF makes it impossible to provide a telephone service for passengers on commercial airlines (although an HF telephony service for private aircraft is available in some countries). Aeronautical satellite communications is thus the only available means for the provision of reliable communication with aircraft which are not within sight of terrestrial stations.

Trials in 1963–64 using the SYNCOM 3 satellite showed that an aeronautical satellite communications service was technically possible. These trials were followed by the AEROSAT programme which was intended to lead to the development of a commercial service. Unfortunately the project was way ahead of its time. The satellites available in the late 1960s and early 1970s would have supported only two or three voice circuits when working to the aircraft antennas available at that time and aeronautical satellite communications would therefore have been ridiculously expensive. The AEROSAT programme made little progress and it was terminated in the late 1970s when airline revenues suffered a temporary decline.

Interest in aeronautical satellite communications was rekindled in the mid-1980s by the success of maritime satellite communications together with rapid growth of the air-to-ground telephone service in the USA and it is now expected that aeronautical satellite communications will become very profitable in the early 1990s. As a consequence the Airlines Electronic Engineering Committee (AEEC) has written an international standard for aeronautical satellite communications, which conforms to requirements laid down by the Future Air Navigation Systems (FANS) Committee of the International Civil Aviation Organisation (ICAO). Two organisations (INMARSAT and AvSat) have announced their intention of providing a worldwide service to the AEEC specification. In addition, Canada and the USA plan to introduce a mobile satellite system in 1993 which will serve all of North America and will provide aeronautical voice and data communications (as well as land mobile and maritime channels), the European Space Agency (ESA) has been carrying out a programme (PRODAT) which includes the development of a low-data-rate aeronautical terminal and Japan plans to conduct experiments in aeronautical communications using its own (ETS 5) satellite.

INMARSAT leads the field. It is already carrying out a programme of tests, demonstrations and trials leading to full global operation of an air-to-ground public telephony service in 1989. The organisation is in the advantageous position of having satellites already in position and many years of practical experience of mobile satellite communications. COMSAT in the USA and BTI in the UK, amongst others, have announced their intention of providing service via INMARSAT satellites.

AvSat (the Aviation Satellite Corporation) which, under its charter, is to be wholly owned by 'members of the international airline community', must start from scratch but it is working with ARINC (Aeronautical Radio Incorporated); the latter is an established US 'not-for-profit' corporation which provides ATC communications under contract to the US Federal Aviation Agency (FAA) and operational and managerial communications for many airlines (most but not all of which are based in the US). AvSat's initial plans envisaged a phase-1 service using leased transponders on four satellites (the first of which is to be launched in 1989) and a phase-2 service using six dedicated satellites launched at intervals of six months starting in 1991; these plans were, however, rejected by the US FCC at the end of 1987, on both financial and technical grounds.

All the technical problems associated with SESs are present, in more acute form, in the design of an aircraft earth station (AES). There will be two basic types of AES, one for low-speed data and the other for high-speed data and voice transmission. The latter will require an antenna with a gain of about 12 dB and one of the most difficult tasks is that of adding such an antenna to an existing aircraft. The antenna must not affect the safety of the aircraft in any way, or increase drag by more than a small fraction of a per cent, and this imposes considerable constraints on the design. The antenna should be mounted so that it has an unimpeded view of the satellite at all normal attitudes of the aircraft but its position may be constrained by the need to mount the LNA and HPA nearby (this can be a particular problem with commercial passenger-carrying aircraft).

The proposed high-gain antennas fall into three categories:

(i) Electronically-steerable antennas comprising linear-phased arrays printed on a flexible base; these 'conformal' antennas are laid flat on the skin of the aircraft.
(ii) Electronically-steerable antennas incorporated in small aerofoils which are mounted on top of the aircraft.
(iii) Mechanically-steerable antennas which are mounted inside small radomes; these antennas are claimed to have better coverage patterns than electronically-steered antennas.

In the case of data-only terminals very small (near-omnidirectional) antennas can be used; these can be blade antennas (i.e. of the aerofoil type) or patch antennas (i.e. printed on a flexible base and attached to the skin of the aircraft).

A number of companies are now developing and producing both high-gain and omnidirectional antennas, and aircraft designs are already being modified to facilitate the fitting of various types of antenna. Estimates of the cost of the first few AESs (uninstalled) range from \$100 000 to over \$250 000 but these costs will undoubtedly fall with time in a similar way to the cost of SESs.

The proposed INMARSAT system is rather like the Standard-C system with the addition of SCPC voice transmission. There are four types of channel:

(i) P-channels, which are packet-mode TDM channels carrying signalling and system management data; one of these channels is transmitted continuously by each ground earth station (GES); GESs in the aeronautical system are analogous to CESs in the maritime system.
(ii) R-channels, which are random-access TDMA (slotted-ALOHA) channels used to carry signalling messages from AESs.
(iii) T-channels, which are reservation-TDMA channels used to transmit traffic.
(iv) C-channels, which are SCPC voice channels.

The basic channel bit rates are 600 bit/s for ground-to-air data transmissions, 600 and 1200 bit/s for air-to-ground data transmissions and 21 kbit/s for voice channels; bit rates up to 12.6 kbit/s will also be available on the data channels after the initial operational phase. Transmissions are in the same frequency bands as the maritime system. The modulation methods will be aviation-BPSK (A-BPSK) for bit rates up to and including 2400 bit/s and aviation-QPSK (A-QPSK) for higher bit rates; A-BPSK is the same as 2,4-QPSK and

A-BPSK is a form of O-QPSK (see Section 5.6.3). The GESs will comprise a C-band antenna with a diameter of about 13 m, an L-band antenna with a diameter of about 3.7 m (or a combined C-band/L-band antenna), RF equipment and ACSE (access, control and signalling equipment).

A problem that is peculiar to airborne public telephones is where to put them in order to avoid any possibility that users will obstruct the aircraft gangways. The present US terrestrial aeronautical service uses cordless telephones in wall-mounted units; passengers insert their credit card into a unit and the telephone is then released so that it can be carried to a seat. Similar equipment will be used for the INMARSAT trials but it seems likely that telephones will, in time, be mounted on the backs of seats (at least in the first-class cabin accommodation). The initial aeronautical satellite-communication public telephony service will provide only air-to-ground calls; this is because of the difficulties of routing calls to airline passengers (for example, aircraft are identified by their tail number and this is not usually known to the public). However it is expected that calls from ground to air will become available in the second stage of the service.

So far we have been talking about the use of aeronautical satellite communications only for public telephony but there are large benefits to be gained from the use of satellite communications for ATC, and for operational and management communications. These benefits cannot be fully realised until most aircraft operating on long-distance routes have been equipped and, in the past, this has been seen as a major obstacle. However, there is no doubt that satellite communications will lead to improved voice and data communications with cockpit and cabin crew and that, in the longer term navigation, automatic monitoring of aircraft performance and many other functions will be integrated into the satellite communications system.

It is expected that aeronautical telecommunications will evolve by the integration of terrestrial systems (providing communication with aircraft over densely-populated areas) and communication-satellite systems (providing communication with aircraft over the oceans and sparsely populated areas).

6.9 Land-mobile systems

Many land vehicles spend most of their time in, or close to, areas with well-developed terrestrial networks. The communication requirements of these vehicles can usually be served by connecting them to the nearest terrestrial network via short-distance radio links (such as the fast-growing cellular networks).

However, in many large countries (even those with advanced communications such as the USA) there are extensive areas where it is uneconomic to provide links to the nearest terrestrial network; moreover, a vehicle equipped to operate to one cellular network may well find that it cannot operate to other networks because of lack of standardisation. There is thus a potential market for land-mobile satellite communications amongst long-distance travellers and recent developments have shown that it should be practicable to produce mobile terminals at a price (say, less than $10 000) which will be attractive to this market.

There has, for some years, been considerable pressure from Canada and the USA for the provision of frequency bands for land-mobile satellite communications and in 1987, at the WARC-MOB conference in Geneva, the aeronautical satellite-communications allocations at 1.5/1.6 GHz were reduced by two 4 MHz bands to provide exclusive allocations to the land-mobile service and this service was also given the right to use two 3 MHz bands in the maritime satellite-communications 1.5/1.6 GHz allocations.

The first comprehensive commercial land-mobile satellite-communications service seems likely to be one serving the whole of North America. This system had its genesis in the MSAT programme, which was initiated in 1980 by Telesat (Canada). It is planned to start operations in 1993 using two geostationary satellites (one for the USA and one for Canada) each of which will provide back-up for the other. Mobile stations will be able to use either low-gain or high-gain antennas; the low-gain antennas will be virtually omnidirectional and will have a gain of about 4 dB, the high-gain antennas will have a gain of about 13 dB and will need to be steerable. The system will provide a range of land-mobile services including:

(i) SSB voice channels spaced at 5 kHz
(ii) Broadcast-quality voice channels at 32 kbit/s
(iii) Standard voice channels at 9·6 and 4·8 kbit/s
(iv) Data and emergency voice channels at 2·4 kbit/s
(v) Packet data channels at 2·4 kbit/s.

The system will also offer aeronautical voice and data (at 8 kbit/s), and maritime voice and data (at 2·4 kbit/s).

The original MSAT studies foresaw the use of the system in connection with:

(i) Transportation of goods and passengers by road, rail, water and air
(ii) The search for and exploitation of oil and minerals
(iii) Forestry
(iv) Law and order (the Royal Canadian Mounted Police at present use about 13 000 terrestrial mobile and portable stations and it was estimated that the use of MSAT would give large savings in operating costs)
(v) Emergencies
(vi) The provision of communications for remote and thinly-populated regions; there are considerable similarities between the requirements of mobile and thin-route services (for example both require simple, reliable earth stations with low power consumption) and there could be advantage in providing the two types of service via one system.

Land-mobile communications-satellite services are also being studied by, amongst others, AUSSAT, ESA, INMARSAT and companies in the USA, Japan and Europe. The first thoughts of INMARSAT are that their land-mobile (Standard-M) service would use amplitude-companded SSB (ACSSB) like the M–SAT system. Other proposed systems are digital systems using low-rate encoding and at least one proposed system uses code-division multiple access (CDMA) with FEC. INMARSAT is trying to foster agreement on standards for a land-mobile satellite service.

In the meantime the European Land Mobile Satellite Trials Working Group is conducting tests and demonstrations using the INMARSAT Standard-C system. A pan-european land-mobile cellular radio system (the GSM system) should start service in 1991; however, even when fully developed this system will cover only the main urban and industrial areas, and major highways of Europe. It is intended to use the Standard-C system to provide a pilot service to those parts of Europe that the GSM system cannot reach. The Satellite Trials Working Group includes more than a dozen countries (including the USSR) and it is expected that about 50 countries will take part in the final phase of the trials (starting in 1988); the service should become fully operational in the second half of 1989.

Trials of paging via satellite are also being conducted (from 1988 onwards) by British Telecom International. The initial phase of the service is intended to enable long-distance lorry drivers to be contacted while travelling in regions within the service area of the INMARSAT Atlantic satellite (e.g. Europe, Africa and the Middle East). Messages will be telephoned, telexed or keyed-in in the usual way and will be routed to the Goonhilly CES where they will be transmitted to the satellite and relayed to small antennas about 5 ins square mounted flush with the roof of the lorry cab. A LNA will be mounted immediately below the antenna and, after down-conversion, the signal is passed direct to a standard paging receiver. A message of up to 90 characters can be displayed on the pager and a paper copy can be provided by a printer. The estimated cost of a mobile paging terminal is about $1000.

The trials of the Standard-C and paging services should establish whether or not, as some experts maintain, two-way voice communications with long-distance lorries is essential to speedy resolution of problems such as breakdown or Customs difficulties.

Most of the technical difficulties relating to maritime and aeronautical communications also apply to land-mobile communications and there is one additional problem; this is that hills, buildings, trees and other obstacles may intrude into the transmission path as the land-mobile station moves. This problem is not so acute for satellite systems as it is for terrestrial land-mobile systems but masking can occur if the satellite is at a low angle of elevation relative to the mobile terminal. In trials conducted by ESA with a geostationary satellite, fades of up to 20 dB were caused by buildings, bridges and trees.

It has been suggested that satellites in highly-elliptical orbits at an inclination of 63° to the equator should be used for service to mobile stations at the higher latitudes (e.g. Europe). There are a number of such orbits, by the use of which a satellite can be made to appear near-stationary over any chosen point for periods of eight to twelve hours a day (the MOLNIYA orbit for example, see Section 1.4). Advantages of these orbits for land-mobile systems would be that:

(i) The satellite could be made to appear at a relatively high angle of elevation over a fairly wide coverage area (irrespective of latitude), thus reducing the shielding effects of hills, buildings etc; this might be of particular advantage for mobiles in areas of high-rise building where terrestrial systems suffer most from obstruction of propagation paths.
(ii) The mobile stations could use a moderately high-gain antenna (which would not require steering because it would always be pointing at or near zenith).
(iii) The movement of the mobile station would be nearly orthogonal to the transmission path and thus Doppler shift would be kept small.

Unfortunately, to give continuous communication it is necessary to use two or three satellites for each coverage area. However, the cost of providing the additional satellites would be partly offset because the reduced fading would allow the use of less satellite power per channel (relative to the equivalent geostationary satellite) and thus reduce space-sector costs per channel. This type of system is being studied by the European Space Agency under its Archimedes programme.

6.10 Navigation and position-fixing systems

A detailed consideration of navigation and position-fixing systems falls outside the scope of this book. On the other hand these systems must be mentioned because their use in conjunction with air, sea and land-mobile communications-satellite services is being extensively studied, and a terrestrial position-fixing system is already being used with a land-mobile communications-satellite system in the USA.

Satellite navigation systems may be either passive or active. In passive systems the mobile terminals calculate their positions by measuring the transmission delays associated with signals received from a number of satellites, the positions of which must be known accurately and transmitted to the mobile terminals. In active systems a mobile terminal, the position of which is to be determined, must transmit a signal which is received by a number of satellites and relayed to a central station where the position of the mobile is calculated. The position information may be used solely by a central office which is keeping track of a number of stations or it may be transmitted back to the mobile station for purposes of navigation.

Active systems are not favoured for military use because radio transmission from mobile terminals can also disclose the position of the terminals to an enemy; thus, since all the navigation systems so far put into operation have been primarily intended for military purposes they have all been passive; the latest of these systems are the US NAVSTAR–GPS system and the Soviet GLONASS system.

There are obvious disadvantages to relying on national military systems for international civil navigation. Organisations which are studying civil navigation-satellite systems include the International Civil Aviation Organisation (ICAO), the International Maritime Organisation (IMO) and the European Space Agency (ESA). In addition a number of commercial systems have been proposed of which the best known are the active systems GEOSTAR (USA) and LOCSTAR (a European system using the same technology as GEOSTAR). Active systems such as GEOSTAR have several advantages for commercial use: firstly that all the equipment for processing the signals and calculating positions can be located at a single earth station and the mobile terminals can therefore be made relatively simple and cheap; secondly that the systems incorporate two-way transmission and can therefore be used to pass messages; and lastly that since a terminal has to transmit when it wants to know its position it is possible to log the transmissions and charge for each position fix.

The establishment of any satellite navigation system is a costly and lengthy business whereas, in the USA at least, customers already see a need for a land-mobile position-locating and messaging service. The GEOSTAR Corporation together with Hughes Network Systems is therefore operating a Radio Determination Satellite Service (RDSS) which uses the LORAN-C terrestrial radiolocation service together with communications satellites; this service enables companies in the USA to keep track of the position of delivery vehicles. Each vehicle carries equipment comprising a LORAN-C position-fixing terminal and a transmit-only satellite terminal which sends the position information to a central earth station which communicates with vehicle fleet-control centres.

Chapter 7

Earth stations

7.1 Introduction

The term 'earth station' is used to mean both an antenna with its associated equipment and a site housing a number of antennas. Thus there are many antennas at the British Telecom International (BTI) Earth Station at Goonhilly and each of these antennas with its associated equipment is an earth station in its own right; for example, one of the antennas at Goonhilly is part of the BTI Coast Earth Station (CES) working to the INMARSAT maritime communications-satellite system. Where there is any danger of confusion, an antenna with its associated equipment is (in this book) called a terminal and the term earth station is used for a terminal (or several colocated terminals) together with any ancillary facilities.

Earth-station equipment such as modems, frequency converters and amplifiers may be modified or replaced at fairly frequent intervals as communications-satellite systems develop but antennas (particularly large antennas) usually survive for long periods; terminals at big earth stations are therefore often identified by their antennas. Thus, for example, the fifth large antenna to be built at Goonhilly is associated with the CES and this station is also therefore referred to as Goonhilly 5.

A terminal always includes an antenna (and large antennas include automatic tracking and steering equipment). Terminals may also include:

(i) Low-noise amplifiers (LNAs) or low-noise down-converters
(ii) High-power amplifiers (HPAs)
(iii) Signal processing equipment, e.g. down-converters, up-converters, modems, codecs, intermediate frequency (IF) and baseband (BB) amplifiers
(iv) Transmission and signalling equipment at the interface between the terminal and the terrestrial network or peripheral equipment
(v) Control and supervisory equipment
(vi) Enclosures to protect the equipment from its environment.

Not all these elements are present in every terminal; for example, a TV receive-only (TVRO) terminal does not include any HPAs. On the other hand a large earth station comprising a number of terminals may require extensive ancillary facilities such as standby power supplies and workshops.

Earth stations and earth-station antennas come in a very wide range of sizes, complexity and cost. Virtually all terminals, other than mobile terminals, use antennas with near-parabolic main reflectors and this is the only type of antenna considered in this Chapter. (Some reference to other types of antenna, used with mobile stations, will be found in Chapter 6.)

Earth-station antennas with reflector diameters of up to 30 m were used in the early days of satellite communications. The main reflectors of modern earth-station antennas range in diameter from about 18 m (for terminals carrying heavy traffic in the international or domestic services) to less than 1 m in diameter (for domestic terminals receiving signals from satellites of the broadcast satellite service, for example). This reduction in the size of antennas reflects first the increased equivalent isotropically-radiated power (EIRP) of modern satellites and secondly the introduction of systems including large numbers (hundreds or even thousands) of earth stations.

A large earth terminal may cost many millions of dollars whereas, at the other end of the scale, a small terminal for receiving television broadcasts direct from satellites must not cost more than about a thousand dollars if it is to sell in large quantities.

7.2 Siting an earth station

7.2.1 Requirements

The site for an earth station should be chosen so that:

(i) The elevation angle to any satellite position to which the station may be required to work is greater than some specified minimum angle (usually 5°). This is not generally a constraint for stations working to national or regional systems but it may restrict the choice of location for stations in countries situated towards the edge of the coverage area of a satellite in an international system. For example, stations in the north-west of the UK would see the INTELSAT satellite over the Indian Ocean at an elevation angle of less than 5° and UK stations working to international systems are therefore in the south of the country.

(ii) There are no objects such as trees or buildings which will obstruct the beam of an antenna when it is pointing at any satellite position to which it may be required to work. In towns and cities it is sometimes difficult to be certain that a site will not be overshadowed by subsequent construction.

(iii) Interference does not exceed the acceptable level.

(iv) There is negligible risk of flooding and the ground (or structure on which the antenna is to be mounted) provides a stable foundation.

(v) There is satisfactory access for purposes of installation and maintenance, and connection can be made to any necessary services (e.g. water and electricity supplies, and drainage).

(vi) The cost of connection to the telecommunications network or the customers equipment is not excessive.

(vii) It does not give rise to serious opposition from local residents or other pressure groups.

Some of these requirements may be difficult to reconcile in practice and finding a site for earth terminals (especially those using large antennas) can be a major problem in many areas of the world. On the other hand there may be virtually no choice of site (e.g. the terminal may have to be located at the customer's premises) and this may put severe constraints on the satellite system.

7.2.2 Interference

7.2.2.1 General

Interference to, from and between satellite systems was discussed briefly in Section 3.6. In Section 7.2.2. we are concerned solely with the problem of interference to and from earth stations.

The flux density at the surface of the earth of signals transmitted by satellites is very low and the signals are therefore susceptible to interference.

Interference to earth stations can arise:

(i) From systems sharing the same frequency band (e.g. terrestrial systems and other satellite systems)

(ii) From equipment operating in other frequency bands which nevertheless radiates energy in the frequency bands allocated for reception by the earth station (e.g. high-power radar transmitters may radiate appreciable amounts of energy at harmonics of their working frequencies)

(iii) From nearby sources such as ignition systems, electric railways and power lines which may radiate at low power levels over a wide range of frequencies.

Interference can reach earth stations in a number of ways, the main ones being line-of-sight propagation, refraction, tropospheric scatter (which is mainly forward scatter), rain scatter (which is more or less isotropic) and propagation via ducts caused by abnormal weather conditions. Some of these mechanisms of propagation are stable (or relatively so) and are associated with interfering signals which are present for all or most of the time; other mechanisms such as scatter from rain or transmission via ducts can result in high levels of interference being received from distant sources for small percentages of the time. It is therefore usual to specify two limits to interference power: one for the interference which can be accepted over the long term (this is usually set at a level which will ensure that its effects on the signal are negligible) and a higher level of interference which can be tolerated for a small percentage of the time.

The maximum allowable interference power at the input to an earth-station receiver depends on the signal power, the amount of interference which is acceptable in the baseband, the characteristics (e.g. modulation method and coding) of both the interfering and interfered-with transmissions and the frequency separation between the two carriers.

In planning an earth station it is, of course, necessary to ensure not only that it does not suffer interference but also that it does not cause interference to other systems.

7.2.2.2 Interference between earth stations and terrestrial stations using the same frequency bands

The avoidance of interference between terrestrial stations and earth stations sharing the same frequency bands can be a major difficulty in regions with well-developed terrestrial microwave services.

The interference is kept to acceptable levels:

(i) By the imposition of internationally-agreed limits on the EIRP which may be radiated at low angles of elevation by earth stations
(ii) By locating earth stations at sites which have good shielding from interference (see Section 7.2.2.4) and which are, as far as practicable, off the line-of-shoot of terrestrial stations
(iii) By careful choice of frequencies
(iv) By using antennas with good radiation patterns (i.e. low sidelobes)
(v) By co-ordination.

Co-ordination

Co-ordination of an earth station is the process by which:

(i) Any potential cases of interference between the station and terrestrial stations or other earth stations are identified
(ii) Each potential case of interference is examined to establish whether or not unacceptable interference would actually result
(iii) In any case where unacceptable interference would result, some method of avoiding the interference is agreed with the other party, or parties, involved.

Co-ordination within national boundaries is a matter for the country concerned but, under some conditions, interference can occur between stations a thousand or more kilometres apart and international co-ordination is thus often required. Internationally-agreed procedures for co-ordination are specified in the Radio Regulations (RRs) of the International Telecommunications Union (ITU).

Not the least of the technical problems of co-ordination, in regions which include well-developed terrestrial systems, is that of identifying all the stations which might cause interference to, or suffer interference from, the earth station to be co-ordinated. An essential step is the preparation of maps showing the receive co-ordination area (i.e. the area within which terrestrial stations might, in the worst circumstances, cause unacceptable interference to the earth station) and the transmit co-ordination area (i.e. the area within which, in the worst circumstances, a terrestrial station might suffer unacceptable interference from the earth station). Appendix 28 of the RRs gives the agreed procedure for determining the contour enclosing a co-ordination area and Fig. 7.1 shows, as an example, the receive (4 GHz) co-ordination contour for a large (32 m diameter) antenna at the BTI Earth Station at Goonhilly.

Two different types of propagation are considered in determining a co-ordination contour, tropospheric propagation (along near-great-circle paths) and scatter by rain (or other hydrometeors); in the RRs these two types of propagation are called Mode 1 and Mode 2, respectively. Because the scatter caused by rain is more or less isotropic, the Mode 2 contour is a circle; only scatter from rain in the main beam of the antenna is significant and the centre of the rain-scatter circle is offset from the earth station in the direction of the azimuth of the main beam (in the case of Fig. 7.1 the offset distance is about 13 km and the rain-scatter circle, shown as a chain-dotted line, has a radius of 390 km). In Mode 1 propagation, a signal travelling over land is attenuated more rapidly than one travelling over sea; this can be seen clearly in Fig. 7.1 where the co-ordination distance over the south-west quadrant is very much greater than that in an easterly direction. Appendix 28 of the RRs gives the (maximum) EIRP E of a terrestrial station working in the 3·4–4·2 GHz band as 55 dBW and suggests that auxiliary contours should be drawn for lower values of EIRP; in Fig. 7.1 the main contour corresponds to $E = 55$ dBW and the auxiliary contours are for $E-5$, $E-10$, $E-15$ and $E-20$ dB. A co-ordination contour is formed by selecting at each azimuth the Mode 1 or Mode 2 contour which is farthest from the earth station. It will be seen that the co-ordination area for the antenna reaches into Eire, Northern Ireland, Belgium, France and Spain.

Accurate estimation of the effects of interference can be a very complex and time-consuming problem; the procedures of Appendix 28 therefore involve a number of simplifying assumptions and approximations. It is important to remember that, once the co-ordination area has been established, more detailed examination will show that most of the stations in the area will not cause or suffer unacceptable interference. For example, the probability of interference between a terrestrial station and an earth station is greatly reduced if the earth station is more than a few degrees off the line of sight of the terrestrial station.

When an administration proposes to build a new station (operating in a shared frequency band) it prepares a diagram of the co-ordination areas and

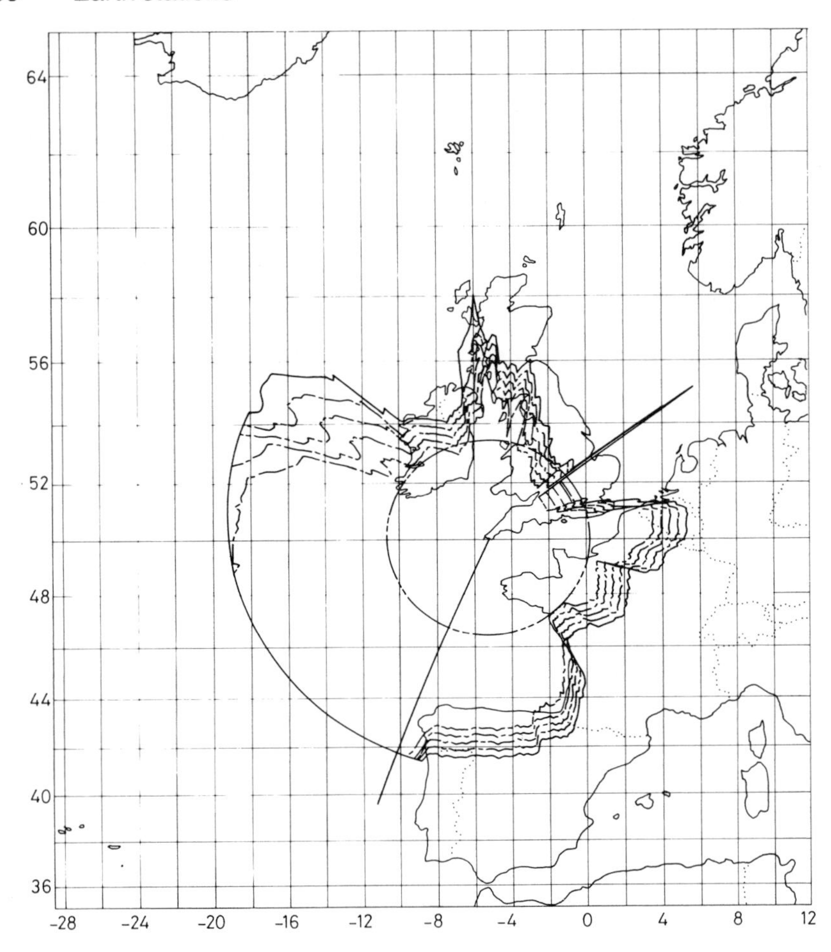

Fig. 7.1 *Co-ordination area (Acknowledgment to British Telecom International)*
———— co-ordination contour
– – – — Mode 2 contour
— – — auxiliary Mode 1 contour

this diagram is published by the International Frequency Registration Board (IFRB) of the ITU. Countries with stations within the co-ordination area provide the characteristics of those stations and the possibility of interference between the earth station and the other stations is now checked, taking account of the actual frequencies in use and proposed, the characteristics of the stations and the transmissions, and the best propagation data available. If unacceptable interference is likely the two administrations concerned must agree on some means of solving the problem before the IFRB is asked to register the frequency assignments (the IFRB will assist in resolving any problems but this is rarely necessary). Further co-ordination is carried out if a new terrestrial station is to be built within the co-ordination area or new frequencies are to be registered. Co-ordination is complete when it has been agreed amongst all the administrations concerned that harmful interference will not occur.

The heavy demand for satellite capacity has made it necessary to reduce the spacing between satellites in certain parts of the geostationary orbit (GSO), particularly the arc used by countries of the American continent. Reducing the spacing between satellites increases the probability of interference between earth terminals and the satellites of other systems; earth-station antennas with improved radiation patterns (see Section 7.3.3) are necessary to compensate for this.

Another potential source of interference is the allocation of the same frequency band to one satellite system for uplinks and to another for downlinks (this is known as 'bi-directional working'). For example the band 17·7–18·1 GHz is allocated for use on downlinks of the Fixed Satellite Service (FSS) and earth-to-space links of the Broadcasting Satellite Service (BSS). There is thus the possibility of interference between earth stations of the two systems. The probability of such interference is small at present but will become greater if the principle of bi-directional working is extended.

7.2.2.3 Interference from radar installations and miscellaneous sources

There are a large number of radar transmitters in use today and, after terrestrial stations, they are probably the most common potential sources of interference to earth stations. In theory no radar installation should be radiating a significant amount of energy in the frequency bands used for satellite communications; nevertheless many of them transmit very high peak powers and it is inevitable that some will have inadequate suppression of harmonics and other out-of-band radiation. In addition some radar installations are sited high up on masts or hills and this increases the area over which they can cause interference.

Once it has been clearly demonstrated that a particular radar would cause unacceptable interference it ought to be possible to persuade the owner to take steps to suppress the out-of-band radiation. However, it can take a lot of time and effort to prove the case and, when that is done, to effect a cure; thus it is often better to avoid the problem by selecting another site if this is practicable.

There are many other possible sources of interference, especially in densely-populated areas, examples are industrial microwave equipment and sources of noise-like radiation such as arc welding, high-voltage power lines, electric railways and ignition systems. The power density of the radiation from these sources is low and none of them (with the possible exception of arc welding) should cause trouble even to the most sensitive earth terminals if they are more than about 200 m distant from the antennas and equipment.

Where an earth station uses an antenna with high gain, it is possible:

(i) that interference to the station could be caused by energy scattered from an aircraft illuminated by a radar transmitter
(ii) that an aircraft flying through the antenna beam could cause partial blockage.

However, as far as is known, both of these effects are negligible for aircraft at altitudes greater than about 5000 ft and that they can be avoided by siting stations so that there is shielding (see Section 7.2.2.4) in the direction of airports or other areas where low-flying aircraft are expected and so that antenna beams do not intersect landing and take-off paths.

7.2.2.4 Site shielding

Interference can be reduced by siting an earth station so that it is shielded by high ground or tall buildings between it and the source(s) of interference. If this is not possible it may be practicable to shield a small antenna by means of a high earth bank, a wire fence or a concrete wall or by locating it in a pit; all of these methods have been used either singly or in combination.

The site-shielding factor (SSF) may be defined as the ratio of the interference power that would be received by an earth-station antenna in the absence of the shielding to the power received in the presence of the shielding. The SSF at a given azimuth is a function of the horizon angle at that azimuth, the distance of the shielding from the antenna, the radiation characteristics of the antenna and the elevation angle at which the antenna is being used.

Fig. 7.2 shows an antenna and a shielding obstacle. If:

(i) The elevation angle E is much greater than the diffraction angle A
(ii) The distance d of the obstacle from the antenna is greater than the Rayleigh distance $D^2/2\lambda$, where D is the diameter of the antenna and λ is the wavelength.

then the SSF should be close to the diffraction loss S associated with the obstacle.*

* For a knife-edge, the diffraction loss S is

$$S(\text{dB}) = 16 + 10\log f + 10\log d + 20\log A$$

where f = frequency GHz
d = distance, km
A = diffraction angle, degrees

For $f = 4$ GHz, $d = 5$ km and A = 2° this gives $S = 35$ dB

The loss over a hill, with a rounded top and vegetation growing on it, is greater (as a rough approximation add 5 dB).

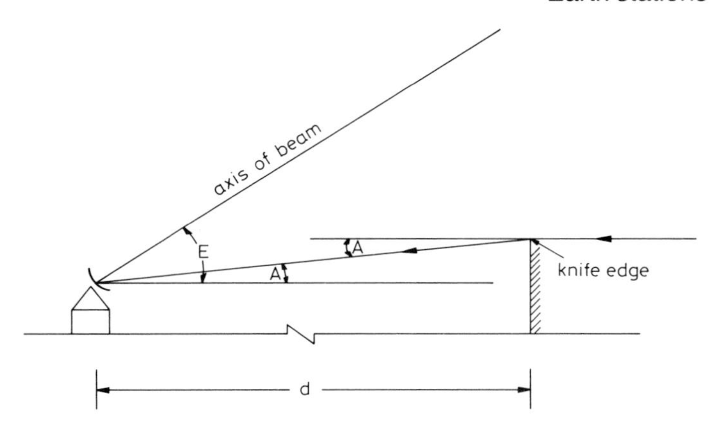

Fig. 7.2 *Site shielding*

As the diffraction angle increases (but the elevation angle is held constant) the SSF rises to a maximum and then decreases again. This can be explained by assuming that the diffracted radiation has an apparent source at the apex of the obstacle and thus, as the diffraction angle A is increased, the angular discrimination $(E - A)$ between the axis of the antenna and the apparent direction of arrival of the interfering signal decreases; as a result, the gain of the antenna in the apparent direction of arrival of the signal increases more and more rapidly until the increase in diffraction loss is outweighed by the increase in antenna gain. For most antennas, maximum shielding should, in theory, be obtained when the diffraction angle is approximately half the elevation angle.

It is often impractical to find sites shielded by natural obstacles outside the Rayleigh distance or, except in the case of quite small antennas, to construct shields outside the Rayleigh distance (for example, the Rayleigh distances for 3 m and 17 m antennas working at 11 GHz are 160 and 5000 m, respectively). It is therefore necessary to consider the effect of shielding which is relatively close to the antenna. The significance of the Rayleigh distance $D^2/2\lambda$ is that up to this distance from the plane of a radiating aperture of diameter D the beam of radiation forms a cylinder based on the aperture and no obstacle outside the cylinder can affect the radiation pattern of the aperture. However, for a practical antenna there are many factors contributing to the sidelobes besides the radiating aperture (see Section 7.3.3) and obstacles closer to the antenna than $D^2/2\lambda$ do therefore provide some shielding. It is however difficult, if not impossible, to calculate the corresponding SSF and the design of artificial shielding is therefore usually more of an art than a science. Shielding factors ranging from 10 to 15 dB for concrete walls to as much as 30 dB for a combination of pit, earth bank and wall have been measured. However, the shielding provided by nearby obstacles can vary sharply with frequency and with the elevation angle of the antenna, especially at sites in built-up areas where an interfering signal is often a composite of components arriving by a number of different paths.

7.2.2.5 *Uncovering potential sources of interference*

It is essential, as soon as a likely site has been located, to seek the co-operation of anyone who might be operating equipment which could be a source of interference. In the case of shared frequency bands:

(i) There are set procedures (co-ordination) for identifying stations in other countries which are potential sources of interference
(ii) The body responsible for national frequency allocation will be able to provide details of authorised radio stations in the home country.

There is no agreed procedure for identifying other likely sources of interference.

Even if there is no evidence on paper to suggest that a terminal at a proposed site might suffer unacceptable interference, it is essential to start making interference measurements at or near the site as soon as possible, as these measurements may disclose unsuspected problems.

It is obviously desirable to detect signals of as low a level as is practicable; measurements from a high point nearby are therefore useful because of the elimination of the effects of site shielding. Measurements from a high point, in conjunction with measurements made at the site, can also provide some reassurance that estimated levels of site shielding will be realised in practice. The collection of statistically-meaningful data on the interference at a specific site likely to be experienced for small percentages of the time can take years; it is therefore usually impractical to rely on measurements alone to give assurance that a site will be free from interference; a combination of calculation, using the best available propagation data, and measurements must be used.

7.3 Antennas

7.3.1 *Gain and noise temperature*

7.3.1.1 *Gain*

The most important characteristic of an earth-station antenna is its effective area, since this determines the power collected by the antenna from a transmission with a given power flux density. The gain G of an antenna is, in theory, proportional both to its area and the square of the operating frequency but, in practice, irregularities of the antenna profile place a limit on the increase in gain with frequency.

As demonstrated in Section 3.1, the carrier-to-noise power ratio C/N at the input to an earth-station receiver is proportional to the figure-of-merit *G/T* of the earth station.

7.3.2.1 Noise temperature

The system noise temperature *T* usually depends almost entirely on the noise temperature of the antenna, the noise temperature of the LNA and the loss of the waveguide or cable connecting the antenna feed to the LNA (see Section 7.4.2).

The main components of the antenna noise power are noise received from the sky via the main beam and noise received from the earth via the sidelobes. The sky noise comprises cosmic (galactic) noise and noise resulting from the absorption and reradiation of energy by water and oxygen molecules in the atmosphere. The cosmic noise decreases rapidly with increasing frequency. The absorption noise is a function of the atmospheric attenuation (see Section 3.5.2) and it increases:

(i) When it is raining or snowing and, to a lesser extent, when it is cloudy
(ii) With decreasing elevation angle
(iii) With frequency

The *G/T* of an earth station also therefore varies with atmospheric conditions, elevation angle and frequency.

The contribution of the clear sky to antenna noise temperature at an elevation of 40° and frequencies of up to 11 GHz is 5 K or less; at an elevation of 5° the noise temperature is about 20 K at 4 GHz and about 30 K at 11 GHz.

The noise power received from the earth depends on the gain of the sidelobes and therefore on the antenna configuration and construction; the corresponding noise temperature can be as little as 10 K for large antennas and as much as 100 K for some small antennas

Because weather conditions affect the propagation of microwave transmissions it is necessary to specify the performance of systems in statistical terms. Thus, for example, CCIR recommends that the 1 min mean noise power at a point of zero relative level in an analogue channel of a satellite system should not exceed

10 000 pW0P for more than 20% of any month
50 000 pW0P for more than 0·3% of any month.

In the 4 GHz band (and lower frequency bands) rain causes relatively little attenuation of the signal or increase in noise temperature; thus, if the system meets the first criterion it usually also meets the second criterion. The required earth-station *G/T* is therefore usually determined by, and specified in terms of, clear-sky conditions.

In the 11 GHz band (and higher frequency bands) attenuation and noise temperature increase rapidly with increasing rain rate (especially at low elevation angles) and the G/T may be determined by the conditions corresponding to small percentages of the time. (A corresponding clear-sky G/T can be derived for purposes of specifying and testing the system.)

7.3.1.3 Antenna sizes

The diameters and uses of some typical antennas are:

(i) *17 m:* Used for stations operating at 6/4 GHz and carrying heavy traffic on international, regional or long national routes, e.g. INTELSAT Standard-A stations with G/T of 35 dB/K. (Note: the original INTELSAT Standard-A specification required a G/T of 40·7 dB/K corresponding to an antenna diameter of around 30 m.)

(ii) *10–13 m:* Used for
  (*a*) Stations operating at 14/11 GHz and carrying heavy traffic on international, regional and national routes, e.g. INTELSAT Standard-C stations with G/T of 37 dB/K. (Note: the original INTELSAT Standard-C specification required a G/T of 39 dB/K corresponding to an antenna diameter of about 18 m.)
  (*b*) Stations operating at 6/4 GHz and serving lightly-loaded routes, e.g. INTELSAT Standard-B stations with G/T of 31·7 dB/K
  (*c*) The international relay of television programmes.

(iii) *2–10m:* Used for a wide range of purposes such as stations serving remote communities, TV receive-only (TVRO) stations serving the head ends of cable-TV systems, stations providing specialised (business) services in international, regional and domestic systems, and hub stations for very-small-aperture terminal (VSAT) systems.

(v) *Less than 2 m:* VSATs, domestic terminals for the reception of direct-broadcast television transmissions, INMARSAT Standard-A SESs.

The size of an earth-station antenna is usually determined by the specified G/T value and the noise temperature of the chosen LNA. In some cases, however, the transmit gain of the antenna is specified and this may require the provision of a larger antenna than is needed to meet the specified G/T. For example, INTELSAT specifes the transmit gain of a Standard-B antenna as not less than 53·2 dB; this ensures that the transmitter power required is not such as to generate EIRPs, in directions off the main beam, which will cause interference to other systems.

7.3.2 Configuration

Many small paraboloidal antennas are front-fed, i.e. the feed is situated at the focus of the reflector (see Fig. 7.3*a*). However, the feeds of larger earth-station antennas may be bulky, especially if they are designed to

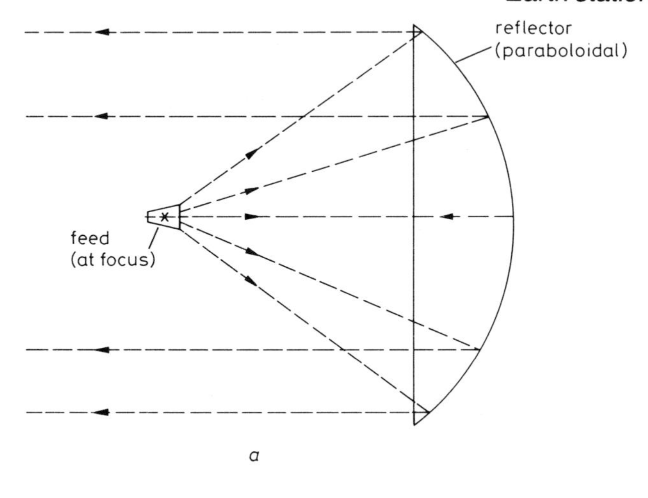

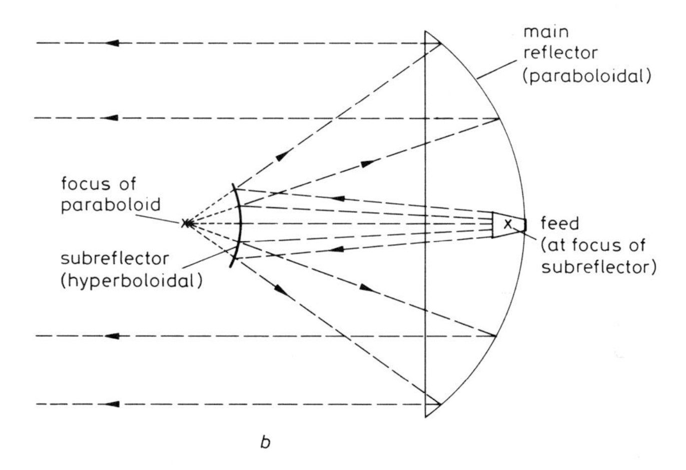

Fig. 7.3 *Axisymmetric antennas*
a Front-fed antenna
b Cassegrain antenna

transmit and receive orthogonally-polarised transmissions, and it is difficult to provide rigid support for such feeds at the focus. Another disadvantage of the front-fed configuration for large antennas is that a long length of waveguide is required to carry signals from the focus to the LNA; the attenuation of this waveguide raises the noise temperature of the antenna (by 7 K for every 0·1 dB loss). Nearly all earth-station antennas with diameters greater than about 4 m are therefore of the dual-reflector Cassegrain type (see Fig. 7.3*b*). In the Cassegrain antenna the feed is at one focus of a hyperboloidal subreflector (the other focus of the hyperboloid being coincident with the focus of the main reflector).

7.3.2.1 Beam waveguide

The feed of a Cassegrain antenna is usually mounted at the hub of the main reflector with the LNA in an enclosure nearby. However, many of the largest Cassegrain antennas use beam-waveguide systems so that their feed horns can be installed at ground level with both the LNAs and HPAs nearby. A beam-waveguide system (see Fig. 7.4) comprises four reflectors which form

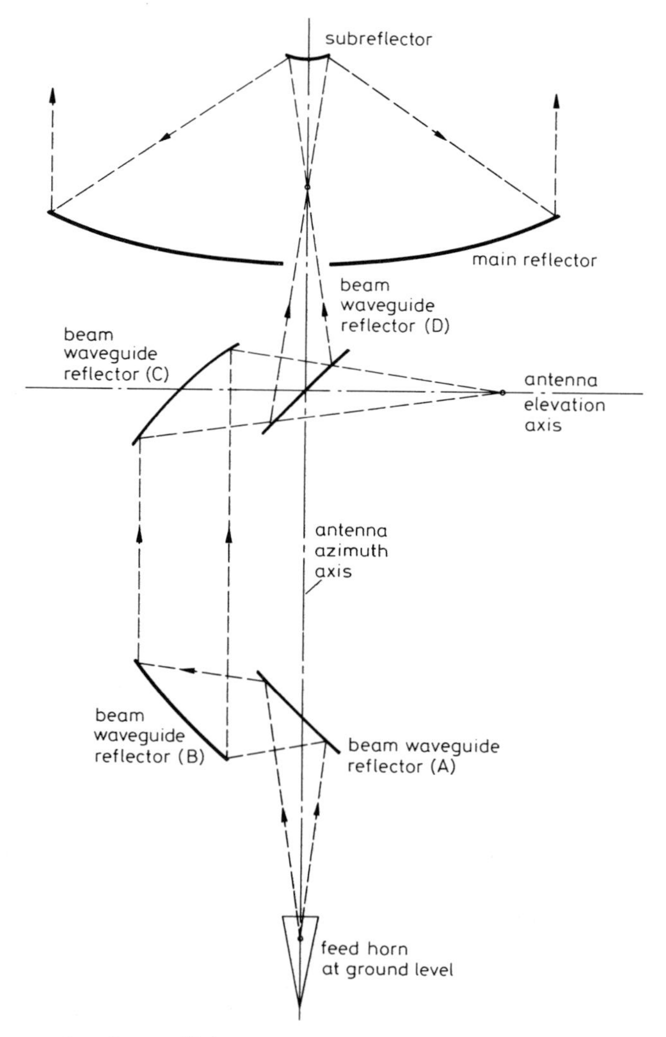

Fig. 7.4 *Cassegrain antenna with beam waveguide*

an image of the feed at the focus of the subreflector. The use of four reflectors avoids the need for physical connections across the elevation and azimuth axes whereas in conventional feed systems flexible waveguide, waveguide rotary joints or cable must be provided.

The beam suffers virtually no divergence between the two curved reflectors (B and C in the diagram) and thus a very low-loss path is provided between ground level and the elevation axis of the antenna. The total attenuation of a beam waveguide system is usually less than 0·1 dB.

7.3.2.2 *Efficiency and sidelobe level*

For maximum efficiency each point in the aperture of a reflector antenna should be illuminated with an RF signal of the same phase and amplitude, and all the energy radiated by the feed should be intercepted by the reflectors. This would require a feed with a radiation pattern with constant gain up to a certain angle off axis and zero gain in all other directions, which is impracticable. With a real feed there is always some spillover which produces sidelobes relatively close to the main beam and thus increases the probability of interference, particularly interference to (and from) adjacent satellites. To limit the spillover it is necessary to reduce the illumination towards the edge of the reflectors and accept some loss of efficiency.

With early Cassegrain antennas, which used truly-paraboloidal main reflectors and truly-hyperboloidal subreflectors, the taper of illumination required to reduce spillover to an acceptable level resulted in a considerable loss of efficiency. Modern Cassegrain antennas achieve greater efficiencies and lower sidelobe levels by using 'shaped' reflectors, i.e. reflectors which deviate from true paraboloids and hyperboloids (these antennas are, however, still usually referred to as paraboloidal antennas).

7.3.2.3 *Offset reflectors*

The antennas shown in Fig. 7.3 use main reflectors whose surfaces are symmetrical about the axis of the paraboloid, i.e. axisymmetric reflectors. It is also possible to use asymmetric paraboloidal surfaces as reflectors (see, for example, Fig. 7.5). Antennas with asymmetric (offset) reflectors can be designed to have better sidelobe performance than antennas with axisymmetric reflectors because it is possible to put the subreflector where it does not block or scatter the energy radiated by, or falling on, the antenna aperture. Offset antennas are also useful for windy sites, such as the roofs of tall buildings, because (as can be seen from the Figure) they can be designed to have a low profile which offers relatively little wind resistance.

Large reflectors are generally assembled from many panels and large axisymmetric reflectors are, in general, cheaper than large asymmetric reflectors because symmetry reduces the number of different panel profiles

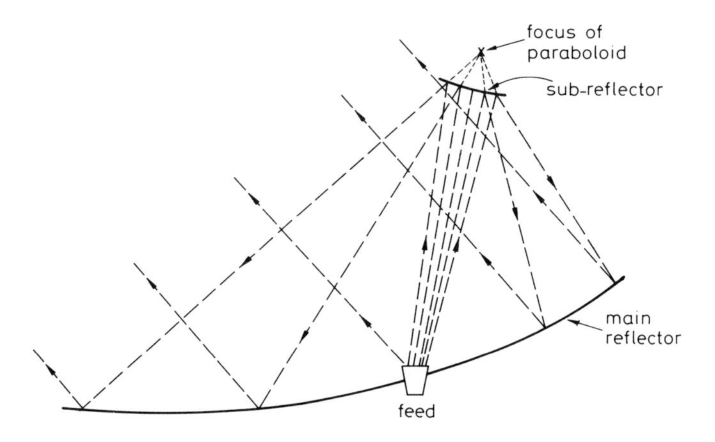

Fig. 7.5 *Offset Cassegrain antenna*

required. Large offset antennas also tend to be expensive because it is more difficult to provide rigid mechanical support for an offset subreflector than for the reflector of an axisymmetric antenna.

However, as the problems posed by interference become worse, the number of offset antennas in use is growing rapidly.

7.3.2.4 Gregorian antennas

Although the Cassegrain configuration is by far the most common for dual-reflector antennas, the Gregorian configuration can also be used. The Gregorian antenna uses a paraboloidal main reflector and an ellipsoidal subreflector; the latter is positioned beyond the focus of the main reflector whereas the subreflector of a Cassegrain antenna is positioned between the focus and the main reflector.

7.3.2.5 The Torus antenna

The Torus antenna (see Fig. 7.6) was developed by COMSAT Laboratories in the early 1970s. The reflector of this type of antenna has a shape which can be generated by rotating an offset sector of a parabola (ABC) about a vertical axis OZ (i.e. the surface is parabolic in plane YOZ and circular in an orthogonal plane parallel to XOY). The Torus has a focal arc instead of a focal point and a feed may be moved along this arc in order to steer the antenna beam or several feeds may be positioned along the arc to give simultaneous access to a number of satellites. For use with geostationary satellites the reflector is tilted so that movement of the feed along the focal arc causes the beam to move along the required arc of the geostationary orbit. Torus antennas with dimensions of about $4{\cdot}5 \times 11$ m, giving access to a 50°

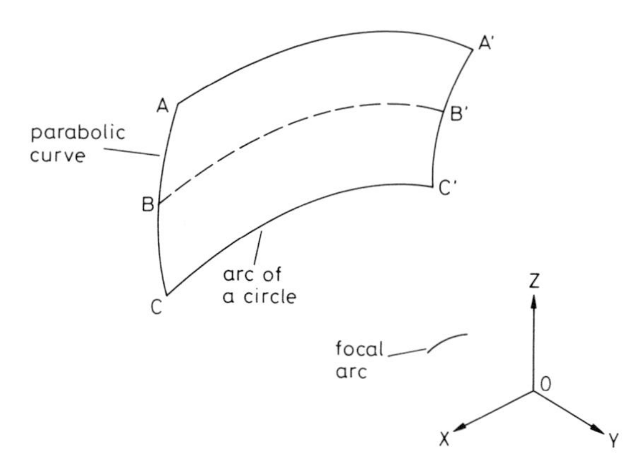

Fig. 7.6 *Torus antenna*

arc of the geostationary orbit, have been used by stations in the USA feeding cable-TV networks. The flexibility of the Torus has to be paid for by a decrease in efficiency and an increase of sidelobe level.

7.3.3 Radiation patterns

The gain of an earth-station antenna must be kept as low as is practicable in all directions outside the main lobe in order to minimise interference from and to other systems. Particular attention must be paid to the near-in sidelobes because they are a major factor in determining the minimum allowable spacing between satellites (and therefore the total traffic capacity of the geostationary orbit).

The main influences on the level of the sidelobes are:

(i) *Aperture Illumination*
Sidelobes are an integral part of the radiation pattern of an aperture. The amplitude of the envelope of these sidelobes decreases with increasing angular separation from the main beam and the larger the aperture the greater is the rate of decrease. Thus, for example, the envelope reaches isotropic level at about 11° and 7° for uniformly-illuminated apertures with diameters of 400 and 100 wavelengths, respectively (these diameters correspond to about 11 and 2·7 m, respectively, at 11 GHz). In practice the illumination of an aperture is usually reduced towards its edge in order to limit spillover and this affects both the amplitude and the shape of the envelope.

(ii) *Scattering and blockage*
Any object which blocks part of the radiation from an aperture causes a disturbance in the wavefront and the formation of additional sidelobes.
The effect of blockage is to add, to the aperture radiation pattern,

the pattern which would be produced by exciting the blocked area with an illumination equal in amplitude but 180° out of phase with the illumination of the aperture.

The blockage pattern of a subreflector is wider than that of the corresponding main reflector (because the subreflector aperture is smaller than that of the main reflector) and the sidelobes produced by the blockage alternately cancel and reinforce the sidelobes of the aperture radiation pattern.

Because of refraction there is a lower practical limit (of about ten wavelengths) to the diameter of a subreflector; thus, as antenna diameter decreases below about 100 wavelengths the subreflector of an axisymmetric antenna begins to block a significant proportion of the main aperture and the near-in sidelobe pattern begins to deteriorate. As a result, the prime-focus configuration may give better sidelobe performance and be cheaper than the Cassegrain configuration for small antennas (provided that only a small simple feed is required).

Struts supporting the subreflector or feed of an axisymmetric antenna cause blockage and scattering of energy. When the antenna is transmitting, some energy is scattered by the struts as the wavefront reflected from the subreflector travels to the main reflector and more energy is scattered after the wavefront has been reflected from the main reflector. It is possible to design the antenna so as to reduce the sidelobes produced by the struts to very low levels in a particular plane (e.g. the plane of the geostationary orbit) at the expense of increasing the corresponding sidelobes in other planes.

The effects of blockage and scattering are a major factor in determining the sidelobe levels of axisymmetric antennas and considerable improvements in performance can therefore be achieved by using offset antennas.

(iii) *Spillover*

Spillover past the subreflector usually makes a significant contribution to the radiation pattern of a Cassegrain antenna, especially at angles between about 10° and 90° off axis. There is also spillover from the subreflector past the main reflector but this is at a lower level.

With a front-fed antenna much of the transmitted energy which spills past the edge of the main reflector is intercepted by the earth. The converse of this is that the earth radiates directly into the feed and raises the noise temperature of the antenna, whereas the feed of a Cassegrain antenna sees only cold sky round the edges of the subreflector. The higher noise temperature of the front-fed antenna is not so important with small antennas which are generally used with simple uncooled LNAs.

(iv) *Profile errors*

The sidelobes caused by errors in the profile of the main reflector are influenced by two factors: the root mean square of the errors (expressed in units of wavelength) and the correlation interval of the errors.

The envelope of the sidelobes caused by profile errors falls off less rapidly with increasing angle off axis if the correlation distance of the errors is reduced. Thus, a repeated pattern of errors in forming a series of panels or assembling them to form a reflector can have a serious effect, especially if the antenna is made up of small panels.

Typical profile accuracies for 'off-the-shelf' antennas with a diameter of less than about 10 m are better than 1 mm RMS for antennas for use in the 6/4 GHz bands and better than 0·7 mm RMS for antennas for the 14/11 GHz bands; greater accuracy, down to about 0·25 mm RMS, can be achieved without too much difficulty or expense.

Radiation patterns for earth-station antennas are included in the reports and recommendations of CCIR Study Group 4 (and in documents issued by other bodies such as INTELSAT and the US FCC). These patterns serve for the determination of co-ordination distance and the assessment of interference when the actual radiation diagram of the antenna to be used is not known. Reference radiation patterns may also be used as specifications or design objectives.

The CCIR radiation patterns for antennas installed before 1987 are defined as follows:

(*a*) *Antennas with a diameter greater than 100 wavelengths*

$$\begin{aligned} G &= 32 - 25\log\theta \quad \text{for} \quad 1^\circ \leqslant \theta < 48^\circ \\ &= -10 \quad\quad\quad\quad\; \text{for} \quad 48^\circ \leqslant \theta < 180^\circ \end{aligned}$$

where G = gain in dB relative to an isotropic antenna
θ = angle off the axis of the main beam

(*b*) *Antennas with a diameter not greater than 100 wavelengths*

$$\begin{aligned} G &= 52 - 10\log(D/\lambda) - 25\log\theta \quad \text{for} \quad 100(\lambda/D) \leqslant \theta < 48^\circ \\ &= 10 - 10\log(D/\lambda) \quad\quad\quad\quad\; \text{for} \quad 48^\circ \leqslant \theta < 180^\circ \end{aligned}$$

where λ = wavelength
D = diameter of the radiating aperture

Because of the increasing risk of interference caused by congestion of the geostationary orbit the design objectives for new antennas are becoming more stringent. For example, for new Standard-A stations INTELSAT specifies that no more than 10% of the transmit sidelobe peaks within ±3° of the plane of the geostationary orbit shall have a gain G greater than:

$$\begin{aligned} G &= 29 - 25\log\theta \text{ dBi for } 1^\circ \leqslant \theta < 20^\circ \\ &= -3{\cdot}5 \text{ dBi for } 20^\circ \leqslant \theta < 26{\cdot}3^\circ \end{aligned}$$

$$\begin{aligned} G &= 32 - 25\log\theta \text{ dBi for } 26{\cdot}3^\circ \leqslant \theta < 48^\circ \\ &= -10 \text{ dBi for } \theta > 48^\circ \end{aligned}$$

Note that the requirement that not more than 10% of the sidelobes shall exceed a specified gain is open to objections. Consider, for example, an antenna which apparently just fails to meet the specification; if more sensitive measuring equipment is used to retest the antenna then the number of low-level sidelobes which can be detected will be increased and the antenna may now be judged satisfactory. It will be seen that specifying the radiation pattern of an antenna is no easy matter; nor is it easy to check that the specification has been met.

7.3.4 *Polarisation*

The vector E in Figure 7.7 represents the electric field of a wave; the vector has constant magnitude and rotates about the axis of propagation at a constant angular speed. The locus of this E-field vector, when projected onto a plane normal to the direction of propagation, is a circle and the wave is therefore described as circularly polarised. The wave is defined as right-hand circularly polarised (RHCP) or left-hand circularly polarised (LHCP) according to whether the vector rotates clockwise or anticlockwise (respectively) when the observer is looking in the direction of propagation.

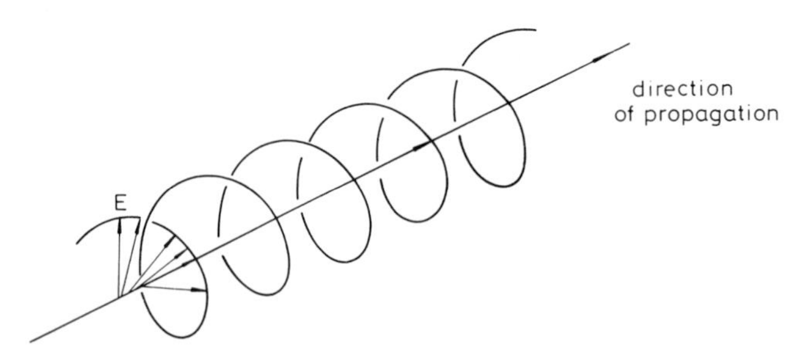

Fig. 7.7 *Circularly-polarised wave*

In the more general case, when the magnitude of the electric-field vector varies periodically, the locus is an ellipse and the wave is described as elliptically polarised. Any elliptically-polarised (or circularly-polarised) wave can be resolved into two orthogonal linearly-polarised waves, one of which is 90° out of phase with the other (see Fig. 7.8).

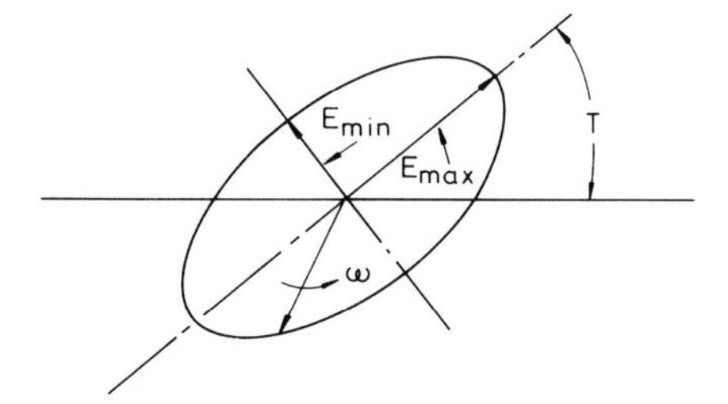

Fig. 7.8 *Elliptically-polarised wave*

7.3.4.1 Polarisation state

The polarisation state of a wave can be completely described by the amplitudes of the major and minor axes of its ellipse E_{max} and E_{min}, the tilt or inclination angle T (which is the angle of the major axis relative to some arbitrary reference axis) and the sense or hand of the polarisation. E_{max}/E_{min} is called the voltage axial ratio (VAR) of the wave. (Note that the inverse definition of VAR is sometimes used, i.e. VAR = E_{min}/E_{max}.)

Strictly speaking a wave is only circularly-polarised if its VAR is equal to 1, but in practice it is usually described as circularly-polarised if the VAR is approximately equal to 1 (say less than 1·5). Similarly, a wave is usually described as linearly-polarised if its VAR is large, although strictly speaking it is linearly-polarised only if E_{min} is zero.

The axial ratio is often expressed in decibels; thus a VAR of 1·06 is the same as an axial ratio of 20log1·06 = 0·5 dB. (If the inverse definition of VAR is used the VAR is 1/1·06 = 0·94 and 20log0·94 = −0·5 dB, but the negative sign is usually ignored.)

An antenna has a polarisation state which can be specified in the same way as that of a wave. The proportion of the power of a wave which is transferred to an antenna depends on the relative senses of their polarisations, the VARs of the two polarisation ellipses and the angle between the major axes of the ellipses (i.e. the difference between the tilt angles). All the power is transferred from a wave to an antenna if the two polarisation states have the same sense of rotation, equal VARs and equal tilt angles. On the other hand, no power is transferred from a wave to an antenna if the two polarisation states have the opposite sense of rotation, the ellipses have equal VARs, and the major axes of the ellipses are at right-angles; in the latter case the wave and the antenna are said to be orthogonally polarised.

7.3.4.2 Dual polarisation

By using a dual-polarised feed with two input ports an antenna can transmit two orthogonally-polarised waves on the same frequency; similarly, by using a dual-polarised feed with two output ports it is possible for another antenna to receive the two orthogonally-polarised waves and separate them out without (in theory) either wave interfering with the other. In practice the waves will be nearly orthogonal rather than truly orthogonal and there will be some interference between them. The departure from true orthogonality arises because of:

(i) imperfections in the antennas and
(ii) changes of polarisation which occur during transmission through the ionosphere and the atmosphere (see Section 3.5.2).

7.3.4.3 XPI and XPD

Consider two orthogonally-polarised waves of equal amplitude represented by E_1 (RHCP) and E_2 (LHCP) transmitted from a satellite and received by a dual-polar earth-station antenna (see Fig. 7.9*a*). If the satellite antenna and the earth-station antenna are correctly designed (and propagation conditions are normal) then the amplitude E_{11} of the co-polar signal (i.e. the component of wave E_1 appearing at the RHCP output port of the antenna) will be very much greater than the amplitude E_{21} of the cross-polar signal (i.e. the component of wave E_2 appearing at the RHCP output port). $20\log(E_{11}/E_{21})$ is called the cross-polar isolation (XPI). (The cross-polar isolation for the other transmission is, of course, $20\log(E_{22}/E_{12})$

When only one polarisation E_1 is transmitted (as is very often the case with polarisation experiments) it is only possible to measure the amplitudes E_{11} and E_{12} (see Fig. 7.9*b*); the ratio $20\log(E_{11}/E_{12})$ is called the cross-polar discrimination (XPD).

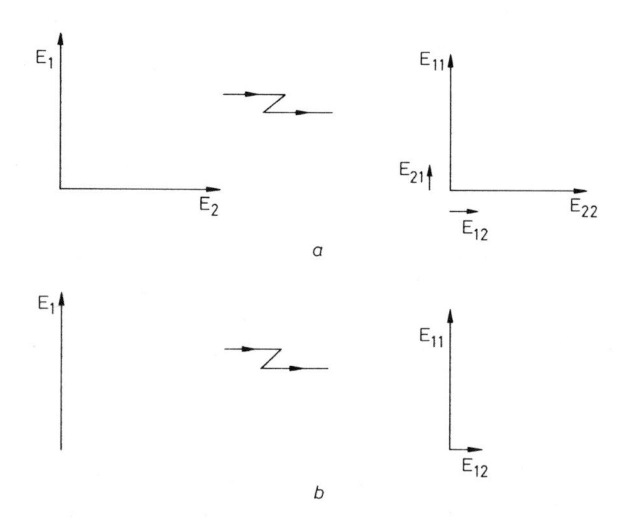

Fig. 7.9 *(a) XPI and (b) XPD*

(Note: These definitions of XPD and XPI are those most commonly used but some books and articles interchange the definitions, i.e. $20\log(E_{11}/E_{12})$ is called cross-polar isolation; as the values of isolation and discrimination are usually very nearly the same this potential source of confusion does not normally matter.)

It can be shown that, for elliptically-polarised waves with axial ratios of less than about 3 dB (which covers most practical cases) a good approximation to the XPI is given by the expression:

$$\mathrm{XPI} = 24{\cdot}8 - 20\log R \text{ decibels}$$

where

$$R_w^{\;2} = R_w^{\;2} + R_a^{\;2} + 2R_wR_a \cos 2d$$

and

R_w = $20\log r_w$ (where r_w is the VAR of the wave)
R_a = $20\log r_a$ (where r_a is the VAR of the antenna)
d = difference between tilt angles

As an example take a wave and an antenna with axial ratios of 0·50 and 0·25 dB respectively, and a difference between the tilt angles of 90°:

$$R^2 = 0{\cdot}5^2 + 0{\cdot}25^2 + (2 \times 0{\cdot}5 \times 0{\cdot}25)(-1)$$
$$= 0{\cdot}0625 \text{ and}$$

$$\mathrm{XPI} = 24{\cdot}8 - 10\log 0{\cdot}0625$$
$$= 36{\cdot}8\,\mathrm{dB}$$

On the other hand if the ellipses are aligned (i.e. $d = 0°$) then the same axial ratios result in an isolation of only 27·3 dB.

Fig. 7.10 shows the limits of XPI for a wave with a VAR of 0·5 and a range of antenna VARs from 0 to 2.

7.3.4.4 Polarisers and feeds

Fig. 7.11*a* shows a linearly-polarised wave ($E\sin\omega t$) incident on a quarter-wave polariser; the angle between the planes of the wave and the polariser being ϕ. The component of the wave in the plane of the polariser is delayed by 90° and, as will be seen from Fig. 7.11*b*, the output from the polariser comprises the components $E\sin\phi \sin\omega t$ and $E\cos\phi \cos\omega t$ which are in time and space quadrature. The locus of the vector formed by adding these two components is an ellipse, and a quarter-wave polariser thus transforms a linearly-polarised wave to an elliptically-polarised wave. If the plane of the polariser is at 45° to the plane of the linear wave then a circularly-polarised wave is generated.

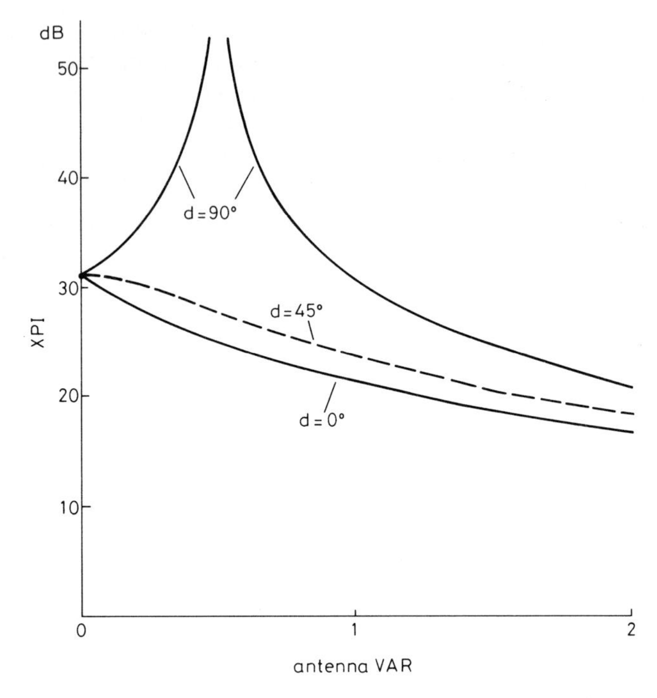

Fig. 7.10 *Variation of XPI with VAR*
Wave VAR = 0·5 dB

If two orthogonal linear waves are applied to a quarter-wave polariser then two elliptically-polarised (or circularly-polarised) waves with opposite hands of polarisation are generated. Conversely a quarter-wave polariser may be used to convert two orthogonal elliptically-polarised waves to two orthogonal linearly-polarised waves.

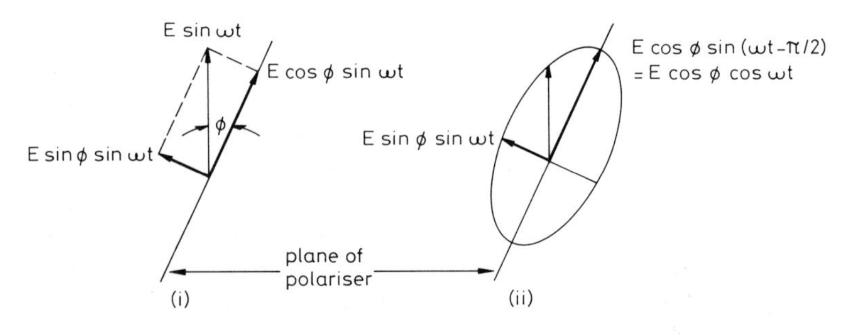

Fig. 7.11 *Action of a quarter-wave polariser*
(i) Input to polariser
(ii) Output from polariser

Fig. 7.12*a* shows a linearly-polarised wave incident on a half-wave polariser, the angle between the plane of the polariser and the plane of the wave being ϕ. The component of the wave in the plane of the polariser is delayed by 180° (i.e. it is reversed in sign). The effect of the polariser is thus, as shown in Fig. 7.12*b*, to shift the plane of polarisation of the wave by 2ϕ. The effect of a half-wave polariser on an elliptically-polarised wave is that the polarisation ellipse is reflected in the plane of the polariser and the sense of rotation is reversed (i.e. clockwise rotation becomes anticlock and vice versa) but the axial ratio is unchanged.

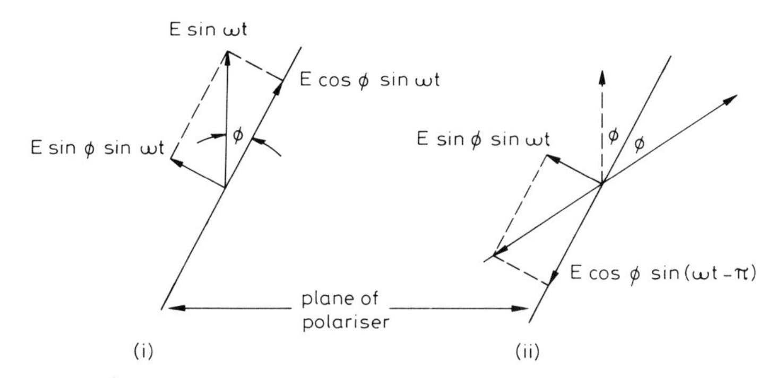

Fig. 7.12 *Action of a half-wave polariser*
(i) Input to polariser
(ii) Output from polariser

With two rotatable polarisers it is possible to convert a wave of any polarisation state to a linear wave with any orientation (or vice versa). For example, a near-circularly-polarised wave, or a near-linearly-polarised wave, can be converted to a linearly-polarised wave by means of a quarter-wave polariser and then rotated into the plane of a waveguide port by means of a half-wave polariser. Such an arrangement of polarisers makes it possible to adjust the polarisation of the earth-station antenna so as to get the maximum practicable discrimination. Polarisers can also be used to provide continuously-adaptable compensation for variations in polarisation state induced by the differential phase shifts caused by rain (see Section 3.5.2) but they cannot compensate for the effect of differential attenuation (which is significant at frequencies of 11 GHz and above). Although polarisers can be used to improve discrimination, they cannot make a pair of non-orthogonal transmissions into orthogonal transmissions.

Good discrimination between cross-polarised transmissions is usually only possible for transmissions received from or transmitted in directions close to the boresight; this discrimination is usually limited mainly by the variation with frequency of the phase shift and attenuation of the polarisers and other components in the dual-polar feed. Off the boresight but within the main beam the discrimination is determined mainly by the characteristics of the feed horn but antenna reflectors and beam waveguides may also reduce discrimination. Outside the main beam it is unwise to rely on significant discrimination.

The cross-polar performance of a horn varies with frequency but corrugated-horn feeds generate only low cross-polar fields over a wide frequency range (and the cross-polar field can be expected, typically, to be at least 35 dB below the co-polar level in the main beam of the antenna). The asymmetry of offset reflectors can be a cause of increased cross-polar radiation but good discrimination can be obtained by designing the feed or shaping the reflectors to compensate for this effect.

Cross-polar discrimination of greater than about 30 dB does not usually result in a significant improvement in system performance (because of the effect of other types of degradation). The axial ratios specified for the INTELSAT V satellite hemi/zone antennas and for earth stations working to the satellite are 0·75 dB and 0·5 dB, respectively. This corresponds to an overall discrimination of 24—28 dB depending on the difference between the satellite and earth-station antenna tilt angles.

Fig. 7.13 is a block diagram of a dual-polar feed for the transmission and reception of orthogonal circularly-polarised signals. Two orthogonal linearly-polarised 6 GHz transmissions (V and H) are combined in an orthomode transducer (OMT) and passed to the polariser. The linearly-polarised signals are converted by the polariser to circularly-polarised signals and the two orthogonal signals (RHCP and LHCP) are passed to the diplexer (which provides isolation between the 4 and 6 GHz signal paths) and thence to the feed horn. Conversely, the two orthogonal circularly-polarised receive signals are passed from the horn via the diplexer to the polariser where they are converted to two orthogonal linearly-polarised signals (v and h). The feed may also include a coupler for a monopulse tracking system (see Section

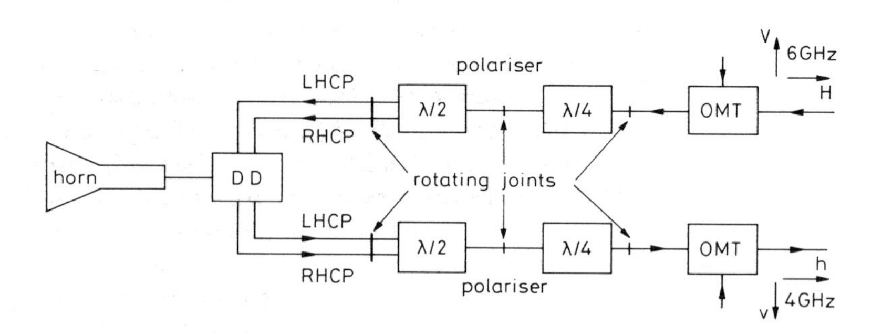

Fig. 7.13 *Dual-polar feed system*
DD = dual-polarised diplexer
OMT = orthomode transducer

7.3.5); this is not shown in the diagram but could be included between the horn and the diplexer.

Feeds for linearly-polarised signals do not require quarter-wave polarisers. It is however necessary, with linear polarisation, to align the polarisation direction of the earth-station antenna with that of the satellite.

7.3.5 Antenna pointing and tracking

As a result of the suspension of Shuttle launches from January 1986 and problems with other launch vehicles which occurred in 1986–87 (see Section 2.3.3) many new satellites are still on the ground long after they should have been in service. This has made it necessary to keep some older satellites in use for longer than was intended and, when these satellites become short of on-board fuel, their positions can no longer be constrained within normal limits. This is a temporary situation and, in what follows, it is assumed that the position of geostationary satellites is closely controlled. Thus, under normal circumstances the position of an INTELSAT V satellite should be maintained within 0·1° N–S and E–W of its nominal position and the station keeping of future satellites, such as INTELSAT VI, should be better than this.

The apparent change in elevation and azimuth of a satellite, as seen from an earth station, is less than the N–S and E–W movement of the satellite (except for an earth station at the sub-satellite point) but, even so, the apparent angular movement of geostationary satellites can be comparable with the 3 dB beamwidth of a large antenna (e.g. the beamwidth of an antenna with a diameter of 13 m at 14 GHz is about 0·1°). Large antennas must therefore be provided with a means of tracking the satellite.

On the other hand, there is usually no point in providing a tracking system for, say, an antenna with a diameter of 1 m working at 14 GHz; such an antenna has a beamwidth of about 1·5° and the loss of gain if the satellite moves 0·1° off axis is less than 0·1 dB.

Somewhere between the two extremes (of large antennas which must track the satellite and small fixed antennas whose gain remains virtually unchanged when the satellite moves slightly off the axis of the main beam) there is a range of antenna beamdwidths where, in the absence of a tracking system, the gain of the antenna does vary significantly with satellite movement but it is cheaper to do without a tracking system and provide a slightly larger fixed antenna to allow for the reduction of gain when the satellite is not in the centre of the antenna beam. When the satellite drift is limited to about 0·1° N–S and E–W, there is evidence to suggest that a fixed antenna may be cheaper than a tracking antenna when a gain of less than about 53 dB is required (a gain of 53 dB corresponds to reflector diameters of about 3·5 m at 14 GHz and 9 m at 6 GHz, respectively); these figures are, however, very approximate since a large number of factors must be taken into account.*

An earth-station antenna may be required to work to satellites in different positions in the geostationary orbit during its lifetime and some antennas may need to switch between satellites quite frequently (for example, an antenna at an earth station which is required to relay programmes from a number of different satellite systems to a cable-TV network or a local broadcasting station). In such cases it should be possible to repoint the antenna simply and rapidly, whether or not it has a tracking system.

7.3.5.1 EL–AZ and HA–DEC

The most commonly used mounts for earth-station antennas are the elevation-over-azimuth (EL–AZ) mount and the hour-angle/declination (HA–DEC) mount (the latter is also called the polar mount).

The EL–AZ mount is widely used for both large and small antennas where full steerability is required. The azimuth axis of an EL–AZ mount is vertical and the elevation axis is horizontal (see Fig. 7.14*a*).

* See, for example, DINWIDDY: 'Performance of fixed-mount earth-station antennas', *ESA Journal*, 1981, **5**, pp. 187–196

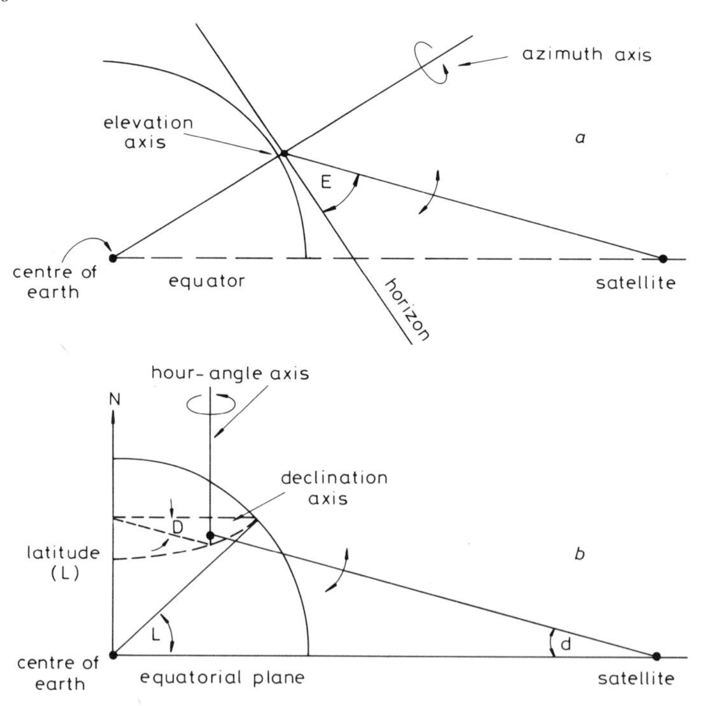

Fig. 7.14 *EL–AZ and HA–DEC axes*
(a) EL–AZ axes
(b) HA–DEC axes

The HA–DEC mount is much used by astronomers. The hour-angle may be equated with the difference in longitude between an observer and a body being observed; the declination is the angular distance of the body from the celestial equator. The hour-angle axis of an HA–DEC mount is parallel to the earth's axis (see Fig. 7.14*b*); thus, as the earth rotates, the beam of a telescope or antenna may be kept pointing at a very distant object (such as a star) by rotating the mount about the HA axis in the opposite direction to the earth but at the same angular speed. If the star is in the equatorial plane then the beam will point parallel to the equator (i.e. the declination is zero). However, if the beam is pointed at an object in the equatorial plane, such as a geostationary satellite, which is not infinitely distant from the earth, then the beam will be at an angle d to the equator (measured in the plane of the HA axis); this angle d is a function of the latitude of the site L and the difference D between the longitudes of the station and the satellite. When an antenna is pointed at the geostationary orbit d varies only slightly with changing D provided that D and L are not too large; it is thus usually possible to direct an antenna at any point in a limited arc of the geostationary orbit with reasonable accuracy by rotating it about the HA axis alone, i.e. it is possible to use a single-axis mount. By slightly tilting the axis of rotation away from the HA axis the single-axis pointing error can be made less than 0·01° for positions anywhere in the arc of the geostationary orbit up to a longitude of about 70° on either side of the earth station, provided that the station latitude is less than about 50°.*

On the other hand there is no possibility of single-axis operation when using an EL–AZ mount because the elevation to the geostationary orbit varies quite rapidly with azimuth.

Single-axis mounts are very popular for small antennas (with diameters of, say, less than 5 m) because they are cheap and because they make it easy to find a satellite in the geostationary orbit once the antenna has been properly aligned. Another advantage of a single-axis mount is that the polarisation of the antenna with respect to the geostationary orbit is virtually independent of its pointing direction. For larger antennas it is difficult to solve the problems associated with supporting a large mass on a non-vertical rotating axis and the EL–AZ mount is almost always preferred.

7.3.5.2 Drives

Many antennas are provided with motorised drives for moving the antenna in elevation and azimuth. The maximum angular speed required for tracking a geostationary satellite is very low indeed and only small motors are required for this purpose. However it is usually specified that it shall be possible to move large antennas (e.g. those for INTELSAT Standard-A and Standard C

* 'Satellite Communications Symposium 1982' (Scientific Atlanta), Section 2B–2)

stations) at speeds of at least 10°/min in the highest wind speed at which the antenna is intended to remain operational. This is necessary for two reasons:

(i) In order that it shall be possible to repoint the antenna reasonably quickly even when the wind is close to the maximum speed at which operation is practicable.
(ii) Because it may be necessary to drive the antenna to stow and lock it in position if exceptionally high wind speeds are forecast. The longer it takes to drive to stow, the more chance there is that the wind forces will rise to a level which the motors cannot overcome before stow is reached. (Note: the stow position is usually that position in which the antenna presents minimum resistance to the wind; for most antennas this is with the reflector horizontal.)

Drive speeds for small and medium-sized antennas cover a wide range from about 0·1°/min to super-speed drives capable of 120°/min; the latter are used for TVRO stations which need to receive successive programmes from different satellites. Most antennas with a diameter of 11 m or more are provided with motorised drives as standard; such drives are often an option for smaller antennas with diameters down to about 5 m, and antennas with a diameter of less than about 5 m are usually repositioned manually.

7.3.5.3 Tracking

When satellite communications was in its experimental phase earth stations were steered manually by operators using joysticks to maximise the signal. Manual steering was soon abandoned in favour of automatic systems which were of two basic types: programme track and autotrack.

With programme track the pointing angles for the earth station are calculated from the track of the satellite. The orbital elements required for the generation of the data must be regularly updated by tracking stations and the pointing angles must be generated as frequently as is necessary to ensure the required pointing accuracy. Programme track only works satisfactorily when the satellite orbit is stable and very accurately known and it requires the generation of large amounts of information. The method is less popular than it was but it is still used by some domestic systems.

Autotrack systems use a signal from the satellite to generate pointing information. The two varieties of autotrack are steptrack and monopulse.

Steptrack systems work by making small changes in the pointing direction of the antenna and noting whether the signal increases or decreases; the information collected from a series of successive samples is used to determine the best direction of pointing.

Monopulse systems originated in the world of radar. There are many different types of monopulse system in use with large earth-station antennas; the thing which they all have in common is that they derive information about the direction of the satellite from continuous processing rather than by the comparison of successive samples. Probably the most common form of monopulse system used with earth-station antennas is that which works by detecting higher-order modes which are generated in the antenna feed by signals received from sources off the centre of the antenna beam.

Monopulse systems give very good accuracy but require complex, and therefore expensive, feeds and receivers. Conventional step-track systems are relatively cheap and simple but the stepping results in a loss of gain and the systems are easily confused by changes in the level of the signal being sampled (such as those caused by rain attenuation or by rapid changes in satellite position when orbital corrections are applied).

Improved steptrack systems are now available, e.g. smoothed step track (SST).* SST combines the advantages of step track and programme track by using the step-track process to build a model of the satellite track rather than using it directly to optimise antenna pointing. Once the SST equipment has established an accurate model of the satellite track it is able to keep this model up to date (under normal circumstances) by taking a sample every 15—30 min. If the signal which is being sampled is subject to random variations SST will ignore any samples which would introduce errors; if the signal disappears altogether, for example as a result of failure of the beacon or a receiver, SST will predict satellite position for up to 24 hours. If SST detects a serious departure from the model such as would be caused by an orbit correction then it restarts the learning process. The accuracy of SST is claimed to approach that of monopulse and the low mechanical duty cycle under normal conditions improves the life of the steering equipment and reduces any modulation of signals resulting from the stepping process.

7.3.6 Structure, stability and safety

The main elements of an earth-station antenna are a reflector with its backing structure and a pedestal or mount which supports the reflector. For large antennas the whole structure must be very rigid in order that the antenna remains undistorted and can be pointed accurately at the satellite, even during high winds. Pointing error is usually approximately proportional to the square of wind speed (over a limited range). Ideally one needs to know the pointing accuracy of large antennas as a function of wind speed up to the maximum operational speed; this, together with the cumulative distribution of mean hourly wind speed at the earth-station site and the gust factor (i.e. ratio of gust speed to mean speed), enables the calculation of the percentage of the time for which the loss of gain from pointing error is likely to result in degraded or unacceptable quality of transmission. Specifications which quote

* EDWARDS, D. J., and TERRELL, P. M.: 'Developments in antenna tracking' in 'Earth stations for the fixed satellite services', IEE Colloqium Digest 1984/78, p. 9/1.

pointing accuracy as, say, 0·04° RMS in a wind of 12 m/s gusting to 18 m/s and 0·1° RMS in a wind of 18 m/s gusting to 24 m/s give about the minimum amount of information that is useful. Specifications which only quote pointing accuracy at one arbitrary wind speed are not much help.

Most reflectors with a diameter greater than 5 m (and many smaller reflectors) are made of aluminium alloy; this has a high ratio of strength to weight and has good resistance to corrosion. Backing structures are also usually made of aluminium alloy or, for some of the larger antennas, of steel. The use of the same material for both reflector and backing structure avoids problems resulting from differential expansion.

Reflectors (except those for very small antennas) are usually made up of a number of panels: from about six panels for an antenna with a diameter of 3 m to over 100 panels for some INTELSAT Standard-A or Standard-C antennas. For the larger antennas the panels are usually stretch-formed by processes similar to those used in the aircraft industry. The panels for smaller reflectors (with diameters less than about 8 m) are often formed in the same way as automobile bodies, by pressing between matched dies. Surprisingly good surfaces can be achieved in this way; reflectors with an accuracy of 0·4 mm RMS have been assembled from pressings and even a dustbin lid, used as a reflector for an amateur earth station, was found to depart from a parabola by less than 0·2 mm over most of its surface.* Other methods of forming smaller reflectors are by spinning (which gives accuracies of about 1 mm RMS) and by casting followed by machining (which gives accuracies of 0·25 mm RMS or better). Reflectors with a diameter of less than about 5 m are sometimes made from moulded plastic with a conductive coating; these had a bad reputation at one time but this method of production is perfectly satisfactory when it is properly carried out. The use of moulded plastics has particular advantages in the fabrication of asymmetrical antennas.

Pressed panels are formed complete with flanges at the edges which provide stiffness. Stretch-formed panels are usually welded or riveted to radial and circumferential stiffeners in special jigs which ensure that the panels are not distorted by the attachment of the stiffeners.

Small reflectors are usually provided with very simple backing structures but the largest reflectors (e.g. those for INTELSAT Standard-A and Standard-C stations) have a complex structure comprising a massive central hub and radial trusses, the latter being interconnected by further beams to give a deep framework with maximum rigidity.

The wheel-and-track mount is commonly used for the largest antennas (with reflector diameters of 16 m or more). Fig. 7.15 shows a wheel-and-track antenna; the reflector is counterbalanced and is moved in elevation by means of the large sector wheel. The whole of the moving structure rests on four flanged wheels at the corners of the base of the mount; these wheels take the

* EVANS, D.: 'A dustbin-lid aerial for 10 GHz', *Radio Communication*, March 1976, pp. 194–198.

Fig. 7.15 *Wheel-and-track antenna (acknowledgment to Marconi Satellite Systems Ltd)*

weight of the antenna and resist the wind forces trying to overturn it, while any drag force is taken by a large central azimuth bearing. The wheels run on a single rail and two of them are driven. One of the main problems in designing a wheel-and-track antenna is ensuring that the friction between the wheels and the rail is sufficient under all conditions to prevent the antenna being blown round in azimuth.

Another way of mounting large or medium-sized antennas (with reflector diameters of, say, 7 m or more) is to use a kingpost (see Fig. 7.16) which rotates in azimuth bearings at the top and bottom of the post; the lower of these bearings usually takes the weight of the rotating structure while the upper bearing resists any overturning moment.

The reflectors of smaller antennas (with limited steerability) may be connected to the mount by means of pivots.

Small antennas (and some large ones) may be driven by means of jackscrews. It is usually only possible to drive an antenna over a limited sector using a jackscrew and various means are used to extend the usable range when this is necessary; for example, a number of anchorage points may be provided for the azimuth jackscrew and extension struts may be inserted in the elevation drive.

The weight of the moving structure of a typical modern wheel-and-track antenna with a reflector diameter of about 18 m is about 150 tonnes and even

Fig. 7.16 *King-post antenna (acknowledgment to Marconi Satellite Systems Ltd)*

a 3 m antenna weighs around 300 kg (0·3 tonnes). A thick coating (say 12 mm) of ice can double the effective weight of an antenna.

The maximum wind force which can be exerted on an antenna is approximately proportional to the square of the wind speed and the area of the antenna. The wind force on even a relatively small antenna can be considerable; for example, the maximum force exerted along the axis of a typical 5 m antenna by a wind of 45 m/s is about 5 tonnes.

'Off-the-shelf' antennas with diameters of, say, 5—10 m are typically designed to survive in winds gusting to about 55 m/s or 125 miles/h; the survival speed for smaller antennas is often lower. For the largest antennas it is usual to specify both the wind speed at which the antenna can survive when it is in the 'stow' position and a maximum operational wind speed. For example, the antenna may be capable of surviving in its operating position up to a wind speed of 46 m/s and in its stow position at a wind speed of up to 60 m/s. Antennas with a diameter of 13 m or less are not usually driven to stow but there may well be some restrictions (e.g. on their pointing direction) associated with the survival wind speed quoted by the manufacturers.

Large antennas and their foundations are usually designed to be capable of withstanding the most adverse combination of forces which may reasonably be expected, including the worst gust of wind likely to be encountered in (at least) 50 years. The value of the once-in-50-years wind speed depends on the area of the world in which the antenna is sited, on topography (e.g. coastal sites are usually particularly windy or the wind may be channelled down valleys) and on the height above local ground level of the centre of the antenna. What adverse combination of forces it is reasonable to expect is a question which can only be answered by an engineer who is conversant with the local meteorological conditions, likelihood of earthquakes and other such factors. Even small antennas may exert large forces on their foundations or the structures to which they are attached, particularly when it is windy, and consideration must be given to the associated civil and mechanical engineering problems before they are installed.

In choosing a site for a large earth station all available geological data on the area should be studied and, before a final decision is made, a thorough geotechnical survey of the site should be made. This survey usually entails the drilling of boreholes, the examination of soil and water taken from the holes and the performance of soil loading tests. At the time of the site survey the resistivity of the soil should be measured as this information is required for the design of a proper earthing (grounding) system which is essential for safety. The result of the survey will determine whether the site is suitable for the proposed antenna and, if so, what type of foundations should be used. A concrete raft is usually a suitable foundation for even the largest antenna but, at some sites it may be necessary to use piles. For smaller antennas steel piers may be placed in holes drilled with augers which are then filled with concrete. For temporary installations the antenna may simply be erected on a heavy metal base.

When an antenna is to be mounted on an existing structure then a thorough survey of the structure must be made to ensure that it can withstand the loadings which will be imposed and a suitable means of mounting the antenna must be devised. With tall buildings it is essential to check that the movement of the building in high winds does not cause the total pointing errors to exceed those allowable for the antenna to be used. One way of mounting an antenna, with a diameter of a few metres or less, on the roof of a building is to attach it to a heavy metal base which serves as an anchor and also spreads the load.

Very small antennas, with diameters of the order of 1 m, may be mounted on poles embedded in concrete footings or attached to the walls of buildings.

7.3.7 Radiation hazards

There is no universal agreement on the power flux density (PFD) below which long-term exposure to microwave radiation causes no effects; the figure in general use in North America and much of western Europe is 100 W/m^2 but lower limits are specified in some countries (notably the USSR).

In considering possible hazards from an earth-station antenna it is useful to divide the space round the antenna into a number of zones:

(i) The zone between the main reflector and the antenna feed or secondary reflector
(ii) The far-field zone, i.e. all points within the main beam of the antenna and at a distance of not less than $2D^2/\lambda$ from the antenna aperture (where D is the diameter of the aperture and λ is the wavelength)
(iii) The near-field zone, i.e. all points at a distance less than $2D^2/\lambda$ from the radiating aperture, and within a cylinder which is based on the radiating aperture and has an axis coincident with that of the main beam and
(iv) The spillover and scatter zone, i.e. all points not included in zones (i) to (iii).

In the far-field the maximum PFD (on the axis of the antenna) is given by the expression:

$$\text{PFD} = \text{EIRP}/4\pi s^2 \qquad \text{watts/m}^2$$

where EIRP = effective isotropically radiated power, watts

s = distance from the radiating aperture, m

It can be shown that the maximum safe PFD of 100 W/m^2 will not usually be exceeded in the far-field zone if the power p radiated by the antenna feed is less than:

$$P = 160\ D^2 \qquad \text{watts}$$

where D = diameter of the antenna aperture, m

The PFD in the near-field zone undulates with increasing distance from the aperture up to a distance of about $0{\cdot}2D^2/\lambda$ at which point it reaches a maximum. It can be shown that a PFD of 100 W/m^2 will not normally be exceeded in the near-field zone if the power delivered by the feed is less than:

$$P = 4\ D^2 \qquad \text{watts}$$

Flux densities can reach very high levels in the zone between the antenna main reflector and the feed or secondary reflector. In particular, the average PFD over the aperture of the feed horn is given by:

$$\text{PFD} = P/A \qquad \text{watts/m}^2$$

where A = area of the feed horn aperture, m^2

The distribution of power in the sidelobe zone depends very much on the characteristics of the antenna. Measurements made by British Telecom International on a large antenna showed that the maximum PFD at ground level in the spillover and scatter zone occurred immediately below the edge of the main reflector. These measurements suggest that a PFD of 100 W/m^2 is unlikely to be exceeded in the spillover and scatter zone if the power delivered by the feed is less than 100 D^2 (where D is the diameter of the antenna in metres).

The foregoing estimates of safe powers allow for total reflection from obstacles except in the case of zone (ii); they should be conservative but provide an indication of the powers at which it would be wise to make PFD measurements in order to be sure that there is no hazard. However, it must be remembered that there are (unusual) circumstances which could lead to the production of high PFDs, e.g. if the radiating antenna is focused in the near field during measurements or if the power radiated by the antenna is focused by a curved metallic surface.

Care must be taken to ensure that no one, on or off site, is at hazard. Usually this means, as a minimum:

(i) That access must be forbidden to certain zones on site (such as those close to the radiating aperture) while the transmitters are on
(ii) That the transmitters must be automatically switched off when the antenna is pointed below the elevation at which the beam is well clear of any buildings or high ground on or off site.

7.4 Electronic equipment

7.4.1 General

Electronic equipment for earth stations can conveniently be considered under four heads:

(i) Radio-frequency (RF) equipment (the main items being LNAs and HPAs).
(ii) Equipment, other than RF equipment, needed for processing FDMA transmissions (e.g. up-converters and down-converters, intermediate-frequency amplifiers and filters, baseband amplifiers and filters, equalisers, modulators and demodulators). This equipment is sometimes known as ground communications equipment (GCE).
(iii) Equipment, other than RF equipment, needed for processing TDMA transmissions.
(iv) Auxiliary equipment, e.g. modules for the provision of engineering service circuits, and for control and supervision.

7.4.2 Earth-station noise temperature and low-noise amplifiers

7.4.2.1 Noise temperatures

The noise temperature and figure of merit G/T of an earth station are usually determined mainly by the LNA and the antenna.

Fig. 7.17 shows an antenna connected via waveguide to a LNA which, in turn, is connected via coaxial cable (or waveguide) to a receiver. In order to calculate the noise temperature of the system we need the following expressions:

$$T_r = (NF - 1) \times 290 \quad \text{K}$$
$$T_p = ((1 - G)/G) \times 290 \quad \text{K}$$
$$T_o = GT_i \quad \text{K}$$

where T_r = noise temperature corresponding to a noise figure NF
T_p = input noise temperature of a passive network (assumed to be at a physical temperature of 290 K)
G = gain of a network, i.e. G = (power out)/(power in), $G > 1$ for amplification and $G < 1$ for attenuation
T_o and T_i = noise temperatures at the input and output of any network with gain G

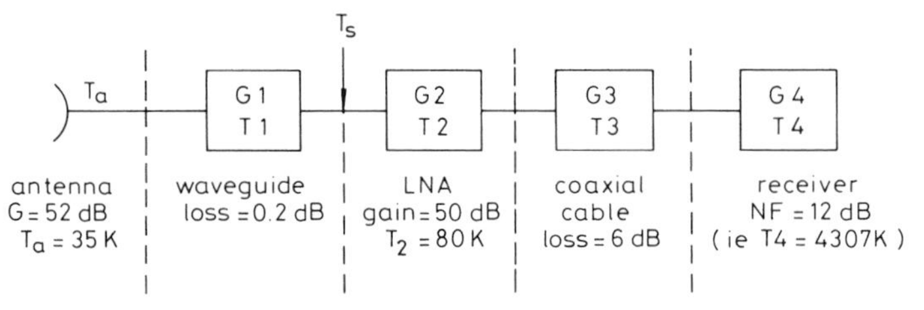

Fig. 7.17 *System noise temperatures*

We can now write:

$$T_e = T_aG_1 + T_1G_1 + T_2 + T_3/G_2 + (NF - 1) \times 290/G_2G_3$$

Where T_e = noise temperature of the earth station referred to the input of the LNA
T_a = noise temperature of the antenna (at its output terminals)
G_1, G_2, G_3 and T_1, T_2, T_3 = gains and input noise temperatures, respectively, of the networks
NF = noise figure of the receiver

Now with the values given in the Figure:

$$T_1 = [(1 - 0{\cdot}955)/0{\cdot}955]290 = 13{\cdot}7\,\text{K}$$
$$T_3 = [(1 - 0{\cdot}25)/0{\cdot}25]290 = 870\,\text{K}$$
$$T_4 = (15{\cdot}85 - 1)290 = 430\,\text{K}$$
$$T_e = 35(0{\cdot}955) + 13{\cdot}7(0{\cdot}955) + 80 + 870/10^6 + 4307/(10^5.0{\cdot}25)$$
$$= 33 + 13 + 80 + 0{\cdot}0087 + 0{\cdot}17$$
$$= 126\,\text{K}$$

From this we see that:

(i) Even a small attenuation between the antenna and the LNA raises the noise temperature of the system by a significant amount; the LNA must therefore be mounted as close as practicable to the antenna feed.

(ii) The receiver, and the cable connecting the receiver to the LNA, have a negligible effect on the system noise temperature provided that the LNA has a sufficiently high gain (50—60 dB is usual); the receiver can therefore be sited quite a long way from the antenna and LNA if this makes for convenience.

The gain of the antenna in Fig. 7.17 is 51·8 dB when referred to the input to the LNA, and the G/T of the system is therefore $(51{\cdot}8 - 10\log 126) = 30{\cdot}8$ dB/K. (Note that the G/T is independent of the point to which the gain and noise temperature are referred, provided that the same reference point is used for both quantities.)

7.4.2.2 LNAs

The two main types of LNA now in use are parametric amplifiers (paramps) and gallium-arsenide field-effect transistor (GaAs FET) amplifiers. Both types may be operated at ambient temperature or cooled thermoelectrically (to about −50° C using Peltier-effect diodes); these types of amplifier are much cheaper and easier to maintain than the LNAs cooled by gaseous helium which were used with large earth stations in the 1960s and early 1970s. Table 7.1 gives typical noise temperatures for various types of LNA (but note that further development of paramps may well result in even lower noise temperatures in the future).

Table 7.1 *Low-noise amplifiers*

Frequency	Type	NT (K)
4 GHz		
	Cryogenic (gaseous helium)	15
	Thermoelectrically-cooled paramp	30— 40
	Thermoelectrically-cooled GaAs FET	50— 60
	Uncooled (ambient) GaAs FET	70—120
11 GHz		
	Thermoelectrically-cooled paramp	80—100
	Thermoelectrically-cooled GaAs FET	120—150
	Uncooled GaAs FET	170—250

The wide range in performance is accompanied by a correspondingly wide range of price and it is necessary to consider carefully what combination of antenna and LNA will give the required G/T at minimum cost.

Bandwidths of at least 500 MHz at 4 GHz and 750 MHz at 11/12 GHz are usual.

Because the LNA has such a wide bandwidth it may receive and amplify a large number of signals (many of which may not be intended for the terminal of which it is part). LNAs must therefore be able to handle a wide range of signal powers without introducing distortion. The specification may typically require the gain compression of the LNA to be less than 1 dB at a total output power of +5 dBm. The LNA is usually protected against the transmissions of the HPAs associated with the terminal by means of a bandstop filter, with a minimum rejection of about 50 dB, at its input port.

Modern LNAs are very reliable. However, if a LNA does fail and the terminal is not equipped with a standby then all traffic is lost; all large stations and many small stations are therefore equipped with a standby LNA which is switched in automatically if the primary LNA fails. The waveguide switch, and the bandstop filter mentioned in the previous paragraph, add about 5 K to the system noise temperature.

7.4.3 High-power amplifiers (HPAs)

7.4.3.1 TWTAs and combiners

Most large earth stations are equipped with HPAs using travelling-wave tubes (TWTs) or klystrons (or both).

TWTs with bandwidths of 500 MHz or more are available and TWT HPAs can therefore be used to amplify a number of carriers of widely differing frequencies; however, when a TWT is used to amplify more than one carrier it must be 'backed off', i.e. used at reduced output power in order to avoid unacceptable intermodulation between the carriers (see Section 3.4).

If the traffic through a station grows to a level where it exceeds the capacity of the original HPA, it would be helpful if further HPAs could simply be added as they are required. Unfortunately it is not easy to combine the outputs of two or more TWT HPAs. A combining network must be placed between the HPAs and the antenna feed in order to provide isolation between the output ports of the TWTs, which must be connected to well-matched loads for correct operation. Ideally this network should have a bandwidth equal to that of the TWTs and it should have a very low loss but, in practice, it is necessary to choose between wide-bandwidth combiners with high losses (e.g. networks using directional couplers, see Fig. 7.18*a*) and low-loss combiners which are frequency conscious (e.g. networks using circulators and bandpass filters, see Fig. 7.18*b*).

If a frequency-conscious combining network is used then the frequency spectrum must be divided into sections (with guard bands between adjacent sections) and each TWT can only be used to amplify signals in one of the sections. This imposes constraints on the carrier frequencies which can be

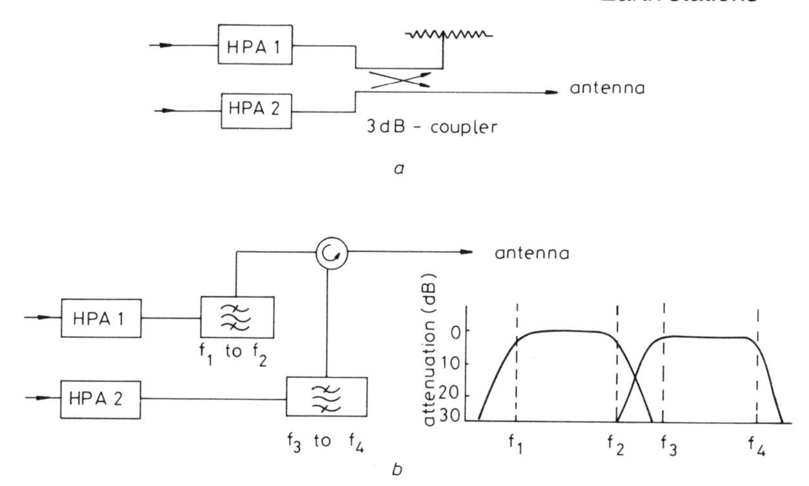

Fig. 7.18 *Combining networks*
a Broadband combiner using a directional coupler
b Combiner using bandpass filters

allocated to the terminal and on the development of new frequency plans for the satellite system. On the other hand, if a broadband combiner is used then a lot of power is dissipated in the combiner and the efficiency of the transmitter system is seriously reduced.

Instead of dividing the signals into sets and amplifying one set in each HPA, it is possible to connect HPAs in parallel; one of the advantages of parallel operation is that, if one of the tubes fails, the terminal may be able to continue transmitting at reduced power until the faulty tube is replaced. If the outputs from the HPAs have identical phase and amplitude they may be combined without loss in a 180° hybrid, but small differences in the amplitude and (particularly) the phase result in significant loss. In practice it is simple to match the phase and amplitude of the outputs of two HPAs at a single frequency but much more difficult to maintain the match over a bandwidth of several hundred megahertz. TWTs with saturated output powers of not more than a few hundred watts use helical slow-wave structures and the characteristics of such tubes are such that it is possible to operate them in parallel, but it has not proved practicable to operate high-power TWTs in parallel because the gain/frequency and phase/frequency responses of these tubes are much more variable than those of low-power TWTs.

7.4.3.2 Klystrons

Klystrons have relatively narrow instantaneous bandwidths (e.g. 40 and 80 MHz at 6 and 14 GHz respectively); they can, however, usually be tuned over a band of 500 MHz or more. Klystron HPAs have the advantage that they are generally simpler and cheaper than TWT HPAs.

A typical transmitter configuration for a large earth terminal constructed in the 1970s or early 1980s comprises a separate klystron for each satellite transponder used by the terminal and a frequency-selective combiner with pass bands aligned with those of the transponders. A standby transmitter is provided for each three or four working transmitters and these standbys are provided with automatic tuning to minimise out-of-service time, whereas the main HPAs are tuned manually. Each HPA, except for those amplifying SCPC transmissions, handles only a single carrier; this has the advantage that there is no requirement for back-off. Air-cooled klystron HPAs with a power rating of up to 3 kW are simple and cheap compared with large liquid-cooled TWT HPAs; thus it is usually possible to provide a reliable system using a separate klystron HPA for each transponder and a complex combining and switching network for considerably less than the cost of the equivalent system using TWT HPAs.

The advent of TDMA systems made it necessary to take a new look at the characteristics of HPAs. Transmissions in the INTELSAT and EUTELSAT TDMA systems occupy a bandwidth of 80 MHz and provision is made for 'transponder hopping' (i.e. switching, in the course of the TDMA frame, between transponders occupying different frequency bands or transponders using opposite polarisations). HPAs used for frequency hopping must employ TWTs in order to have adequate bandwidth.

In systems relying solely on TDMA most earth stations need to transmit only one carrier. However, where earth stations need to transmit both TDMA and FDMA carriers it is usual to employ a separate HPA for the TDMA carrier; this avoids the phase modulation of the FDMA carriers which might otherwise be caused by the on/off keying of the TDMA carrier and AM/PM conversion.

7.4.3.3 Linearisation

It is possible to improve the linearity of an amplifier by pre-distortion, i.e. the insertion before the HPA of a nonlinear network with AM/AM and AM/PM characteristics which are, as far as practicable, the inverse of those of the HPA.

Pre-distortion can be used with amplifiers of any power rating, from satellite TWTAs with output powers of a few watts to the most powerful earth-station HPAs with saturated output powers of many kilowatts. The use of a lineariser may make it possible either

(i) To reduce the output backoff required to limit intermodulation between FDMA carriers by 2—3 dB, or

(ii) To reduce the output backoff required to limit the spectrum spreading of TDMA transmissions by 1 or 2 dB

However, some linearisers give a significant improvement only over a fairly narrow range of characteristics; before buying a wideband pre-distortion network it is therefore wise to test it with the particular type of tube and with transmissions of the specific powers and frequencies with which it is to be used.

7.4.3.4 Power supplies and protection circuits

The structure and mode of operation of high-power TWT and Klystron tubes is such that they must be provided with complex power supplies and protection circuits. In both types of tube the signal is propagated along a slow-wave structure and interacts with an electron beam; amplification results from the transfer of energy from the electron beam to the signal.

The most common slow-wave structure for TWTs is a helix. This helix, the anode (which accelerates the electrons) and the collector (see Fig. 7.19) all require different high-voltage supplies (typically 5—10 kV) which must be regulated to about one part in a thousand and must be free from any spurious frequencies which could modulate the electron beam and thus the signals being amplified.

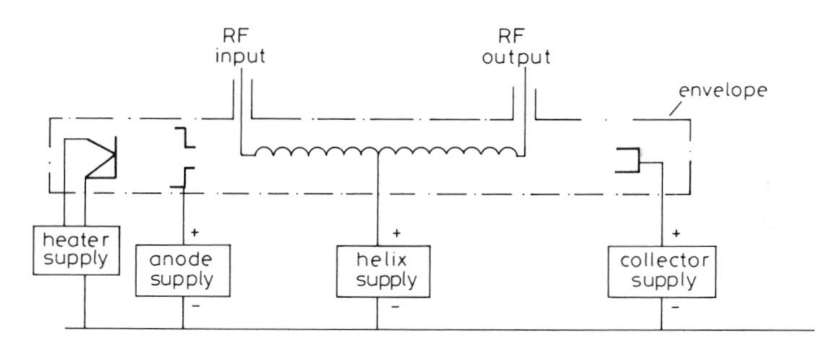

Fig. 7.19 *TWT and power supplies*

The electrons in the beam of a TWT are accelerated to very high velocities and still have high kinetic energy after they have interacted with the signal. This residual energy is dissipated as heat when they strike the collector. If the maximum rated output of a TWT is (say) 1 kW then several kilowatts of power will usually be dissipated inside the tube. For TWTs with a maximum rated output power of up to two or three hundred watts the heat may be removed by conduction; for tubes with higher powers, up to about 3 kW, it is necessary to use forced-air cooling. Tubes with still higher power outputs usually require the use of a liquid coolant and this adds considerably to the complexity and cost of the HPA; however, although tubes with output powers of over 3 kW were commonly used in the early earth stations, they are rarely required nowadays.

TWTs need protection from a number of hazards. In particular, the helix can be destroyed by local overheating arising from:

(i) Excessive helix current (caused, for example, by defocusing of the electron beam) or
(ii) A bad mismatch at the output of the tube which causes power to be reflected back and dissipated in the helix.

Methods of protection include:

(*a*) Circuits which ensure that supplies to the tube are switched on in the correct order
(*b*) Devices such as helix current monitors, temperature monitors and (in tubes with output powers of over 1 kW) arc detectors, all of which are arranged to switch the tube off if there is any danger that it would suffer damage.

Klystron HPAs are rather more simple, robust and efficient than TWT HPAs; for example, klystrons usually need only a single high-voltage supply. Klystron HPAs are nevertheless still relatively complex devices and it is not surprising that HPAs (whether klystron or TWT) often account for a high proportion of faults at earth stations.

7.4.3.5 Solid-state amplifiers

Solid-state wide-band amplifiers using GaAs FETs with outputs of up to 100 W at 4 GHz or tens of watts at 14 GHz are now available. These amplifiers are cheap, reliable and efficient and are more linear than klystrons or TWTs. They are suitable for use with most types of station but the output power available may not be sufficient to allow their use as output stages for terminals using combining networks with high losses.

Low-power TWT HPAs (which are simple devices compared with the medium-power or high-power versions) are usually the best choice where there is a requirement for rather higher powers than can be delivered by solid-state amplifiers.

7.4.3.6 Characteristics of HPAs

The most important characteristics of output amplifiers apart from operating frequency, bandwidth and output power, are:

(i) *Gain:* The gains of TWTs and klystrons range from about 35 to 50 dB. Intermediate-power amplifiers (IPAs) are usually needed between the modulator and HPAs. IPAs may use low-power TWTs or solid-state devices (the latter being more common in modern equipment). Equalisers are usually provided to compensate for the variation of gain with frequency of the HPAs and other units in transmit chains.

(ii) *Linearity:* The effects of AM to AM (power transfer) non-linearity and AM to PM conversion are dealt with in Sections 3.4 and 4.3.3, respectively.

(iii) *Variation of group delay with frequency:* See Section 4.3.2.

(iv) *Noise and spurious outputs:* Any amplifier generates thermal noise. The noise power delivered to the antenna by an output stage is proportional to $[G_p G_o(\mathrm{NP}-1) + \mathrm{G_o}(N_o-1)]$ where G_p and G_o are the gains of the preamplifier and output amplifier, respectively, and N_p, N_o are the corresponding noise figures (NFs); the NF of the preamplifier is therefore the major influence on the noise power delivered to the antenna. The NF of a medium-power or high-power microwave tube is often of the order of 30 dB but special low-noise TWTs or solid-state devices must be used for preamplifiers. Solid-state amplifiers with NFs of about 6 dB at 6 GHz, and slightly higher at 14 GHz, are available. Other unwanted effects of HPAs are the generation of spurious tones and bands of high-level noise, and amplitude modulation of output signals as a result of ripple on the power supplies. It is usual for the space-sector operator to specify (explictly or implicitly) limits on all unwanted outputs from earth-terminal output stages; for example, the EIRP outside the bandwidth allocated to the wanted transmission may be limited to not more than 4 dBW in any 4 kHz unit of bandwidth.

7.4.4 *Signal processing equipment*

7.4.4.1 *Introduction*

Voice and data signals arriving at an earth station via the public switched telephone network (PSTN) are usually in the form of FDM or TDM assemblies.

Terrestrial FDM assemblies arriving at an earth station working to an FDM–FDMA satellite system usually have to be demultiplexed for one or more of the following reasons:

(i) Not all the channels in the assemblies are intended for transmission by the earth station.

(ii) The forms of the terrestrial and satellite multiplexes are not identical (e.g. the multiplex assemblies used in the INTELSAT system are not the same as the standard terrestrial multiplexes.

(iii) The station is transmitting more than one carrier, and channels (or groups of channels) from a multiplex have to be divided between these carriers.

The extent to which the multiplexes have to be disassembled varies considerably. If the terrestrial network and the earth station belong to the same organisation then most or all of the traffic in the FDM assemblies may be intended for transmission by satellite and only limited reconfiguration of the groups may be necessary. On the other hand, if the earth station is accepting traffic from a number of independent terrestrial carriers it may have to extract a few channels from each of a number of large assemblies and it may be necessary to disassemble to channel level; in this case the amount of demultiplexing equipment required at the earth station may be large.

An increasing proportion of the traffic arriving at earth stations is in the form of TDM assemblies. Once again, not all the channels in the assemblies may be intended for transmission by satellite, in which case they will have to be demultiplexed. On the other hand the TDM multiplexes may be intended for modulation directly onto carriers for transmission in TDM–PSK–FDMA form or they may be carrying traffic for a TDMA system. Whatever the method of transmission via the satellite, the line coding must be removed since earth-station equipment will usually accept only binary signals.

Where signals for transmission by TDMA or TDM–FDMA arrive at an earth station in FDM form it is necessary to convert them either:

(i) By demultiplexing to 4 kHz channels, encoding the channels in digital form and remultiplexing them as a TDM assembly, or
(ii) By using a transmultiplexer which converts the signals directly from FDM to TDM (or vice versa). This is less complicated and cheaper than demultiplexing and remultiplexing.

Signals for transmission in SCPC form must, of course, be completely demultiplexed.

All signals for transmission to the satellite, whatever their form, must be modulated onto a carrier at IF and up-converted to the frequency of transmission. The inverse processes must be performed on signals received by an earth station; i.e. the signals must be down-converted and demodulated and the channels must, if necessary, be demultiplexed and remultiplexed in the form required by the terrestrial system.

The major elements of the various types of signal-processing equipment at earth stations are described briefly in the following Sections.

7.4.4.2 FDM–FDMA

Fig. 7.20 comprises block diagrams of transmit and receive chains for processing a FDM–FDMA carrier.

(*a*) *Transmit chain*

At the input to the transmit chain, the engineering service circuits (ESCs) and a pilot carrier are combined with the baseband traffic signal in a baseband

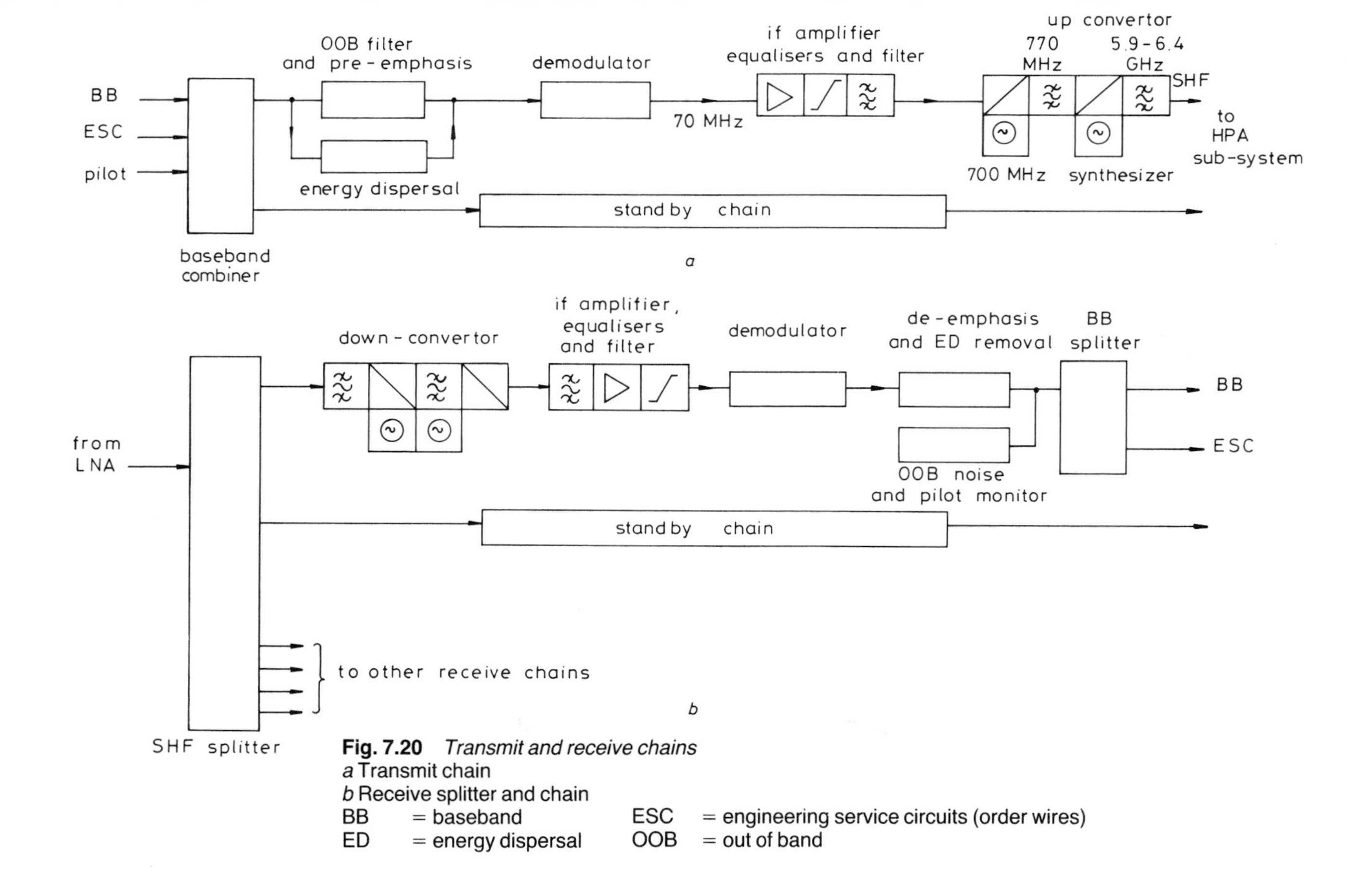

Fig. 7.20 *Transmit and receive chains*
a Transmit chain
b Receive splitter and chain
BB = baseband
ED = energy dispersal
ESC = engineering service circuits (order wires)
OOB = out of band

combiner unit. The pilot (usually a 60 kHz carrier modulated with a continuous tone) is monitored at the receiving station and if the carrier is not received the absence of the pilot operates an alarm.

The baseband signal is passed through an out-of-band (OOB) noise filter; this band-stop filter blocks a narrow band of frequencies just above the top baseband channel. At the receiving station a monitor circuit measures the noise present in the band; this noise must have been generated after the OOB filter and it can therefore be used to give warning of deteriorating or unacceptable performance of the satellite link.

The signal is now passed through a pre-emphasis network and, when necessary, an energy dispersal signal is added (see Sections 4.2.3.1 and 4.4, respectively).

The next stage in the transmit chain is the modulator, which converts the baseband signal to a frequency-modulated carrier at the intermediate frequency (IF). The IF is chosen to enable the processes of amplification, filtering and equalisation of the signal to be carried out as simply and economically as is practicable. For transmissions with bandwidths of up to 36 MHz the usual IF is 70 MHz; for transmissions with a bandwidth of over 36 MHz an IF of 140 MHz is customary (but FDM–FDMA signals with a bandwidth greater than 36 MHz are rare). A bandpass filter limits the bandwidth of the transmission and prevents it interfering with transmissions in adjacent frequency bands; the INTELSAT system uses bandwidths of 1·25 to 36 MHz (corresponding to assemblies of 12 to around 1000 channels).

Equalisers are provided to compensate for the variation with frequency of the gain and group delay of the transmit chain and, in the INTELSAT system, additional equalisers are provided on the transmit side of the earth station to compensate for the variations of group delay introduced by the satellite.

The carrier is usually translated from IF to the transmission frequency by means of a double up-converter (DUC). The reason for using a double up-converter rather than a single-stage up-converter can best be illustrated by a simple example. Consider a system using the up-path frequency band 5925—6425 MHz and suppose that there are two carriers at an IF of 70 MHz which are to be transmitted at frequencies of 6100 and 6400 MHz. If we use single-stage up-converters, one with a local-oscillator (LO) frequency of 6030 MHz and one with a LO frequency of 6330 MHz, we get image signals at 5960 and 6260 MHz as well as the wanted signals at 6100 and 6400 MHz. Both pairs of signals fall within the band 5925—6425 MHz and the wanted signals must be separated from the unwanted by means of filters, a separate design of filter being required for each transmission. Now suppose that we use double up-converters in which the first-stage local oscillators have a frequency of 700 MHz (as shown in Figure 7.20); the IF signals are converted to 770 MHz (signals at 630 MHz are also generated but are rejected by a filter). If we now up-convert again using local-oscillator frequencies of 5430 and 5630 MHz we get our wanted transmissions at 6100 and 6400 MHz. We still get image signals but this time they are at frequencies of 4730 and 4930 MHz, i.e. they fall well outside the wanted frequency band; it is thus possible to reject all image signals by means of a single design of filter with a passband corresponding to the wanted frequencies of 5925—6425 MHz. The same result could, of course, be achieved by using an IF of 770 MHz instead of 70 MHz but components and circuits designed for this frequency are much more expensive and difficult to adjust than those working at 70 MHz. In a large earth station the second local oscillator is usually a frequency synthesiser but a smaller station may be equipped with units offering a choice from a limited number of frequencies.

The output of the up-converter is passed to the output amplifier which, in the larger stations, comprises an intermediate-power amplifier (IPA) followed by a high-power amplifier (HPA).

(*b*) *Receive chain*

A receive chain receives signals from a LNA and performs the inverse processes to a transmit chain. Where there is more than one receive chain a splitter must be inserted between the LNA and the receive chains to provide an input to each chain. The required carrier is selected by tuning the first local oscillator of a double down-converter (DDC). A bandpass filter at IF limits the thermal noise and protects the transmission against adjacent-channel interference. Equalisation and amplification at IF are provided before the signal is passed to the demodulator.

After demodulation the baseband signal is amplified, the emphasis is removed, the pilot and OOB noise are monitored and finally the traffic channels are separated from the ESCs in a baseband splitter unit.

7.4.4.3 SCPC equipment

(a) SCPC–FM

The basic characteristics of SCPC-FM transmissions are described in Section 4.6. The functions required of FDM–FM transmit and receive chains (i.e. baseband processing, IF processing, up-conversion and down-conversion) are also required of SCPC–FM equipment, but SCPC equipment usually provides additional functions such as companding, voice-activated carrier switching and automatic frequency control (AFC).

SCPC–FM equipment comprises common equipment and channel equipment (see Fig. 7.21).

A typical channel unit is able to generate or select any one of, say, 1200 carriers spaced by 30 kHz across an IF band of 52—88 MHz; each unit must therefore include a frequency synthesiser.

The transmit side of the common equipment combines the outputs from

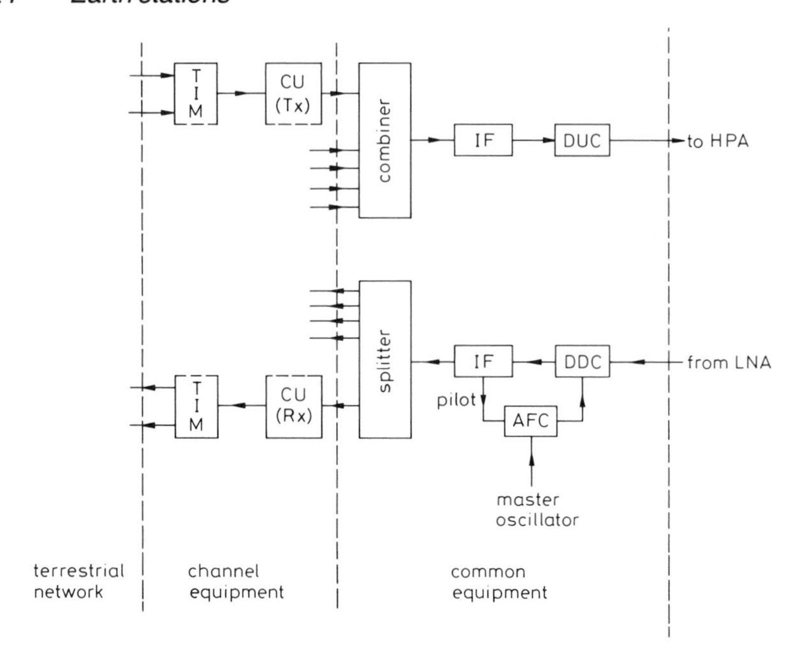

Fig. 7.21 *Block diagram of SCPC–FM equipment*
TIM = terrestrial interface module
CU(Rx) = channel unit (receive)
CU(Tx) = channel unit (transmit)
DUC & DDC = double up-converter & down-converter
AFC = automatic frequency control

a number of channel units, provides IF processing and up-conversion, and passes the composite output signal to an HPA. On the receive side the common equipment takes a composite signal from the LNA, performs down-conversion and IF processing, and passes the signal to a number of channel units, each of which selects a wanted channel for further processing. The frequency of the local oscillator in the second stage of the double down-converter is controlled by an AFC signal which is derived by comparing the frequency of a stable oscillator at the earth station with the frequency of a pilot signal transmitted by a reference station and received at the earth station; this compensates for errors in frequency translation in the satellite as well as virtually eliminating frequency errors resulting from down-conversion at the earth station. All reference frequencies required by the equipment (e.g. those required by the channel-unit frequency synthesisers) are also generated in the common equipment.

A voice-channel unit includes:

(i) A compandor

(ii) Pre-emphasis and de-emphasis circuits
(iii) A modem (the demodulator section of this is usually a threshold-extension demodulator)
(iv) A frequency synthesiser
(v) A speech detector, the output of which is used to operate an IF switch which suppresses the carrier when no speech is present; as the speech detector cannot operate instantaneously, the signal is passed through a delay line with a delay of the order of 10 ms in order to avoid clipping of the initial syllable of each speech burst.

In some SCPC equipment echo suppressors are provided in the channel units thus avoiding the need for the provision of suppression or cancellation at the international switching centre.

A terrestrial interface module may be incorporated in the channel unit or provided separately. The main purpose of this module is to convert signalling on the terrestrial system to that required by the satellite system and vice versa. (Terrestrial signalling systems may depend on an unbroken connection between subscribers whereas an SCPC voice channel suffers frequent breaks because the carrier is voice-switched.)

(b) SCPC–PSK

The common equipment for SCPC systems transmitting digital channels is much the same as that required for SCPC–FM operation. Voice channel units include PCM, ADPCM or delta-modulation codecs with voice-switched carriers. The usual method of modulating digital channels onto the carriers is QPSK. Voice-channel units must include synchronising circuits which provide a preamble for each speech burst and thus enable the receiving units to recover the carrier and the clock timing.

In the INTELSAT system, channel units are available for the transmission of voice, voice-band data and data at 48, 50 or 56 kbit/s. FEC is applied to all data transmissions at rates greater than 4·8 kbit/s.

(c) SCPC–DA

SPADE (see Section 1.10.2.1) is SCPC equipment which assign channels on demand. Each earth station includes a demand-assignment signalling and switching unit (DASS), which incorporates a digital processor, and all stations exchange information via a TDMA common signalling channel (CSC). The DASS keeps an up-to-date list of the carrier frequencies in use and when demand for a new call arrives at the station a terrestrial interface unit informs the DASS and the latter chooses a pair of unused frequencies and supervises the setting up of the call.

SPADE is a DA system with distributed control since all earth stations keep a list of the frequencies in use and select their own carriers from the common pool. If the control functions are centralised at a single station the amount of equipment required at all the other stations is reduced; DA systems with centralised control are therefore cheaper and a number of domestic satellite-communications systems use this type of system.

7.4.4.4 TDMA

The basic principles of TDMA are described in Section 5.8.3.

A TDMA terminal (see Fig. 7.22) comprises terrestrial interface modules (TIMs) and a common TDMA terminal equipment (CTTE).

The CTTE controls the transmission to, and reception from, the satellite of bursts of data. These bursts of data are assembled by the CTTE from sub-bursts generated by the TIMs. The CTTE stores the burst time plan (i.e. the start times and durations of the bursts to be transmitted and received by the terminal) and provides all timing functions for the terminal.

In the transmit direction the TIMs accept signals from the terrestrial network and convert them to sub-bursts of data which are stored in buffers and read out, under the control of the CTTE, at the satellite transmission rate (this is approximately 120 Mbit/s for the INTELSAT and EUTELSAT systems).

There is a wide range of TDMA systems from simple domestic systems, in which much of the logic may be wired into the hardware, to very complex systems (such as those used by INTELSAT and EUTELSAT) which require the provision of extensive data-processing facilities and software. Some TDMA systems will accept and deliver only TDM multiplex assemblies

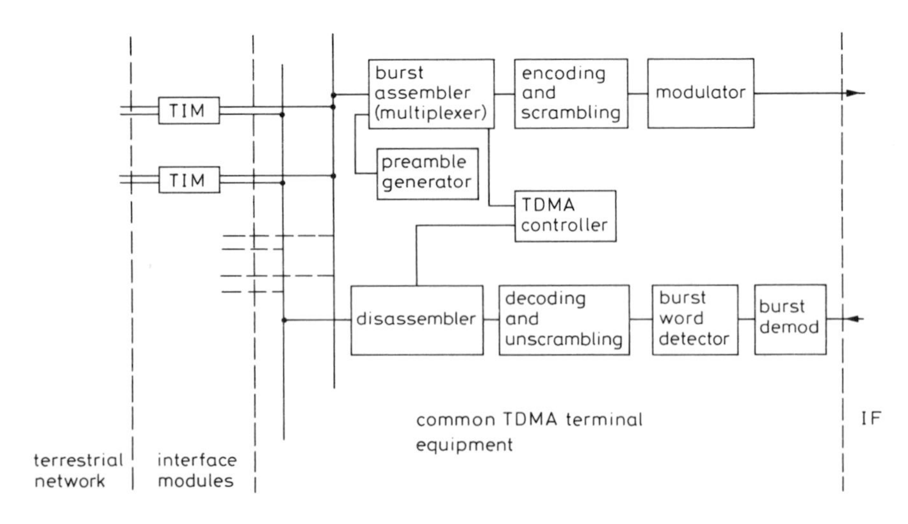

Fig. 7.22 *Block diagram of TDMA terminal*

whereas others will accept traffic from a variety of digital and analogue sources and adapt to changing traffic demands.

An earth station working to a TDMA system must, like most other stations, include up-converters and down-converters, IF amplifiers and equalisers, HPAs and LNAs. The IF bandwidth required for a 120 Mbit/s QPSK transmission is about 60 MHz and an IF frequency of 140 MHz is typical for this type of system.

A problem peculiar to TDMA operation is that time delay between alternative paths through the earth station must be equalised so that differences are small compared with the guard time between consecutive bursts at the satellite (this means typically that the differences must be much less than 1 μs); the paths to be equalised may be main and standby paths or they may be the alternative paths used when hopping between transponders using the same frequency band but orthogonal polarisations, or between transponders using different frequency bands (the latter is usually achieved by the use of frequency-agile up-converters and down-converters).

7.4.5 Supervisory and control equipment

The purpose of supervisory (surveillance or monitoring) equipment is to ensure that any failure or deterioration of the services provided by an earth station (and, if possible, any condition which might lead to failure) is brought to notice immediately. The main parameters which need to be monitored are baseband noise power in an analogue system or bit error ratio (BER) in a digital system, and transmitter output power. The main control function required is usually the switching of traffic from a path in which a unit has failed to a standby path; this may be done either automatically, under the control of the supervisory signals, or manually. Thus a simple receive-only terminal might comprise main and standby LNAs and receivers with monitoring of the main receiver output and automatic switching to the standby equipment in case of a failure.

At large earth stations it is usual to monitor, sample and record a number of terminal characteristics in order to compile a record of the system and equipment performance and to assist in the diagnosis of faults. Most items of earth-station equipment are provided with a number of monitor or test points which may be used to check that units are functioning correctly and to assist in adjustment or repair; the more important of these test points are usually provided with built-in measuring equipment and associated alarms, and these may be used as the basis of a monitor and control system. Alarms may be sub-divided into major (or primary) alarms which indicate failure of a unit or a serious deterioration in performance, and minor (or secondary) alarms which show that there has been a change in performance which gives cause for concern and needs to be investigated as soon as possible; some systems also include advisory alarms which indicate slight deterioration of performance.

Other information which may be collected by the monitoring system includes the state of coaxial and waveguide switches, antenna pointing direction, and weather conditions such as wind speed and temperature. Control functions at large stations usually include remote steering of the antenna, adjustment of transmitter power and retuning of receivers.

Alarm signals and all the data collected by the supervisory system are usually relayed to a central monitor and control position. It would be very confusing if all the information available were to be displayed simultaneously and it is essential that the signals and data are processed in a way that gives the operator as much help as possible in understanding the cause of any problems and taking appropriate action; this is particularly necessary in the case of faults which would otherwise cause multiple alarms. Monitor and control systems using microprocessors are now available from several manufacturers; systems suitable for use with relatively simple single-antenna stations are available as well as those designed for complex multi-antenna stations. Microprocessor systems can:

(i) Monitor a large number of points
(ii) Check that the monitored values are within acceptable limits
(iii) Raise alarms and give operators assistance in determining the cause of problems
(iv) Exercise a wide range of control functions including the reconfiguring and retuning of equipment at set times
(v) Keep a log of system performance and faults

A particularly valuable feature of microprocessor systems is that, unlike earlier monitor and control systems with hard-wired logic, they are flexible and can be adapted without difficulty to expansion or modification of an earth station. There may be difficulties in connecting a proprietary monitor and control system with earth-station equipment from a different manufacturer but some operators of very large earth stations have introduced custom-made microprocessor-based systems with advanced facilities such as interactive diagnostics, statistical analysis of performance, automatic preparation of reports, computer-based equipment manuals, maintenance schedules and spares holdings, and integration with electronic mail and message systems. Microprocessor systems are well suited to the provision of centralised control of a number of earth stations and this may prove of increasing importance in an era when the number of earth stations in satellite systems is increasing rapidly.

Three essential characteristics of any monitor and control system are first that it shall be simple to use, secondly that it shall be very reliable and thirdly that, if it does develop a fault, it shall fail safe and return the system to manual control.

Short Bibliography

1 CCIR: 'Handbook on satellite communications (fixed satellite service)' (International Telecommunications Union, Geneva, 1985). This book, prepared by an *ad hoc* group of experts from CCIR Study Group 4 primarily to assist developing countries, contains a wealth of information and is essential reading for anyone new to satellite communications

2 'Recommendations and Reports of the CCIR: Vol. IV, Fixed Satellite Service'. XVIth Plenary Assembly, Dubrovnik 1986 (International Telecommunications Union). This volume is essential to the system designer for reference

3 EVANS, G. G. (Ed.): 'Satellite communication systems' (Peter Peregrinus, 1987)

4 MARAL, G., and BOUSQUET, M.: 'Satellite communications systems' (John Wiley, 1986)

5 ALPER, J. R., and PELTON, J. N. (Ed.): 'The INTELSAT global satellite system' (American Institute of Aeronautics and Astronautics, 1984). This book was produced to mark the twentieth anniversary of INTELSAT. It has no index and is becoming rather out of date but some of the chapters are well worth reading (e.g. Chap. 6 'INTELSAT satellites; Chap. 11, 'INTELSAT and the ITU')

6 IPPOLITO, L. J.: 'Radio propagation for space communication systems', *Proc. IEEE*, 1981, **69**, pp. 697–727. An excellent summary of the subject

7 IEE Conference Publication No. 294: 'Fourth international conference on satellite systems for mobile communications and navigation, 1988'. The proceedings of this conference include a number of useful general surveys of mobile satellite communications.

List of abbreviations

Items in the following list of abbreviations will be found in the index under the full title, not in the abbreviated form (for example, there is no reference in the index to 'ACI' but there is an entry under 'adjacent channel interference')

ACI — adjacent channel interference
ACSE — access, control and signalling equipment
ADE — above-decks equipment
ADPCM — adaptive differential PCM
AFC — automatic frequency control
AKM — apogeee kick motor
ARQ — automatic repeat request
ASK — amplitude-shift keying

BAPTA — bearing and power-transfer assembly
BDE — below-deck equipment
BER — bit error ratio
BPSK — binary PSK
BSS — broadcasting satellite service
BTP — burst time plan

CBTR — carrier-and-bit-timing recovery
CCIR — International Radio Consultative Committee
CDMA — code-division multiple access
CES — coast earth station
CFDM — companded FDM
CTTE — common TDMA terminal equipment

DA — demand assignment
DBS — direct broadcast satellite
DCME — digital channel multiplication equipment
DM — delta modulation
DNI — digital non-interpolation
DSI — digital speech interpolation

EIRP — equivalent isotropically radiated power
EL-AZ — elevation-over-azimuth (antenna mount)
EPIRB — emergency position-indicating radio beacon
ESA — European Space Agency

FCC — Federal Communication Commission
FDM — frequency-division multiplex
FDMA — frequency-division multiple access
FEC — forward error correction
FGMDSS — future global maritime distress and safety system
FM — frequency modulation
FSK — frequency-shift keying
FSS — fixed satellite service

GCE — ground communications equipment

HA-DEC — hour-angle/declination (mount)
HPA — high-power amplifier

HRC hypothetical reference circuit

IBS INTELSAT business service
IDR Intermediate Data Rate (an INTELSAT service)
IFRB International Frequency Registration Board
IMO International Maritime Organisation
INMARSAT International Maritime Satellite Organisation
INTELSAT International Satellite Communications Organisation
ISI intersymbol interference
ITU International Telecommunications Union

LNA low-noise amplifier
LRE low-rate encoding

MAC multiplexed analogue components
MCS Maritime Communications Subsystem
MSK Minimum shift keying

NASA National Aeronautics and Space Administration (US)
NBFM narrow-band FM
NCS network co-ordination station
NPR noise power ratio

O-QPSK offset QPSK

PCM pulse code modulation
PEP packet error probability
PKM perigee kick motor
PSK phase-shift keying

QAM quadrature amplitude modulation
QPSK quadrature phase-shift keying

RCC rescue co-ordination centre
RDSS radio determination satellite service
RRs Radio Regulations

SCC spacecraft control centre
SCPC single channel per carrier
SES ship earth station
SOLAS safety of life at sea
SPEC speech predictive encoded communication
STS Space Transportation System
SS-TDMA satellite-switched TDMA
SSB single-sideband modulation

TDM time-division multiplex
TDMA time-division multiple access
TIM terrestrial interface module
TOCC technical and operational control centre
TRMA time-random multiple access
TT & C tracking, telemetry and command
TVRO TV receive only
TWT travelling-wave tube

VAR voltage axial ratio
VSAT very-small-aperture terminal

WARC World Administrative Radio Conference

XPI cross-polarisation isolation
XPD cross-polarisation discrimination

Index

Above-deck equipment, 247–248
Access, control and signalling equipment, 252
Acquisition, 229
Active satellite, 3
Adaptive differential PCM, 44, 187–188
Adjacent channel interference, 207
Aeronautical service, 31–32, 267–270
Albedo, 58
ALOHA, 231
AM/PM conversion, 158–159
Amplitude shift keying, 198
Analogue modulation, 36–39
Angular momentum, 77
ANIK, 24
Antenna
  beamwidth of, 103
  Cassegrain, 287–289
  despun, 11
  directional, 15
  drive, 303–305
  earth-station, 19, 27–30, 284–294
  front-fed, 286
  gain, 98–99, 102–103, 284
  Gregorian, 290
  kingpost, 307
  noise temperature, 99, 285
  offset, 289–290
  pointing, 301–303
  radiation pattern, 291–294
  safety, 305–311
  satellite, 8, 11–14, 15–16, 18, 47–48, 87, 104
  sidelobes, 289
  structure, 87, 305–309
  Torus, 290–291
  tracking, 301–305
  wheel-and-track, 306–307
Apogee, 3
Apogee kick motor, 59
ARABSAT, 22, 26
Ariane, 55, 59, 61
ASC, 24
Attenuation
  atmospheric, 96
  clear-sky, 113–115
  free space, 94
  rain, 115–120
  snow, 119
Attitude control, 73–74, 76–80
AUSSAT, 25
Automatic frequency control, 251–252
Automatic repeat request, 44, 214–215, 261
Azimuth, 51

Backoff, 37, 87, 105, 110
Bandwidth
  of FM signal, 148–150, 174–175
  of PSK signal, 208–209
Batteries, 72
Baud rate, 199
Beam waveguide, 288–289
Beamwidth, 11, 12, 103, 258–259, 301
Bearing and power-transfer assembly, 14
Below-deck equipment, 247
Binary PSK, 197–198
Bit error ratio, 44, 196, 201, 233–235
Bit-timing recovery, 226
Block codes, 215–217
Body stabilisation, 66, 77–80
Boosters, 55
BRAZILSAT, 25
Broadcasting satellite service, 33–35, 170
Burst time plan, 43, 229
Bus, 65

Business systems (see specialised services)

Capture effect, 147
Carrier-and-bit-time recovery, 226
Carrier power to noise power density ratio, 97
Carrier to intermodulation noise power ratio, 96, 111–112
Carrier-to-noise power ratio, 33–35, 151–152, 201
Carson's rule, 148
Channel quality (see Performance)
Characteristics
  earth terminal, 98–104
  satellite, 104–106
Clarke, Arthur C., 1, 9
Closed network, 29
Coast earth station, 249–252
Code-division multiple access, 221–222
Coding, 182, 184–190
  gain, 217
Command (see telemetry, tracking and command)
Common TDMA terminal equipment, 42, 231
Communications module, 17, 80–81
Companded FDM, 37, 166–167
Companding, 163–166, 255
COMSAT, 7, 238
COMSTAR, 24
control, 64, 80–81
Convolutional encoding, 217–221, 261
Co-ordination, 124–127, 278–281
Costs, 34, 54, 91
COURIER, 2
Coverage area, 51
Cross-polar discrimination (see orthogonal polarisation)
Cross-polar isolation (see orthogonal polarisation)
Cross-strapping, 19
C/T, 97

Decoding, 182
De-emphasis, 144–145
Delay (see group delay or propagation delay)
Delta modulation, 186–187
Delta-I method, 125
Demand assignment, 40
Demodulation
  coherent, 203
  differential, 204
Depolarisation, 119
Differential encoding, 204
Differential PCM, 185–186
Digital-channel multiplication equipment, 44, 195
Digital modulation, 195–214
Digital pipes, 223–224
Digital speech interpolation, 43, 192–195
Digital transmission, 39, 180–235
Direct broadcast satellite, 33–35, 170, 178–179
Direct sequence, 221
Distortion, 155–159
Diversity (see site diversity)
Digitally non-interpolated, 193–194
Domestic system, 23–25
Doppler shift, 233, 251
Drift, 74
Dual polarisation (see orthogonal polarisation)

EARLY BIRD, 5
Earth station, 275–327
  antenna, 284–311
  electronic equipment, 311–327
  INTELSAT
    standard-A, 19, 286
    standard-B, 19, 27, 286
    standard-C, 19, 286
    standard-E, and standard-F, 28
  sites, 276–284
  supervisory and control equipment, 327–328
Eb/No, 97, 201
Eccentricity, 52–53
ECHO, 2
Eclipse, 71–72
Elevation, 51
Elevation-over-azimuth mount, 302–303
Emergency position-indicating radio beacon, 32, 266–267
Encoding
  adaptive, 187–188
  block, 215–217
  convolutional, 217–220
  differential PCM, 185–187
  hybrid, 189
  low-rate, 184–190
  pulse code, 181–184
  sub-band, 188
  television, 189–190
  vocoders, 188–189
Energy dispersal, 159–163, 175

Energy-per-bit, 97
Equivalent isotropically-radiated power, 94, 98, 102, 105
Error correction, 214–221
Error performance objectives, 233–235
European Space Agency, 268
EUTELSAT, 22, 26
Exhaust velocity, 56–57
EXPLORER, 2
Eye diagram, 196–197

Faraday rotation, 103
FDM–FM–FDMA, 11, 36–37, 129, 136–138, 319–323
Federal Communications Commission, 23
Feed, 98, 286–289
Figure of merit, 95, 98, 100, 104–105, 311, 313
Filtering, 207–210
Fixed satellite service, 33
FORDSAT, 24
Forward error correction, 44, 215–220
Frame
  TDM, 191
  TDMA, 224
Frequency
  allocations, 121
  bands
    6/4 GHz, 4, 5, 10, 15, 16–19, 21
    14/11 GHz, 16–18, 21
    20/30 GHz, 48
    shared, 122
  modulation, 4, 36, 136–179
    bandwidth required, 36
    frequency deviation, 141, 148, 171–174
    narrow-band, 157, 244
    threshold, 146–148
  register, 124
  reuse, 15–18
Frequency-division multiple access, 10, 136, 221
Frequency-division multiplex, 11
Frequency-shift keying, 156
Future Global Maritime Distress and Safety at Sea System, 32, 265–266

GALAXY, 24
Gateway, 28
GEOSTAR, 274
Geostationary
  orbit, 1, 49–53
  satellite, 1, 5
Global beam, 11, 16–17, 83, 87
GLONASS, 274
Ground communications equipment, 311
Group delay, 155–158
GSTAR, 24
G/T, 95, 98, 100, 104–105, 311, 313
Gyroscopic stiffness, 79

Half-power beamwidth, 103, 301
Hemispheric beam, 16–17, 83, 87
High-power amplifiers, 314–319
Hohmann transfer ellipse, 58
Hour-angle/declination mount, 302–303
Hydrometeors, 119
Hypothetical reference circuit, 139

Inertia, 77–88
INMARSAT, 1, 31, 236–273
  satellites, 252–258
  standard-A system, 239–255
  standard-B system, 237
  standard-C system, 258–265
  standard-M system, 272
INTELNET, 30
INTELSAT, 1, 5, 7–23
INTELSAT I, 5, 7–9
INTELSAT II, 9
INTELSAT III, 9–12
INTELSAT IV, 12–14
INTELSAT IVA, 15–16
INTELSAT V, 16–22, 83–87
INTELSAT VA, 20
INTELSAT VB, 20, 28
INTELSAT VI, 20–22, 87–88
INTELSAT VII, 22
INTELSAT business services, 28
Interference, 102, 120–127, 277–284
Interleaving, 220, 261
Intermediate Data Rate Service, 42, 223
Intermodulation, 36–37, 96, 106–110
International Frequency Registration Board, 124–126
International Maritime Organisation, 32, 237–238, 274
International Maritime Satellite Organisation, 32, 237–238, 239
International Radio Consultative Committee, 293
International Telecommunications Union, 124, 278
Intersymbol interference, 207

Land mobile service, 6, 32, 270–273
Launchers and launching, 54–65
Leased transponders, 26
Lifetime (satellite), 9, 22
Linearisation, 316–317
Link
  budgets, 97, 127–135
  equations, 95
Low-noise amplifiers, 311–314
Low-rate encoding, 44, 184–190

Major-path satellite, 14–15
MARECS, 240, 254–256
MARISAT, 31, 238–240
Maritime Communications Subsystem, 20, 31, 239, 252–254
Maritime service, 31–32, 229–267
Mass ratio, 55
Minimum-shift keying, 213
Mobile satellite service, 31–32, 42–47
Modulation index, 141
MOLNIYA, 5–6, 273
Momentum
  angular, 77
  bias, 79
  wheel, 19, 79
Monitoring, 64
Monopulse tracking, 304–305
M-SAT, 271–272
Multiple access, 5, 221–232
Multiplexed analogue components, 169, 178–179
Multiplexer, 86

National Aeronautics and Space Administration, 2, 8
Narrow-band FM, 167, 244
Navigation and position-finding system, 273–274
NAVSTAR-GPS, 274
Network co-ordination station, 240, 243–246, 262–265
Noise
  power, 94
    allowable, 140–141
    ratio, 152–155
  temperature
    earth station, 311–313
    satellite, 104–105
Non-linearity, 106–110
NTSC, 169–172

Offset reflector antenna, 289–290
OLYMPUS, 48
On-board processing, 82, 89
Open network, 29
Orbit
  control, 73
  elliptical, 3, 6, 273
  geostationary, 1, 49–53
  inclined, 3, 6, 273
  MOLNIYA, 5–6
  synchronous, 49–50
Orbital errors, 52–53, 301
Orthogonal polarisation, 16–19, 295–301
  discrimination and isolation, 16–18, 119–120, 296–297
Overdeviation, 174–175

Packet error probability, 261
PAL, 169–172
PALAPA, 25
Parabolic antenna, 276, 286–290
Parking orbit, 58, 60
Passive satellite, 3
Path loss, 93–94
Payload, 61–62, 65
PCM–QPSK–SCPC, 41
Performance, 138–141, 172, 175, 183–184, 196, 255
Perigee, 3
  kick motor, 60
Phase-shift keying, 196–200
  O–QPSK, 211–212
  2,4–PSK, 212–213
Pitch, 76
Platform (despun), 14
Plesiochronous, 191
Pointing accuracy, 305–306
Polarisation, 103–104, 294–301
  state, 294–295
Positioning systems, 273–274
Power flux density, 12
Power subsystem, 69–73
Preamble, 228
Pre-emphasis, 144–145, 172–173
Pre-emptible capacity, 26
Primary satellite, 14–15
Private-user systems, 27–30
Programme track, 304
Propagation, 112–120
  atmospheric, 113–120
  delay, 5, 7, 9, 45
  ionospheric, 113
  tropospheric, 279
Propellant, 54–55, 57

Psophometric weighting, 140–141
Pulse code modulation, 180

Quadrature amplitude modulation, 201–203
Quadrature PSK, 199–200

Radiation hazards, 309–311
Radiation patterns, 291–294
Radio-determination satellite service, 274
Radio Regulations, 121–122
Rain
  attenuation, 115–119
  scatter, 279
Raised-cosine filter, 208
Random-access TDMA, 231, 242
Range (of satellite), 50–51
Reaction wheel, 77–78
Receive chain, 321, 323
Redundancy, 92
Reference burst, 224
Regeneration, 88
Regional system, 26
RELAY, 3, 5
Reliability, 90–92
Rescue co-ordination centre, 32, 241
Roll, 76
Roll-off, 208

Safety, 64, 307–311
Safety of life at sea, 237, 241, 265–267
Satellite-switched TDMA, 230–231
Saturation power, 107, 110–111
SBS, 24
SCORE, 2
Scrambling, 211
SECAM, 169–172
Sensors, 76
Service module, 65–81
Ship earth station, 31, 240–242, 247–249, 258–259
Side lobes, 291–294
Signal-to-noise power ratio
  of FM transmission, 142–143
  in FDM–FM channel, 143–144
  of TV (FM) channel, 170–176
Single channel per carrier, 37–38, 133, 138, 167–169, 222, 323–326
Single-side-band modulation, 38–39
Site diversity, 118
Site shielding, 282–284
Small earth-station antennas, 27–28, 286, 306–309
Soft decisions, 220
Solar array and solar cell, 8, 19, 22, 69–71
Solar interference, 100–101
Spacecraft control centre, 80, 256
Space environment, 67–68
SPACENET, 24
Space platform, 65–81
Space Transportation System, 60
Space Shuttle (see Space Transportation System)
SPADE, 40
Specialised services, 27–30
Specific impulse, 56–57
Spectrum
  FM transmission, 141–142
  pulse train, 206–207
Speech predictive encoded communiation, 195
Spot beam, 12–14, 16–18, 83, 87
Spreading area, 93
Spread-spectrum multiple access, 221
SPUTNIK, 2
Stabilisation, 8, 11, 19, 66, 77–80
Station keeping, 74–76
Step-track system, 304
Stiffness (gyroscopic), 79
Structural subsystem, 66–68
Supervisory and control equipment, 327–328
Switching matrix, 86
Syllabic companders, 163–166
Synchronisation, 229, 232
SYNCOM, 5

TDF, 25
TDM–PSK–FDMA, 41, 223–224, 320
Technical and operational control centre, 80
Telemetry, tracking and command, 14, 64, 80–81
Teleport, 29
TELESAT, 25
Television, 4, 6, 24, 45–46
  FM transmission, 169–179
  link budget, 131, 138
  overdeviation, 174–175
  receive-only stations, 25
  signal-to-noise power ratio, 171–176
  sound, 177–178
  unified system, 172–173
TELSTAR, 3–5, 24

Terrestrial interfaces, 232–233
Terrestrial interface module, 42, 231
Test-tone deviation, 143, 173
Thermal control, 68–69
Three-axis stabilisation, 77
Threshold (FM), 146–148
Thrusters, 53, 74–75
Time-division multiple access, 21, 40, 42, 132, 224–232, 244
  satellite-switched, 21, 87–88, 230–231
Time-division multiplex, 190–192, 221, 243–244
Tracking, 64, 248
Transfer orbit, 58–59
Transmit chain, 320–323
Transmultiplexer, 232
Transponder, 3, 8, 82–90, 104–106
  hopping, 229
Travelling-wave tube, 314–318
Truncation distortion, 148
TVRO, 25
TV–SAT, 25

Unified system (TV), 172–173
Unique word, 226–227

VANGUARD, 2
Velocity increment, 55–56
Very-Small-Aperture Terminal, 27, 29–30, 134, 309
VISTA, 167
Vocoders, 188–189
Voice activation, 40–41, 133, 167, 194, 325
Voltage axial ratio, 103–104
VSAT, 27–30

Weighting
  psophometric, 140–141
  TV, 172–173
World Administrative Radio Conference, 122

Yaw, 76

Zero relative level, 140
Zone beam, 16, 83, 87